T0073158

The Story *of* Genetics, Development *and* Evolution

A Historical Dialogue

The Story *of* Genetics, Development *and* Evolution

A Historical Dialogue

GÁSPÁR JÉKELY

Max Planck Institute for Developmental Biology, Germany

World Scientific

NEW JERSEY · LONDON · SINGAPORE · BEIJING · SHANGHAI · HONG KONG · TAIPEI · CHENNAI · TOKYO

Published by

World Scientific Publishing Europe Ltd.

57 Shelton Street, Covent Garden, London WC2H 9HE

Head office: 5 Toh Tuck Link, Singapore 596224

USA office: 27 Warren Street, Suite 401-402, Hackensack, NJ 07601

Library of Congress Cataloging-in-Publication Data

Names: Jékely, Gáspár, author.
Title: The story of genetics, development, and evolution : a historical dialogue / by Gáspár Jékely
(Max Planck Institute for Developmental Biology, Germany).
Other titles: Mester, ébren vagy? English
Description: Hackensack, New Jersey : World Scientific, 2017.
Identifiers: LCCN 2016054154| ISBN 9781786342522 (hc : alk. paper) |
ISBN 9781786342539 (pbk : alk. paper)
Subjects: LCSH: Genetics--Popular works. | Developmental biology--Popular works. |
Evolution (Biology)--Popular works.
Classification: LCC QH437 .J4513 2017 | DDC 572.8/6--dc23
LC record available at https://lccn.loc.gov/2016054154

British Library Cataloguing-in-Publication Data
A catalogue record for this book is available from the British Library.

Copyright © 2018 by World Scientific Publishing Europe Ltd.

All rights reserved. This book, or parts thereof, may not be reproduced in any form or by any means, electronic or mechanical, including photocopying, recording or any information storage and retrieval system now known or to be invented, without written permission from the Publisher.

For photocopying of material in this volume, please pay a copying fee through the Copyright Clearance Center, Inc., 222 Rosewood Drive, Danvers, MA 01923, USA. In this case permission to photocopy is not required from the publisher.

Desk Editors: Dr. Sree Meenakshi Sajani/Mary Simpson/Shi Ying Koe

Typeset by Stallion Press
Email: enquiries@stallionpress.com

Printed in Singapore

"Translated from Hungarian by Lara Strong"

Contents

Prologue

At the Harbour

ALKIMOS: Democritus, Democritus, I found you!

DEMOCRITUS: Is that you, Alkimos?

ALKIMOS: I roamed the entire market in search of you until I heard you were down at the harbour.

DEMOCRITUS: Ah, but you know it's here that I do my best thinking. The wind off the ocean inspires me to greater thoughts.

ALKIMOS: Master, have you read my poem?

DEMOCRITUS: I have, my friend.

ALKIMOS: I intend to submit it to the poetry contest. A poem in praise of Democritus is sure to gain distinction. Finally, thank Zeus, I've found the right profession for me. Poetry will be my lot. I'm prepared to write a heroic epic on the war between Titans and the Gods. I shall sing the War of the Centaurs and Prometheus.

DEMOCRITUS: Your commitment is admirable, and although I'm flattered you've written a poem about me, I should note that the oft-repeated phrase—"And the wise Democritus, in answer to his question, said," violates the rule of dactylic hexameter.

ALKIMOS: Master, that's not such a serious offence.

DEMOCRITUS: Yes, but if you remove the "rosy-fingered dawn" or "looking askance upwards" and the other tropes from your poem, not much remains. Show me the heart of your art. Show me the lines deserving of the name *true poetry*.

ALKIMOS: Master, listen! I swear to Zeus, Apollo himself could not have composed a better distich:

> *In your spacious skull, intellect takes flight*
> *Your every thought eternal and bright*

DEMOCRITUS: I'm sorry to say, but it lacks spontaneity.

ALKIMOS: And this masterfully composed hexameter?

> *With a murmur, the gentle wind of Aphrodite gained force.*
> *Scattering the magic, as Eos did the morning mist.*

DEMOCRITUS: Trust me, it's rather insipid. Were those lines really in the poem? Interesting.

ALKIMOS: Oh, master, please listen to my favourite part:

> *What sublimity pours from the pupils of the wise and clever Democritus*
> *The light of a new sunrise greets the ungracious world.*

DEMOCRITUS: The content isn't bad, but the structure is a bit clumsy.

ALKIMOS: Clumsy? This beautiful verse? Oh, if only I could see it.

DEMOCRITUS: One of the greatest virtues a person can have is the ability to see his own mistakes.

ALKIMOS: Oh master, can't you spare me a reassuring word? Your severity dampens my spirits; now all I see are my weaknesses. You force me to concede—poetry is nothing but a vain, childish fantasy—another pitfall of my youth. My poetic inspiration doesn't soar. No, at the most it wallows in the mud; the toxic air of my intellectual swamp deadens my non-existent genius.

DEMOCRITUS: Now, Alkimos, don't be so hard on yourself.

ALKIMOS: Oh master, as you see, my undisciplined mind wanders nonstop, never arriving at shore. Tame it, restrain it. It's not as useless as my body is pitiable, as my legs are crooked, my belly soft, my chest withered. Lead me down the path of rigor; let me be lord of my tangled thoughts.

DEMOCRITUS: Your determination is impressive, Alkimos. You have quickly given up poetry, and now, if I understand, you'd like to be my apprentice and learn to be a philosopher?

ALKIMOS: Master, give me a chance!

DEMOCRITUS: How can I guide a three-wheeled cart down the right path? I'm suited for contemplation not teaching. But I must admit, I've become fond of you. Yes, I think perhaps I'd enjoy helping you exercise your intellect.

ALKIMOS: Master, thank you! I'm going to be a philosopher! Oh thank Zeus, I've found my path!

DEMOCRITUS: I hope so, son.

At the Market

DEMOCRITUS: Come on, Alkimos, let's walk through the market and give ourselves over to the joy of thinking.

ALKIMOS: Yes, master. I'd also like to give myself over to the joy of a crispy, fried flounder and maybe some of that Thracian wine I've got in my leather flask.

DEMOCRITUS: Yes, yes, sometimes we have to lower ourselves and respond to our physical needs.

ALKIMOS: I swear to Zeus, lowering myself certainly feels good.

DEMOCRITUS: Son, if we pay too much attention to our stomachs, we have less time for our thoughts.

ALKIMOS: But Master, as an impassioned poet ... or rather philosopher, I can pay attention to two things at once. I'm anxious for my first lesson.

DEMOCRITUS: All right. If you recall, yesterday we talked about the creation of the world.

ALKIMOS: Yes.

DEMOCRITUS: I explained to you how atoms of all shapes and composition, as they collide, can form larger and larger bodies. These bodies — thanks to the variety of materials they're made of—can also take on countless forms and colours. The infinite assortment results in an infinite number of structures. Indeed, if we look around, we'll discover the unbounded richness of our world. Look at the abundance of sea creatures caught in the fisherman's nets, or the huge variety of birds in the sky. If we count the grass in a meadow and the game in the forest, the number will approach that of the stars.

ALKIMOS: Yes, there are millions of forms; the variations are endless. An inestimable wealth unfolds before me.

DEMOCRITUS: Son, don't let me fool you. Take a closer look at the fishmonger's display or the flowers in a field. What do you see? You've just eaten a flounder with the voracity of a child. But does the fishmonger have any more of the same to sell? If you pick a blade of grass, are there countless more of the same blades in the meadow?

ALKIMOS: Well … not exactly the same.

DEMOCRITUS: That's right. So how do you know if a flounder is a flounder, or a blade of grass is a blade of grass if they're not exactly the same as the others?

ALKIMOS: Because they're similar.

DEMOCRITUS: Yes, indeed. Flounders more closely resemble each other than they do other fish. The same is true of grass and flowers. And of people, too, regardless of how dissimilar we seem. But of course, we have souls made of fire atoms, which distinguish us from all other things.

ALKIMOS: Ah, now I get it. Similar things are more similar to each other than to things that are different. It doesn't take wisdom to understand that.

DEMOCRITUS: No, but we have to travel a long road to understand why that's true. Why do baby birds resemble their parents? Why do only partridges emerge from partridge eggs? Why not larks? If the farmer plants wheat seeds, why does only wheat sprout from the ground and not rye? Why don't grapevines produce figs?

ALKIMOS: Excuse for saying so master, but that seems a bit crazy. I mean, who thinks a fig would grow on a grapevine? Everyone knows—even those of us who aren't very clever—that that's impossible.

DEMOCRITUS: Don't be fooled by what you're accustomed to. Just because we observe a phenomenon every day, and it's obvious to everyone, doesn't mean we don't need to explain it. It's always the easiest questions that are the most difficult to answer. That's because they touch upon the very essence of things—the secrets our minds can't grasp, but yet we believe we know them, because they're so familiar. Our never-ending task is to explore these questions.

ALKIMOS: I'm not objecting, but I don't understand why we have to examine questions of that sort. Why not be satisfied that things happen a certain way because they do. The gods know how to run the world. If Demeter sees a peasant sow wheat, she won't trick him by making rye sprout.

DEMOCRITUS: But I'm not satisfied, son. My goal is to understand things at the atomic level.

ALKIMOS: But how?

DEMOCRITUS: Let's take our thoughts a step further. How can the endless crashing of an endless number of atoms produce similar looking things? We must presume that something prompts some atoms to stick more closely together. Let's imagine that crooked atoms would rather join other crooked ones and not smooth ones. The smooth ones likewise prefer the company of smooth ones. This results in the large variety of things that resemble each other. This is how slithering worms, winged and four-legged creatures, the throngs of slippery fish, and the rank grasses that cover the earth came to exist. After birth, similar creatures mated and reproduced more creatures like themselves, so their great variety remained, the order of their atoms and the pulsating essence of their lives were established at the moment the atoms crashed.

ALKIMOS: That's exactly how I would have put it! I understand the point you're trying to make. Your thoughts are so beautiful I can't resist a sip from my wine flask. Ah, such a heavenly drink!

DEMOCRITUS: Give it here, son. I haven't had such fine wine in a long time. Ah, consuming wine in moderation helps the mind to generate ideas and put them into words. As the pure wine circulates in your veins, you'll be able to imagine the minuscule and the gargantuan. Your thoughts will propel you through the endless oceans. You'll reach the bottom of the bottomless, the limits of the limitless. But Alkimos, one day the moment will come when you'll realise you know nothing, even when you think you know everything. In every era, we think we've solved one or two mysteries, but the unknown is without end, a boundless territory. Still, our task is to conquer as much of this land as we can. I have travelled to Hellas, Thrace, and Persia, and met with many wise men, but the wisest were always those who acknowledged they knew nothing. Perhaps over the coming centuries, thousands of wise philosophers may turn this nothing into something. My most fervent desire is to be present at the time the great secrets are revealed, and we'll be able to say that our knowledge, as rudimentary as it is, is something rather than nothing. Now give me some wine. Let's drink up and sleep well.

ALKIMOS: My flask was full to the brim, but now only a few drops remain. Allow me to finish it, and then let's walk to the sandy shore and rest before nightfall.

DEMOCRITUS: Come, Alkimos, lead me there. My head has grown heavy. I wouldn't mind a short nap. I see my atoms clearly, the world I've created. They crash, they dissolve, they join, in an endless current. Oh, if I had the strength to watch them, to examine their special world, to understand the infinite secrets of their essence, their forms, their functions, and their roles. It's my only dream, my stubborn, burning desire.

ALKIMOS: We can rest here, master. The breeze is gentle, the sea calm.

DEMOCRITUS: Give me your hand. Help me to lie down. I can put my head on the soft sand. No, give me the flask. I'll put it under my head; there that's better.

ALKIMOS: Drat the sand. It's getting in my ears …

Part One

The Mystery (and Sperm) of Life's Origins

Deus ex Machina

EXTON: Ah, Exton, you've done well. They're making their way. Now, let's see. A little more chronic turbulence and here they are. The old codger, too.

DEMOCRITUS: Alkimos, Alkimos, where are you? I can't see a thing in this thick fog.

ALKIMOS: Here I am, master, over here!

DEMOCRITUS: I don't know what's happened to us. I've never been swept into such a vortex, not even when I visited the Oracle of Delphi. My head is spinning.

ALKIMOS: Hold on to me. I see a gate over there. Let's see what lies beyond it.

DEMOCRITUS: I don't understand. Maybe it was the wine. Alkimos, you'll have to lead the way.

ALKIMOS: Here we are, but the gate doesn't want to open. I fear the Heavens are playing tricks on us. I'll have to force it. Kronides! Master, look! I've never seen the likes of this! Is this Olympus? Have we arrived at the home of the Cloud-Gatherer? Such lustre, such brilliance!

DEMOCRITUS: I'm blinded. Alkimos, I see nothing but glitter!

EXTON: Greetings.

3

ALKIMOS: Who's speaking? Oh dear Zeus, is it the Cloud Gatherer himself? How my heart pounds!

DEMOCRITUS: Alkimos, you heard something too? There's someone here!

EXTON: Have no fear, my friends. Come in, come in!

ALKIMOS: Oh, the immortal Zeus! We'll sacrifice one hundred oxen, but please spare our lives. We didn't mean to intrude. We drank too much wine and lost our way.

EXTON: You have nothing to fear. I'm not Zeus. I'm only a mortal being such as yourselves. My name is Exton. You were brought here because of your great desire for knowledge.

DEMOCRITUS: Where are we?

EXTON: Alexandria. This is the Deus ex Machina Centre for the Teleportation and Training of Philosophers.

ALKIMOS: Master, did you hear that?

DEMOCRITUS: I don't understand, son.

EXTON: You were brought here because of your deep and enduring desire for knowledge.

DEMOCRITUS: By whom?

EXTON: Let's just say the gods. I'm here to acquaint you with all the revolutionary discoveries that have been made about the living world during the past centuries.

ALKIMOS: What are you talking about? The past centuries? But my master knows everything about the earlier developments of philosophy.

DEMOCRITUS: This is true, stranger. What revolutions are you talking about? Some new nonsense devised by the Athenian philosophers? My master Leucippus and I have developed the most complete theory about the origin and properties of all living things. Of course in Athens they don't read my treatises. Such abominable insolence.

EXTON: Perhaps it's difficult for you to understand, but with the help of our newly developed Deus ex Machina you've been sleeping for 2400 years. We are now in the year 2015, after Christ. Since you fell asleep in the sands of Abdera, the Earth has circled the Sun 2400 times.

DEMOCRITUS: What gibberish. The only thing circling is the stupidity in your mind.

ALKIMOS: Certainly not the Earth.

DEMOCRITUS: And who is Christ? Is he that troublemaker—that Athenian philosopher who refuses to acknowledge my work?

EXTON: I see you don't understand. You're in Alexandria. About three thousand years have passed since the Trojan War. You have a unique opportunity to witness the awe-inspiring changes that have taken place in the study of the living world over the centuries. I'll be your guide, but you'll need to work hard to understand. So what do you say—would you like to undertake this journey? If not, I'll return you to the sands of Abdera and provide you with a full flask of wine to compensate for your everlasting ignorance. Now decide, because there are several Athenians on the waiting list.

ALKIMOS: Master, let's leave this place. We're being fooled with. We shouldn't believe a word he says. He must be a god, or demigod—his words are full of scorn. Let's take our full flask of wine and depart, quickly!

DEMOCRITUS: Alkimos, there's no point. You cannot escape the power of the gods. They'll find us, no matter where we hide. Let's try instead to overcome our suspicions and have a look around in this strange, but interesting world. We might even gather a few crumbs of this knowledge we crave. Yes, I'm going to stay, and I advise you do the same.

ALKIMOS: Of course, master. I have no other desire than to learn. But I'm not willing to sacrifice my life for this knowledge. I can do without crossing the River Styx for a while.

EXTON: So you've decided? If you don't want to stay, grab your flask and leave quickly through the gate. Now let's see ... the next name on the list is Diogenes.

DEMOCRITUS: Who? That Sinopean fool? You mean he'll gain that knowledge instead of us? Alkimos, we can't allow that!

EXTON: Ah, I've persuaded you. My assistant Hermes will spread the news: the visitors have arrived! Now my friends, bathe and get dressed. Hermes, bring them soap and a change of clothes! Oh, and a lice comb!

DEMOCRITUS: Diogenes, hah. That worm of a simpleton. If I ever tell the others about this in Abdera …

ALKIMOS: Master, the water is foaming!

DEMOCRITUS: Ah, so it is. How odd! But we must accept what the gods provide.

ALKIMOS: And look at these clothes. They're splendid!

EXTON: Are you ready? Can we go?

DEMOCRITUS: Wait a moment, Exton. You have to understand, we're overwhelmed. Our loins are covered in clothes with metal teeth; we're surrounded by materials and smells unknown to us; in the glass on the wall, we see our own shaggy faces. The image is so clear, so unlike the reflections we see in a bronze plate or water. I'm afraid the Deus ex Machina has blinded us with its brilliance and left us a bit worn out.

ALKIMOS: Yes, my master is right. Please understand, dear Exton, at the moment we're too exhausted to face the knowledge of your age. Maybe you need to show us first how the distant future can satisfy our hunger and thirst. Nothing special, of course, just a pitcher of honey wine, roasted capon, and some partridge eggs, or some other simple food. After all, we haven't eaten a morsel for almost three thousand years.

EXTON: I'll see what I can do.

ALKIMOS: Thank you for your generosity. We can see you have a noble heart.

EXTON: I'll be right back.

ALKIMOS: Master, if we're staying, let's have a look around. What is knowledge good for if we don't know anything about the foods, the people, and the cities of this future world? Let's see what the women of Alexandria are like three thousand years after the Trojan War! When we return home, you'll need to have good stories to tell at the marketplace. Trust me, nobody's going to care about your profound knowledge if it turns out you had no dalliances with a woman not yet born.

DEMOCRITUS: Son, quiet. What guides wisdom is not what guides our human companions. Anyone can admire food and women. We were brought here because of our powerful desire for knowledge, and we must not diverge from that path if we are to experience this miracle of the gods. People and food don't change, but our knowledge does, increasing with the passage of time. We need to find out what the philosophers have discovered about the living world. What makes living things move? What gives them their unique shapes and their ability to survive and reproduce? What secrets reside in a plucked flower or a ripe fruit? Yes, Alkimos, we must remain true to our task. This is the only way we can learn about the wonders of life.

ALKIMOS: I understand, master. We must forget about Earthly pleasures and imbibe only knowledge. Philosophy may not be able to alleviate out hunger, but it's good for something.

EXTON: Here I am, my famished guests. I'm afraid it's a modest meal, but it should fill you up.

DEMOCRITUS: Thank you, dear Exton. This will keep our hunger pangs at bay.

ALKIMOS: Where are the partridge eggs and wine? What kind of mouldy pie is this?

The World Egg

DEMOCRITUS: Dear Exton, thank you for the fine food. But now, in keeping with our true mission, let us turn to wisdom. I believe your knowledge is so vast and varied that we shall only absorb it if we approach it slowly and with the appropriate humility. One day will scarcely be enough for us to comprehend a thousand years of knowledge.

EXTON: Yes, that's true. But you won't be here for just one day.

ALKIMOS: Master, you mean we'll have more time to uncover the mysteries of life?

DEMOCRITUS: It appears so. How fortunate for us. You see, son, life is a complicated tangle. All around us it clatters. However, I worked out a coherent theory to explain it. All life was created from the clustering of atoms with like properties; it is a consequence of the atoms' most inner traits.

ALKIMOS: Yes, master. Crooked with the crooked and smooth with the smooth.

EXTON: Indeed, Democritus' mechanistic atom theory provided a far better explanation than the alternatives, such as the ancient myth of the world egg. Of course, the idea of the world egg has its beauty as well. Let's read what Aristophanes had to say about the origins of the world in his comedy *The Birds*:

> *In the beginning there existed only Chaos, Night, Black Erebus and Dreary Tartarus: there was no Earth, no Air, no Sky. It was in the boundless womb of Erebus that the first egg was laid by black-winged Night; and from this egg, in due season, sprang Eros the deeply-desired, Eros the bright, the golden-winged. And it was he, mingling in Tartarus with murky Chaos, who begat our race and hatched us out and led us up to the light.*

ALKIMOS: The egg emerged from nothing, and from it hatched the bird, life, love.

EXTON: Indeed, we'll need to talk a lot about whether life sprang from non-living or living material. The first is known as *abiogenesis* and the other reproduction.

DEMOCRITUS: The world egg was produced through abiogenesis, but it created further lives through reproduction.

ALKIMOS: In other words, it laid eggs.

DEMOCRITUS: Probably.

ALKIMOS:

> *With a grunt the Cosmos spontaneously laid a speckled egg,*
> *From which hatched Life, light radiating its heat.*

DEMOCRITUS: With the mixing of the atoms in the eggs, life is born again and again. From a formless mass sprang a feeling-moving creature.

EXTON: This is the process by which an organism forms. It's the subject of embryology and developmental biology. I'll have a lot to say about that.

ALKIMOS: Exton, if we solve all these problems we will have uncovered the key that unlocks the mystery of life.

EXTON: Perhaps.

ALKIMOS: The origins of the world egg.

EXTON: The Greek pioneers in embryology probably had similar dreams. The first detailed embryological observations were made by Hippocrates and his students on the island of Kos. Often we don't know if the books produced by the Hippocratic School were written by the master or his students. Of course, this doesn't diminish the value of this wonderful institution. The Hippocratic School formulated detailed mechanistic explanations of embryonic development. For example, here is a quote by Hippocrates, or one of his students:

> [*The embryo*] *is set in motion, being humid, by fire, and thus it extracts its nourishment from the food and breath introduced into the mother. First of all this attraction is the same throughout because the body is porous but by the motion and the fire it dries up and solidifies; as it solidifies, a dense outer crust is formed, and then the fire inside cannot any more draw in sufficient nourishment and does not expel the air because of the density of the surrounding surface. It therefore consumes the interior humidity. In this way parts naturally solid being up to a point hard and dry are not consumed to feed the fire but fortify and condense themselves the more the humidity disappears—these are called bones and nerves.*

DEMOCRITUS: That corresponds to my beliefs. For some time I have stated that the external shape of the embryo develops before the internal organs form. This is the *crust* according to Hippocrates.

ALKIMOS: And what do you think, master, about the fire?

DEMOCRITUS: Hm, the fire … I suppose the fire animates the atoms.

ALKIMOS: But what is this fire, master? What is the flame that burns in all living things and is extinguished at the moment of death?

DEMOCRITUS: I don't know, son. That's something we'll have to investigate.

EXTON: The spark of life or *vis vitalis*. The core idea of vitalism.

ALKIMOS: This fire can only come from Eros. The flames of Eros send us into the arms of a woman, so we may light the fire of a new life.

DEMOCRITUS: But it was Hermes' flame and seed that gave life to Eros in Aphrodite's womb.

ALKIMOS: Whether the flame comes from Eros or not, the union of a man and woman is necessary for birth. It's no different for animals. Otherwise why would there be ewes and rams?

EXTON: They've thought that since the world came into existence.

ALKIMOS: Yes, but what unites with what?

EXTON: According to the Hippocratic School the male seed unites with the female's fluids. After mating the male and female seeds play an equal role in forming their offspring. This is the doctrine of "two seeds." However, according to Aristotle, one of the greatest philosophers and biologists of ancient times, the male seed doesn't unite with female fluids, it doesn't contribute anything material. It only directs, shapes and sets in motion the foetus. The basic material of the foetus is the egg, or in the case of people, a woman's menstrual blood. These two views competed with each other for centuries. We should keep this problem in mind, since we'll encounter it again many times.

ALKIMOS: Exton, you won't be disappointed in my memory.

EXTON: According to Aristotle the developing foetus, under the direction of the seed, doesn't inherit properties, but only the ability to create them. Its properties, the various organs, and shapes develop gradually from a formless mass. The heart, lungs, eyes and other organs do not appear simultaneously, but in succession. During this change, a given state is the starting point for the creations of the next, more differentiated state. This gradual development of the organism is known as *epigenesis*.

ALKIMOS: *Epi*, as in upon, and *genesis*, as in creation.

EXTON: Aristotle based this on his own observations. He studied the development of numerous animals. He collected his samples in the sands of the Pyrrha lagoon. He received his zoological information from fishermen of Mytilene. He cracked open chicken eggs at various stages of incubation. He also dissected mammals and creepy crawlers. The representatives of the Hippocratic School also made observations of chicken embryos, but they were not nearly as thorough as Aristotle. They believed every organ in the embryo started to form and grow at the same time. We may consider their conclusions the forerunners of the doctrine of *preformation*.

ALKIMOS: *Pre* meaning prior; so "prior formation."

EXTON: Exactly. According to this theory, every part of the embryo is present from the start, and the development of the organism simply means growing larger. The competition between these two theories is the history of embryology.

DEMOCRITUS: This is incredibly exciting!

EXTON: We shall devote plenty of time to it.

ALKIMOS: And according to the great scientists, after the union, why does the child resemble his parents?

EXTON: Hippocrates thought that during procreation, tiny particles of every body part flow into the parents' fluids. Through the melding of these particles, an offspring is created that bears the mark of both parents. This is known as *pangenesis*.

ALKIMOS: *Pan* meaning everything, *genesis* meaning creation. So creation from everything. In other words, the entire body contributes tiny particles containing the traits of the parents.

DEMOCRITUS: But there's a problem. Hippocrates made a mistake.

ALKIMOS: Master!

DEMOCRITUS: According to my atom theory, the pangenetic particles, being indestructible and indivisible, cannot fuse in the offspring. At the most they can settle beside each other. Thus the offspring can receive the same number of atoms from each parent. This explains why a child resembles both of its parents.

EXTON: This leads us to the theory of *particulate inheritance* or inheritance of particles. Aristotle wrote an extensive criticism of Hippocrates' theory of inheritance. He demonstrated that offspring sometimes resemble distant relatives, and don't inherit the traits of their parents. Therefore, these traits cannot originate with the fluids that collect during procreation. What's more Aristotle believed that if the male and female seeds had an equal role in fertilisation, then two embryos would develop. This dispute continued until the arrival of Charles Darwin and Gregor Mendel. As we follow this debate, we'll attain a clearer understanding of the laws of inheritance.

ALKIMOS: Do you hear that, master? No matter what I ask, Exton can provide us with an answer.

DEMOCRITUS: I'm more interested in the nature of life atoms.

EXTON: I see you are starting to formulate the essential questions. Without understanding atoms, we cannot understand the development of organisms, inheritance, and biological diversity. We need to devote all our energy to investigating the secrets of atoms.

ALKIMOS: There's hope!

DEMOCRITUS: Let's see what the wise men of the generations following Aristotle had to say about these questions.

EXTON: Interestingly, for one thousand eight hundred years no one came up with anything new. With Aristotle, biology became an independent science. He created something lasting in the area of comparative embryology and in the study of biological diversity. He was the first expert curator of the living world. For someone to outdo Aristotle in terms of his significance and influence in the study of the Earth's diversity, we had to wait two thousand two hundred years, until Darwin. His observations on embryology were also unsurpassed for one thousand eight hundred years.

ALKIMOS: But why?

EXTON: For the most part, because no one made any observations until the Renaissance. In 1554, a city surgeon called Jacob Rüff, relying on the work of Aristotle and the prominent 2nd century Greek–Roman surgeon and philosopher Galen, sketched the development of the human foetus, as you see here:

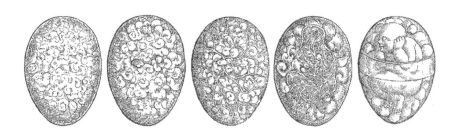

Although his imaginative drawings relied partly on his dissections, they probably have more artistic value than scientific.

DEMOCRITUS: How is that possible, Exton? Aristotle showed us the path. All we had to do was continue on it.

EXTON: Surprisingly, we can partly blame Aristotle himself. Never before has such a comprehensive work been written, and there didn't seem to be anything to add to it. Aristotle's theories gradually solidified, and their author garnered ever greater respect. But respect for authority is the death of science. In later periods, instead of continuing or questioning Aristotle's work, scholars referred to him as if he were a certain, infallible source. As a result, his mistakes were never discovered and nothing new was developed.

ALKIMOS: See Master, that's why I never allow the power of your authority to persuade me.

DEMOCRITUS: That's right, son. Only good reasoning matters.

EXTON: I should mention another thing interesting about Aristotle: his theory of the "final cause." This theory, which was reinterpreted by Plato, also contributed somewhat to the thousand-year delay. Aristotle used four causes to explain the phenomena of nature. For the chicken embryo to develop it needs the "moving" cause—the warmth of the brooding hen. But that warmth would not give life to anything, which is why the "material cause" is also needed, the egg. This is the guarantee that the egg, affected by the warmth, will hatch. But these two causes are only enough for something to hatch from the egg, but not necessarily a chicken. For a chicken to hatch, you need the "final cause." This cause determines that the shape of a hen or rooster guides the development of the egg, assuring that it takes the shape of its parents. The fourth, "formal cause," ensures that a particular chicken will hatch in a particular chicken yard.

ALKIMOS: Well, Aristotle really wanted to make sure it was a chicken that developed from a chicken egg, and not a quail.

EXTON: In the Middle Ages, the "final cause" meant the same things as God. Slowly the other causes grew obscure, as the "final cause" became the one and only explanation. Creatures move because God moves them.

The existence of the "final mover" explains every movement, and thus it's unnecessary to muse on the nature of motion. Aristotle was not around to object. The revival of the study of natural science occurred during the 17th century with Galileo Galilei and Isaac Newton. They developed a mechanistic worldview, which was necessary for scientists to pick up where Aristotle left off. You see, Newton's laws of motion demonstrated that motion exists even without an apparent exterior propeller. This is the basis of the mechanistic world view.

ALKIMOS: So we've returned to Democritus.

DEMOCRITUS: But with a two thousand-year break.

EXTON: Well, I wouldn't use the word *break*. During this time architecture, painting, and sculpture thrived. The masterpieces of Romanesque and Gothic art could only have been created with the kind of spiritual focus of the Middle Ages. During this period ancient knowledge was transmitted, although largely in the Arabic world—thanks to the flourishing manuscript culture. But if our interest is natural science, it's true, we haven't missed much if Galilei has just arrived. Well … there were a lot of original, 13th century contributions to mechanics that Galileo could rely on. And of course the work of Copernicus in 1543… But we will concentrate on the living world.

DEMOCRITUS: I can hardly wait, Exton, to learn about the work of later scientists.

EXTON: To do so, we'll need to travel a lot, to many places. But our well-designed Deus ex Machina can fly us from Alexandria to wherever these preeminent scholars reside. Then you will be able to question them in person.

ALKIMOS: My Zeus! What an adventure! Master, I'm so glad we stayed.

DEMOCRITUS: I told you, son …

Spontaneous Generation and Meat Broth—*Lazzaro Spallanzani*

EXTON: Let's get started, my friends! Let's see what other theories, after the world egg, were developed to explain the origins of life.

ALKIMOS: Master, you've told me a lot about this …

DEMOCRITUS: Yes, son. My atomic theory deals with genesis and renewal. In my view, all different kinds of atoms have all different kinds of relationships in the darkness of the cosmos. Life is constantly being created, as it is in forest soil, in rich muddy humus, or in the foam of a turbulent sea. The morass produces maggots and mosquitoes. From the decaying forest litter arise black flies and beetles; the endless depths of the sea give birth to snails and shells and thousands of other creatures. This eternal mixing is the source of life, but its sustenance depends on the internal makeup of these beings. Nothing can disappear completely. Birth and Death are the continual coming together and breaking apart of atoms. Renewal is a part of decay. The eternal mixing is the essence of this game played by indivisible atoms.

ALKIMOS: I share my master's opinion, but what would happen if we compared it to the knowledge of this future world? Exton, how does this idea of atoms mixing and combining hold up?

EXTON: We'll get to atoms soon enough. First, let's address the theory that maggots arise from mud and black flies from forest litter. What are scientists' thoughts on this two thousand years after Aristotle and Democritus? I think it's best if we find Doctor Lazzaro Spallanzani.

ALKIMOS: Excellent! I don't know who Spallanzani is, but I assume he works here in Alexandria and is a great maggot expert.

EXTON: No, Lazzaro Spallanzani doesn't work in Alexandria. He worked mostly in Reggio, Modena, and Pavia in the 18th century. He's famous for a lot of amazing work, including ingenious experiments that disproved the theory of spontaneous generation.

ALKIMOS: Spontaneous generation? Italy? I have a feeling we have to take another ride in that horrid contraption.

EXTON: That's right! Hermes, prepare the Deus ex Machine! Direction Modena, 1768!

HERMES: Right away, Sir Extonos.

ALKIMOS: But it's cramped and dark in there.

DEMOCRITUS: Squeeze in. Come now, make room, son!

ALKIMOS: Ohhh, here we go …

SPALLANZANI: *Ecco, un po' d'acqua e prosciutto ed è pronto*! My experiments have resolved everything! Needham and Buffon have no idea what they're talking about. The theory of spontaneous generation is done with! *La teoria della generazione spontanea è stata abbattuta finalmente*!

EXTON: Buon giorno, or rather my greatest respects, Signor Spallanzani. May we have a moment of your time in your laboratory? *Non vi disturbiamo*? We heard you speaking about spontaneous generation. We'd be delighted if you'd share with us your ideas.

SPALLANZANI: *D'accordo*. Please come into my modest laboratory. Although I have very little equipment, I've produced ideas and results that have done much to enhance my reputation. I've just begun an important test. Countless fools have believed and disseminated ridiculous nonsense about spontaneous generation, but this and my earlier experiments show the falseness of their claims. I'm confident that science shall illuminate the minds of the deceived.

ALKIMOS: So many shiny flasks and tubes—how overwhelming! Why have scientists taken their work indoors, into such a clutter? What happened to the great colonnades and the reverence they inspired? A person can't think straight in this mess of equipment. Ah, but what a tantalizing smell! Meat broth? Master Spallanzani, if you don't mind, I'd love a taste …

SPALLANZANI: *Corpo di bacco*, get your hands off my test substance! *Che insulto*!

DEMOCRITUS: Alkimos, stop! Please forgive us, Master Spallanzani. Could you explain to us the purpose of this brew? What problems are you trying to solve? You've used words that we barely understand. Spontaneous generation, experiments, test substance.

SPALLANZANI: *Che*? I thought you were scientists. How bewildering — *Massimo stupor*! What university are you from?

EXTON: Signor Spallanzani, I'm afraid there's been a misunderstanding. Please, don't listen to my apprentices. I'm a Professor at the University of Mantua. My name is Francesco di Estoni. I'm a fervent opponent of the preposterous theory of spontaneous generation, especially since I had the opportunity to read your work. I've brought my pupils here so they may stand in the presence of your greatness and hear directly from you why the theory is so outrageous.

SPALLANZANI: Ah, now I see, *ho capito*. Your pupils are a bit strange, especially the one with the crooked nose. *E' indifferent. La generazione spontanea ...*

EXTON: We're listening, professor.

SPALLANZANI: I have little hope of convincing the lowly and the uneducated masses that the theory is absurd. After all, who isn't persuaded by a tried-and-tested recipe? Put some sweaty underwear and a handful of chaff in an uncovered jug, and wait twenty-one days. On the twenty-first day, mice pop out of the jug. It's that easy to create life. Similarly, you can put out a piece of meat for a day or a bundle of hay in a basin of water. In the meat you'll find maggots and in the basin you'll find critters you can only see with a magnifying glass.

ALKIMOS: Oh yes, I've observed that, especially the hatching of maggots and mice in the gutter around the Abdera market. It seems my master is correct—to use his words—all kinds of relationships between all kinds of atoms continually give rise to living creatures.

EXTON: And that's not all. The Swiss doctor and alchemist Paracelsus was not satisfied with the creation of maggots and mice. In the first half of the 16th century he wrote a recipe that would help to artificially create people, *homunculi*. To do this you need male seed, a pumpkin, a horse's stomach, and blood. Paracelsus believed that if this concoction worked, the *homunculi* would transform into dwarves or giants. But let's hear Signor Spallanzani. Perhaps he should first tell us about the work of Francesco Redi, who called Paracelsus a charlatan.

SPALLANZANI: Redi? Yes, he was an excellent poet. The dithyrambs of *Bacco in Toscana* are unrivalled ...

EXTON: Yes, Redi was a poet. But he was also a naturalist. Signor Spallanzani, haven't you read Redi's work, *Esperienze intorno alla generazione degl'insetti*, published in 1688?

SPALLANZANI: Of course! Why, that was the work that inspired me to experiment with spontaneous generation.

EXTON: Let me tell my pupils about those remarkable experiments!

SPALLANZANI: Please do.

EXTON: It was a widely-held view that maggots spontaneously appeared in rotten meat. Redi, however, believed that maggots emerged from the eggs laid by flies landing on the meat. In order to prove this, he took several pieces of meat and placed them in different jars. Some jars were plugged, others were covered in thin gauze, and a third set were left uncovered. According to the theory of spontaneous generation the rotting of the meat is what produced the maggots; therefore, maggots would appear in all the meat, even in those pieces in the plugged jars. Redi, however, contended that maggots would appear only in the jars accessible to flies.

DEMOCRITUS: Ingenious! This is not something you can conclude using reason alone. You need to experience how things actually work by carrying out a certain procedure. Is that what you mean by experiment?

EXTON: Yes, something like that. Redi left the jars for a couple of days. Flies came and went, buzzing loudly, and the *vis vitalis* may have hummed in the flasks, too. In the end, what happened?

ALKIMOS: Obviously, all the jars were full of maggots, since the meat itself carried the maggot atoms.

EXTON: No. Maggots did not appear in the plugged jars or the ones covered in gauze. Only the open jars, which the flies could enter to lay their eggs, contained maggots. Without flies, there are no maggots.

DEMOCRITUS: Hah … astounding. Decades of contemplation would not have helped me to achieve such results. Is that the secret of science? We need to ask nature itself and not our shallow grey matter? Perhaps I should remain silent and reject my self-contained theories.

EXTON: It's enough for now if you question just the part of your atomic theory that deals with spontaneous generation. Even Redi, after his experiment, was not prepared to completely reject the theory of spontaneous generation. He still thought it could take place under certain circumstances. His meat experiments, however, supported the statement made by the famous English biologist William Harvey: *omne vivum ex ovo*, all life is from an egg.

SPALLANZANI: Who wouldn't be persuaded by a poet's meat experiment? *Poeta con carne*. Anyone can believe maggots come from flies.

ALKIMOS: And what about mice? And mosquitoes?

SPALLANZANI: Ah, my friend, that's not such a big problem. It's easy to make similar experiments. But what about the microscopic *animalcula*? The tiny critters that abound in puddles and mud patches?

EXTON: Master Spallanzani is of course referring to the minuscule creatures Antonie Van Leeuwenhoek discovered and named *animalcula*. Van Leeuwenhoek developed the best microscope of his day. The following drawings are based on his observation:

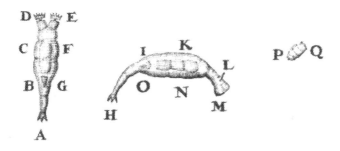

ALKIMOS: What did you call it? *Mikros*—that means small. *Skopein*—that means to look at. In other words: looking at what's small.

EXTON: Even if a microscope can't help us to see atoms, it can still vastly increase the power of our eyes to see. The wonderful invention gave the study of biology a huge push forward. Leeuwenhoek's microscope has just one small lens, unlike Spallanzani's. Leeuwenhoek's entire microscope would fit in the palm of my hand.

DEMOCRITUS: Tell us more about this contraption and how it works.

SPALLANZANI: *Ecco*, my friends. I see that you have not come far in your study of science. However, you will be rewarded for your efforts. Sooner or later the gates of your mind will open and be flooded with the light of knowledge. *L'illuminazione della scienza*. Now, let's have a look at this indispensable tool of science. We would call it diabolical, if we didn't know this was the work of just two diligent glass engravers. Its magic lies in the finely crafted lenses. Because of the exact curvature and juxtaposition of the lenses, whatever is placed under them appears larger to our eyes. *Guardatelo*! Have a look at a drop of pond water!

ALKIMOS: Ah ... what huge worms wiggle. This tube is full of leeches! Master, have a look. A snake with two enormous wheels on its head. What an appalling sight—I can't look anymore!

DEMOCRITUS: You fool, there's no snake there—it's just a drop of water. Now, let me have a look. Oh my, a nest of vipers! In a drop of water! Is Zeus playing a trick on us? Ah, but wait a moment. I remember something from my younger days: after a rain shower, I often observed that a large drop of water seemed to enlarge the part of the leaf beneath it. I can only conclude that this is the same phenomenon.

EXTON: It's just as you say. But instead of a drop of water the lenses do the job of enlarging what's under them. What your eyes see is the real world magnified a hundred fold.

DEMOCRITUS: Astounding. I always knew that life had a dimension imperceptible to our senses. But I never thought there would come a time when we could step into that dimension. With the help of such an ingenious contraption we can expand the limits of what we can know about the world. This is the second great weapon of science after experimentation.

SPALLANZANI: *Ecco*, my friend. Although you seem to know little of science, you quickly grasp the significance of things; clearly you have a sharp mind. But I should remind you that you came to hear about my latest experiments. In all your enthusiasm you've barely let me speak! Don't worry; I'm not angry since that kind of eagerness is what opens minds. But, *scusate*, I'd like to return to a discussion of spontaneous generation.

ALKIMOS: Please go on!

SPALLANZANI: Where do the *animalcula* come from if not the puddle water? *Ecco*, it's that simple. At least, John Turberville Needham claimed it was. He boiled some meat broth, sealed it, and a few days later found that it was teeming with tiny creatures. But his procedure was faulty: I'm certain he only boiled the water for a few minutes and perhaps sealed the jars in a careless manner. Sadly, though, a badly-conducted experiment can do much to expand the range of nonsense, and a well-conducted, but perhaps less spectacular experiment, has little power to counteract it. So you see, Needham's shoddy work helped to spread foolish notions. When I repeated his experiment I did not find a single living specimen, and I let my jars sit for weeks.

ALKIMOS: Now I see why you need so much meat broth.

SPALLANZANI: *Finalmente*, it's important to note that my experiment's success is due to my very careful work. I boiled the broth for at least an hour, then sucked the air from the jars and heated them over a flame. This prevented organisms in the air from contaminating the brew.

DEMOCRITUS: So life cannot emerge from nothing? There is no amount of stirring that can give rise to a tiny worm? Either life has existed since time immemorial or it has been created by divine intervention.

ALKIMOS: Master, how can we know what might emerge from the soup if we let it sit for thousands of years? Maybe it takes that much time. Perhaps once the world was nothing but soup and only after several millennia did enough of whatever bears life manage to accumulate in this primeval slop?

SPALLANZANI: I don't know either. I can set aside a few of my jars for posterity. Someone else might uncover the secrets of life's origins.

ALKIMOS: Perhaps we can heat up whatever Master Spallanzani doesn't leave for posterity. A little soup would hit the spot! After all, it hasn't got any maggots in it.

SPALLANZANI: It's time to continue my work. Let's heat up my alembic and test tubes. I'll rekindle the fire.

DEMOCRITUS: The millennia may blend the tissues of life, if we had endless soil. It's time that is key. If only I could wait.

SPALLANZANI: Boil already! I need a larger flame. Bring the bellows! Look at it froth!

ALKIMOS: Master, look at the glow! Isn't the world of experiments wonderful?

SPALLANZANI: Quickly! The tongs! I need to fix this andiron!

EXTON: Come, the test tubes are about to explode!

ALKIMOS: So much power contained in the glass! And the smell of soup!

EXTON: Hurry! The fire has spilled from the stove! The Deus ex Machina is about to catch fire! Quickly, inside! It's time to go!

DEMOCRITUS: Perhaps the boiling, the uncontrolled seething of the brew, and the endless heat will give life to this primeval soup. I don't dare think about it …

ALKIMOS: Exton, wait! Let's at least take one jar with us. Quick a rag! It's hot!

EXTON: Master Spallanzani, you must leave you laboratory!

SPALLANZANI: *Dio mio!* My equipment! The building is in flames! Where are my notes? Where is my offer of a professorship in Pavia signed by Maria Theresa? Heavens!

ALKIMOS: Master Spallanzani, thank you for the soup. Oh, here we go again …

Preformation or Epigenesis?

ALKIMOS: Master, master! Are you awake?

DEMOCRITUS: Yes, I'm just a bit dizzy. The trip has really worn me out.

ALKIMOS: Me too, but it's worth it. We've learned so much. Who would have thought that life doesn't just arise from nothing, that it actually takes life to create life?

EXTON: The truth is that even after Spallanzani's experiments, many people still refused to accept this. They remained steadfast believers in abiogenesis, or spontaneous generation. They claimed that because Spallanzani completely removed the air from his flasks, all he was proving was that spontaneous generation cannot occur in an airless environment. It took almost one hundred years before the theory of abiogenesis was successfully refuted—at least to the satisfaction of those in scientific circles. The French Academy of Sciences offered a prize to the scientist who could give a definitive answer to this question of the origin of life. The winner was the young Louis Pasteur. The experiments he carried out in 1859 were similar to Spallanzani's. The main difference was that instead of plugging or sealing the opening of the glass, Pasteur extended it into a long tube, creating an S shape, as you can see in this figure.

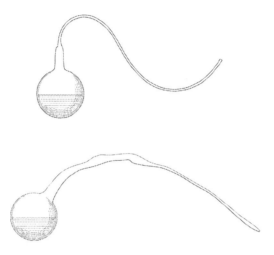

The air could reach the soup Pasteur had placed in the flask through the so-called swan-neck, but the microorganisms floating in the air were trapped in the S-shaped curve. As Pasteur expected the soup remained sterile. But if he slightly tipped the glass and the contents spilled into the neck where dust and other particles had settled, within a short time rapid growth of microorganisms took place. In one fell swoop he both brought down the theory of abiogenesis and proved that the air is full of microorganisms.

ALKIMOS: Oh dear Zeus! I don't dare take a breath. Such disgusting little creatures are swarming about in this seemingly clean air? What a swamp!

DEMOCRITUS: All right, Alkimos, take it easy. There's an interesting side to this: if every breath is full of life, then the seeds of life can make their way anywhere and settle. There isn't a desert in the world that's not teeming with life. And if life can be found anywhere, then life can arise anywhere, but we must accept that life only arises from other living matter. This law of biogenesis may seem in one respect to limit our thinking, but it can also inspire us to ask different questions. If only life can give rise to life, then we now have to ask, how?

ALKIMOS: But Hippocrates already said that it's through the fusion of seeds.

EXTON: Yes, but once again through careful observations and experiments we have managed to move beyond this point.

ALKIMOS: By experimenting with seeds?

EXTON: Yes, and eggs. This is what William Harvey did. He is best known for his detailed descriptions of circulation and the workings of the heart, but it's his book *Exercitationes de Generatione Animalium,* published in 1651, that interests us here.

ALKIMOS: We've already heard of him! He's the one who said *omne vivum ex ovo,* or all life from eggs. I'm sure this is true of poultry, but who has ever seen a woman lay an egg!

EXTON: Alkimos, Harvey didn't state this without any kind of proof; it was based on a great deal of observations. He had little problem discovering the presence of eggs or ova in fish, amphibians, reptiles, birds and even insects. But it's true that mammals did present a greater challenge. As court physician he had access to the royal grounds, and thus used deer as the subject of his studies. Although he didn't find anything in the reproductive organs of does after mating, a couple of weeks later he found a small sack full of liquid covered in a membrane. He thought he'd found the ovum. The truth is that Harvey had not found the ovum, but rather the developing embryo in the blastocyst stage. The mammalian ovum was first described by Karl Ernst von Baer only in 1827. But let's not jump ahead! Instead let's see what Harvey had to say about bird eggs: "the egg is none other than a transitional phase between parents and offspring. A transition between those who have existed and those who will exist. The egg is the stage from which all hens and cocks arise, and the stage which it is their purpose in life to achieve. Nature has created them for this aim only."

ALKIMOS: How poetic ....

EXTON: Harvey also made detailed observations of the chicken embryo developing in the egg. These examinations led Harvey to conclude, like Aristotle, that the parts of an embryo developed gradually from a shapeless mass—that is by epigenesis. Marcello Malpighi, a professor of medicine from Bologna, used a microscope to carry out his detailed studies of the development of a chicken embryo, publishing the results in 1673. This is a picture from his book *De ovo incubato*.

Malphighi was also deeply convinced of the fundamental importance of the egg.

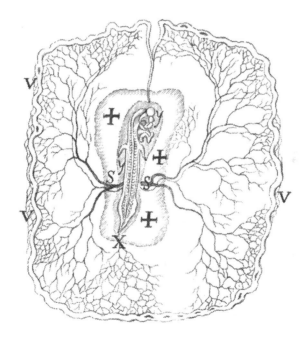

ALKIMOS: I understand all this, but if the eggs are so important, then what role do the males play?

EXTON: This wasn't clear for a long time. Many thought that the male seed was ethereal rather than material in nature—a spirit, vapour, aroma or ray, which penetrated the ovum and stimulated and guided its development. Nehemiah Grew, an English botanist and one of the founders of plant anatomy, wrote that the semen was 'only the vehicle of a certain highly ethereal and volatile spirit, which makes vital contact with the site of conception, that is, the female egg.'

ALKIMOS: That's all they thought the male's role was? Why, Helios could do the same with his rays or Zephyr with his warm breath. So what happened next?

EXTON: Along came a merchant from Delft, who had never been to university but was the best microscope maker of his time and he turned everything on its head! His name was Antonie Van Leeuwenhoek. He worked in the textile trade, in which microscopes were used to ascertain the quality of cloth. He was also interested in the living world and was the first to describe *animalcula*, or microscopic organisms.

ALKIMOS: Where did he point his microscope?

EXTON: Good question, Alkimos. Around 1677 Van Leeuwenhoek started studying the male seed of various animals, including his own. Because of his own strict morals, he felt it necessary to emphasise that he collect his samples "not by sinfully defiling himself but as a natural consequence of conjugal coitus." His animal samples were removed from the reproductive organs of females that had just mated. He had great reservations about publishing his findings.

DEMOCRITUS: I assume sooner or later he did publish his observations.

EXTON: Yes, but his results, published in 1678 in *Philosophical Transactions of the Royal Society*, the world's first and longest-running scientific periodical, were quite surprising. No matter what sample he examined, he found the same thing. The seemingly homogenous material was full of what he called *Samentierchen*. He described them as, 'less than one millionth the size of a coarse grain of sand, with thick, undulating, transparent tails.'

ALKIMOS: Oh, that sounds disgusting!

EXTON: Von Baer later called these little animals *spermatozoa*.

ALKIMOS: You mean the animals living in the seed?

EXTON: Yes, we know them now as sperm. These are Van Leeuwenhoek's illustrations.

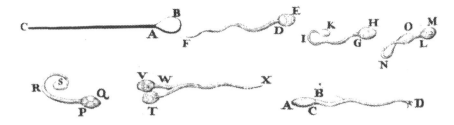

ALKIMOS: If we're full of cavorting little parasites, how come we aren't scratching all the time?

EXTON: Some did think that sperms were parasites typically found in the male seed. Others thought they were a part of the seed that did not play a role in fertility. But Van Leeuwenhoek put forth perhaps the most staggering suggestion. His microscopic observations left him convinced that an entire person lay concealed in every sperm. He believed that only those sperms that came into contact with an ovum would develop, with the ovum providing the nutrients. For Van Leeuwenhoek, development meant the growth of already present, but very small parts. He called the tiny dormant people in each sperm *figura animalculorum* or *animalcula seminalia*, and this theory of an already formed person is called preformation.

ALKIMOS: Wow! Is that what Hippocrates thought?

EXTON: Hippocrates never expounded his views on preformation in great detail. The earliest detailed description of preformation was given by Seneca, a Roman stoic philosopher. In his work *Quaestiones Naturales* he wrote: 'In the seed are enclosed all the parts of the body of the man that shall be formed. The infant that is borne in his mother's wombe hath the rootes of the beard and hair that he shall wear one day. In this little masse likewise are all the lineaments of the bodies and all that which Posterity shall discover in him'

ALKIMOS: Oh, now I understand! The development of an embryo isn't such a great problem!

EXTON: The idea of preformation, which rested on serious philosophical arguments, was promoted in 1674 by the French theologian Nicolas Malebranche, who had been quite impressed by the microscopic observations of the excellent Dutch naturalist Jan Swammerdam. Like most other scholars at the time, Malebranche was also under the influence of the Italian physicist Galileo Galilei's and the French philosopher René Descartes' mechanistic world views. It's no surprise that the notion of an already formed person just waiting to grow fit in well with the idea of a perfect order of the universe created by God. Curiously enough, however, Descartes himself was not a preformationist.

ALKIMOS: So maybe Malebranche misunderstood Descartes?

EXTON: Maybe, I would have to look that up. What I know is that Malebranche was a theologian and he did not make any observations himself. But he expressed his preformationist views very clearly: 'We may say that all parts are in a smaller form in their germs. By examining the germ of a tulip bulb with a simple magnifying glass or even with the naked eye, we discover very easily the different parts of the tulip.' Swammerdam, on the other hand, was very methodical in his work. He was one of the most careful observers of his time. His work *Biblia naturae* from 1737 to 1738, published after Swammerdam's death, is one of the most splendid books on microscopic studies ever published. This drawing from the book shows the reproductive organs of the bee, including the ovaries with strings of developing eggs.

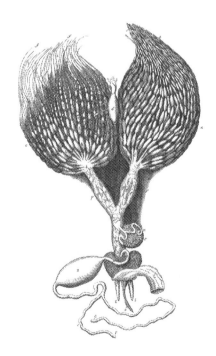

DEMOCRITUS: Amazing! No one could question the power of the microscope now!

EXTON: Swammerdam and the first generation of preformationists, unlike Van Leeuwenhoek, uniformly believed that the ovum was the carrier of the miniature offspring to be. They were *ovists*. Van Leeuwenhoek's peculiar notion set him in opposition to the entire medical community. But he didn't back down. He remained a devoted *spermist*, and continued his research into the *animalculum* nestled in the male seed. He tried to observe the already formed person by opening the membranes of the sperm dried on the slide using a damp brush, but without success. After a series of failures, he determined it impossible that 'human ingenuity will penetrate so deeply into that great secret that, by chance or by the dissection of the animalcula in the semen, we will come to see the entire man.' There was no doubt in his mind, however, that the entire person, in whatever form, was there. These drawings from Van Leeuwenhoek illustrate how he imagined the entire person could fit into a sperm.

DEMOCRITUS: It's wonderful to hear about the efforts of such determined scientists. But with all due respect, isn't it possible that the preformationists hadn't thought everything through? If indeed the entire person were present, whether in the ovum or the sperm, doesn't it follow that in the tiny sperm and ovum of this tiny *animalculum* there should also be more of these tiny *animalcula*? And we could continue this way for generations. So my question is: How could all these smaller and smaller people fit into the smaller and smaller genital organs? Sooner or later we would reach the undividable atoms. This would have to be the last generation, and would mean the extinction of mankind.

EXTON: Indeed the idea of *emboîtement* or the successive encasement of all subsequent generations was the most controversial aspect of the preformation theory. Charles Bonnet, Swiss natural scientist and philosopher and one of the most influential proponents of the theory of preformation, felt that 'we should not suppose an infinite emboîtement, which would be absurd. The divisibility of Matter to the infinite, through which we could support this type of emboîtement, is a geometric truth and a physical error. Any Body is necessarily finite; all of its parts are necessarily determined: but this determination is unknown to us.'

DEMOCRITUS: But my theory of atomism says the same thing. Based on this we can reject this notion of encasement.

EXTON: Not so fast! The preformationists were much more imaginative than that! Bonnet continued: 'We absolutely do not know what are the last terms of Matter's division; and it is this very ignorance that must prevent us from regarding the encasement of the Germs inside each other as impossible. We have only to open our eyes, and to walk around, to see that Matter is prodigiously divided.'

ALKIMOS: And that's just what we see when we look through a microscope. A new world unfolds before our eyes, with its own never-before-seen minute creatures. Why couldn't there be other worlds concealed within this world?

EXTON: This notion was the boldest aspect of the preformation theory. At first the theory appeared to be far more fruitful than the theory of epigenesis. Its power to explain was also much greater, while epigenesis faced irresolvable problems. How could a formless mass give rise to forms and ultimately a complex organism? According to the mechanistic view of the world introduced by Descartes, living matter could be viewed as well-organised machinery. The vitalists' explanations, relying on the concept of a vital spark that would ignite the developmental process, seemed superfluous. Preformationists handled growth as more or less a mechanistic process, which is what made it especially attractive. Additionally, if God created the world, why couldn't he have created all the generations until the end of the world encased in a series of ever smaller capsules. And in contrast what did the theory of epigenesis have to say?

ALKIMOS:

> *While a little person waits for life, slumbering in the seed,*
> *Spontaneous generation arises anew, from formless mass indeed!*

EXTON: Spontaneous generation? Yes, that's abiogenesis, as we've said. Arising from formless mass … why, Alkimos! I think your marvellous couplet has stumbled upon a very important connection!

ALKIMOS: Master! Finally someone likes my poem!

EXTON: You've hit on the connection between abiogenesis and epigenesis. The essence of both theories is creation from formlessness, from nothing.

DEMOCRITUS: If someone is a preformationist, doesn't that mean he must deny the possibility of spontaneous generation?

ALKIMOS: After all, didn't Spallanzani make all his concoctions to prove the theory of preformation?

EXTON: This may have been one motivating factor. At any rate, we know that Spallanzani was an ardent preformationist.

ALKIMOS: Then it's decided! Maybe Spallanzani didn't convince everyone, but Pasteur's experiments left no doubt. Spontaneous generation does not exist; therefore we must reject epigenesis too!

DEMOCRITUS: Now, just wait a minute, son. It seems to me there are some problems with preformationism. If we are spermists, we must insist that offspring can only resemble their fathers. Ovists on the other hand clearly maintain that progeny can only possess characteristics of the mother. But yet we all know of children who look like their father and others who look like their mother. And many that look like both!

EXTON: Excellent! Indeed, nothing is ever that simple. Nevertheless, the epigenesists faced quite a struggle in trying to gain primacy for their theory! Let's see what they were up against! In the 1740s, a Swiss naturalist, Abraham Trembley began to study freshwater hydras. This tiny creature of just a few millimetres in length consists of nothing more than a sack adhered to a base with several tentacles, as you can see in Trembley's drawing.

Trembley wanted to determine if hydras were plants or animals. They at first appear to be plants, but they use their tentacles to capture small worms and other animals swimming by, and cram them into their stomachs. Trembley excitedly told his friend, Ferchault de Réaumur, about his discovery, 'They are carnivorous, and certainly very avid!'

ALKIMOS: If they're meat eaters, then they must be animals.

EXTON: And the fact that hydras respond to touch further supported this. In fact, they can even creep along by turning somersaults. But the number of tentacles was always changing, which Trembley noted was typical of the branches and roots of plants.

ALKIMOS: So, are they plants or animals?

EXTON: Trembley set about to perform the definitive experiment. He cut a hydra in half lengthwise. If the hydra was an animal it wouldn't survive. But a plant would. The result was astonishing. Not only did the hydra survive the operation, but both halves regenerated to form two new creatures. If he cut the hydra in half crosswise, the same thing happened: he got two new hydras. Finally Trembley took the bold step of chopping the hydra up into little bits. And wonder of wonders, each little shred developed into its own hydra. While he was trying to make sense of these puzzling observations, his hydras were busy preparing another surprise. They slowly grew tiny buds on their sides, which eventually separated from the mother and grew into new hydras. Trembley discovered with great surprise that his hydras are also able to proliferate through budding.

ALKIMOS: It's too bad it doesn't work that way with people. If we carved up my master we could make lots of little wise masters, and we'd solve the problem of generation a whole lot quicker!

DEMOCRITUS: Keep quiet until I say otherwise!

EXTON: Later it was established that hydras were indeed animals. But of course Trembley's experiments revealed something far more earth-shattering about hydras—that is their incredible powers of regeneration and ability to reproduce asexually. John Turberville Needham was so inspired by Trembley's findings that he declared the creation of the first woman, Eve, explained. He stated that 'The body of the first woman was not formed from earth, like the body of her husband, but she was rather generated from him through an accelerated vegetative process, nourishing on his substance during his sleep, until she separated in a state of perfection, like what is observed among the young polyps and the organised bodies of the same kind.'

ALKIMOS: Wow, he really has a convincing argument there. I am sure Master Spallanzani had a few words to say about that. But who is Eve?

DEMOCRITUS: Alright Exton, give me a moment to think this over again. We are well aware of budding in plants, but it was really surprising to discover this occurring in the animal world. Nevertheless, this wasn't the real problem faced by preformationists. Instead they had to fit regeneration and budding into their theory.

EXTON: It's not impossible. You just had to presume that the next generation of small preformed *animalcula* could be found scattered throughout the body of the hydra. When the hydra is cut up into pieces, each one still contains plenty of little *animalcula*. So many preformationists readily accepted the results of Trembley's experiments. There were others, however, who felt regeneration and the replacement of lost parts could not be explained within the theory of preformation. The crab's ability to re-grow its claws led Nicolaas Hartsoeker, one of the earliest and most devoted proponents of the spermist view, to finally abandon the theory. He wrote: "That kind of intelligence which is capable of replacing a crab's lost claw is capable of building an entire individual."

ALKIMOS: But it's possible that an already formed little claw lies at the base of the existing claw, isn't it?

EXTON: Hartsoeker did not accept this argument. He felt that the regeneration phenomenon overturned every simple theory of preformation. But of course theories of preformation were anything but simple during that time. Now, let's see how preformationists handled some other challenges to their theory. In 1682, Nehemiah Grew proclaimed in his book *The Anatomy of Plants* that the stamen and pistil are actually male and female reproductive organs. This emboldened many botanists and they began a scholarly examination of hybridisation in plants. Camerarius' experiments showed that if the pollen produced in the stamen did not reach the pistil, the flower remained infertile and did not produce seed. But this was just the beginning. Joseph Gottlieb Kölreuter, a pioneer of plant hybridisation, carried out extensive experiments with cross-fertilisation. Kölreuter's experiments were inspired by a hybridisation experiment the Swedish naturalist Carl Linnaeus had carried out in 1759. Extreme preformationist views were contradicted by the observation that crosspollination of different varieties could produce plant hybrids that had mixed characteristics. Therefore neither the stamen nor the pistil could contain a tiny preformed version of the plant.

DEMOCRITUS: Features of the parents mix in the progeny. It's just like I said before. We have seen this with people. If there is mixing, both parents must contribute, just as Hippocrates stated.

EXTON: Exactly! But keep in mind, that just because someone possesses his father's nose and his mother's leg, this is not conclusive proof. We have to find some trait that is clearly recognisable and does not become less distinct over generations. If both the mother and the father can pass this on, then the theory of preformation would become untenable. We were helped in this endeavour by the French natural philosopher Pierre-Louis

Moreau de Maupertuis. He suggested that little particles that are responsible for inheritance are present in the reproductive fluids of the parents. During mating these mix together, which is why the offspring bear traits of both the parents. Maupertius began to study the inheritance of a very strange defect, *polydactyly*.

ALKIMOS: *Poly* means many and *dactyl* means digits, or fingers and toes, so polydactyl means having many fingers and toes. I've heard of Hekatonkheires, the giants with one hundred hands, but I've never heard of anyone with one hundred fingers.

EXTON: We're not talking about one hundred fingers or toes, just six or seven on the hands or feet of the affected individual. Maupertuis demonstrated that this unusual trait can be passed on by both the mother and the father. He explained the appearance of polydactyly by a change in the minute particles responsible for inheritance. But he made other observations too. In 1743—to the great wonderment of the Parisian public—a five-year-old albino African child arrived in the French capital. His parents were clearly black. Maupertuis wrote: "…certain traits could remain hidden through several generations, only to reappear as the result of some fortuitous mating. Some apparently new kinds may have arisen in this way; if the white Negro were an example of this phenomenon, it need not have either a white mother or father… it is enough that this child had some white Negro amongst its ancestors." According to Maupertuis' theory of particulate inheritance "…elements which represent the condition of an ancestor rather than the immediate parent may enter into the union forming the embryo and so produce a resemblance to that ancestor rather than to the parent."

DEMOCRITUS: I was never able to come up with such a profound explanation of how atoms work! Maupertius' results take us beyond every other theory. He should be called the founder of the laws of inheritance.

EXTON: There's some truth to that. In his work *Vénus Physique* published in 1745 he lists all the evidence against the theory of preformation. One would think as overwhelming as the evidence was the preformationists would give up their struggle. But quite the opposite happened. At about the same time as Maupertuis was making his investigations into inheritance, Charles Bonnet started on some unusual studies of his own. As his results became widely known, the theory of preformation increased in potency, dominating scholarly opinion until the end of the 18th century. Bonnet's studies were on the reproduction of aphids.

ALKIMOS: He must have had strange tastes …

EXTON: In his work *Traité d'Insectologie* published in Paris in 1744 he described an unusual feature of aphids. Female aphids that were separated at birth and had thus never encountered a male were capable of laying eggs that developed without fertilisation—in other words virgin birth, or *parthenogenesis*!

ALKIMOS: So mating is not necessary in reproduction in an animal? Who ever heard of such a thing!

EXTON: Well, you, for example! Have you forgotten about hydras and how they reproduce? They don't need to mate either; they can reproduce by budding. With aphids, the difference is that the offspring develops from an egg, which for ovists is the source of preformed life.

DEMOCRITUS: If an offspring can develop from an egg without the help of a male, then all the secrets of birth and inheritance must lie in the egg. This strongly supports the ovist's version of preformation. Swammerdam must have gained new impetus.

EXTON: Except the poor guy died in 1680. But Bonnet, our friend Spallanzani, and a Swiss poet, anatomist, botanist, and embryologist named Albrecht von Haller must surely have welcomed this result. The ovist theory promoted by these three intellectual giants had a tight hold on scientific thought in the second half of the 18th century.

DEMOCRITUS: What explanation did the preformationists offer for the mixing of traits?

EXTON: Bonnet believed that the features of a preformed offspring could be influenced by the male seed. He wrote that, 'The germ is of a small-scale man, horse, or bull but not a particular man horse or bull. The individualizing effect comes from the environment provided by the mother and the seminal fluid of the father.'

DEMOCRITUS: And what did he say about *emboîtement*?

EXTON: "Nature can work in as small a scale as it wants," Bonnet wrote. If there is no lower limit to the division of matter into smaller particles, then *emboîtement* can continue *ad infinitum*. Haller even calculated that there were 200 billion tiny people encased in succession in the ovary of Eve, the mother of all living.

ALKIMOS: Pre-fabricated generations until the end of time.

EXTON: Despite all this sophistication, preformationist ideas were constantly challenged, foremost by Buffon and Carl Linnaeus. As the number of observations of embryos in their early stages grew and it became increasingly obvious that an egg could not contain a completely formed offspring, preformationists had to concede. By the end of the century preformationists' arguments sounded like this: "Whilst the chick is still in the form of a germ all its parts have a form, proportion and position which differ considerably from those they will ultimately assume. Consequently if we enlarge it at this stage it would not be possible to recognise it as a chick." As embryologists delved deeper into the secrets of the ovum and early development, the *animalcula seminalia* slowly disappeared. They waned, like the moon, only to return in such a completely different guise as to be unrecognisable.

DEMOCRITUS: That poor *animalculum seminalium*! It lost its parts and proportions, and was slowly reduced to a formless mass. What would Haller say about his mass of 200 billion little individuals all melted together? And what do we think of it all? What law or force activates this mass? How does it give rise to form? This is the true mystery of generation!

EXTON: This is how scientists at the end of the 18th century felt too. This question leads us from a century of confusion into the heroic age of embryology!

ALKIMOS: I'm still not sure I understand. So is it preformation or epigenesis?

Types and Rhythms of Embryos—*Karl Ernst von Baer*

DEMOCRITUS: How odd. If we reject the seemingly plausible theory of preformation, we are left with no other explanation of the origins of life. The theory of epigenesis, although based on very careful observations, only disproves preformation, but offers no alternative. Either we suppose that the development of an organism is guided by some power we humans are incapable of comprehending, or we need to work out a mechanistic theory of self-organizing atoms.

EXTON: Ah, the time has not come for such a revolutionary thought as self-organizing atoms. But embryologists quickly arrived at an unfathomable power that guides development: *vis vitalis* or *vis essentialis*. The shaping of an organism from nothing, *morphogenesis*, was now considered indisputable. Caspar Friedrich Wolff renewed the study of chicken embryos. In his dissertation *Theoria Generationis*, published in 1759, and his treatise *De Formatione Intestinorum*, published in 1769, Wolff argued that although it was possible that preformed body parts lay hidden in regions that couldn't be seen with a microscope, once the parts were viewable, they did not in any way resemble the final forms. For example, the intestines of the chicken embryo at first look like flat membranes that slowly fold into a tube shape. The entire embryo initially looks like a flat disk that begins to fold in the middle when an embryonic formation known as the *primitive streak* forms. Wolff could only explain the formation of these body parts with the concept of *vis vitalis*.

ALKIMOS: But what is *vis vitalis*? If we don't understand something, it seems we immediately develop a new concept, a secret creating force, and then we're content. But it doesn't explain anything!

EXTON: For the true nature of development and the guiding force of *vis vitalis* to be understood, we need to wait a while. We're only at the end of the 18th century. Nearly two hundred years need to pass before we find a satisfying explanation. Until then, let's enter the world of Wolff and his fellow vitalists. Without preformation, shape emerges from shapelessness. Some kind of force or law must be driving this process. But it's impossible at this point to know anything about it; but we should call it something. The nature of *vis vitalis*, however, is pure speculation. We can call it spiritual, material, ethereal, or mechanistic, it's all the same.

DEMOCRITUS: Okay, so we'll have to wait to learn about the true causes of organism development. An essential force, *vis essentialis*, is simply a way of explaining what we can't explain but we know must be there. The smartest thing we can do is observe in great detail the stages of development from an amorphous shape into an organism. Perhaps we'll stumble upon the laws of inheritance.

ALKIMOS: You're right! We have a microscope. Let's use it to examine a variety of embryos and see what we discover.

EXTON: Jean-Louis Prévost and Jean-Baptiste Dumas, French embryologists, did just that. In their treatise of 1824, they described the early development of a frog embryo. They had an advantage over Harvey and Wolff because they weren't studying mammalian or bird embryos, which are hidden within eggs or bodies, but amphibian embryos, whose development is in the open. What they discovered was staggering. The spherical egg cleaved in two.

ALKIMOS: Oh, poor *animalculum*! How did it survive?

EXTON: Later another furrow appeared perpendicular to the first. Then more and more, dividing the embryo into a greater and greater number of parts. This process is known as cleavage. While this is happening, the embryo doesn't grow. So the compartments created by the cleavages are smaller and smaller.

DEMOCRITUS: This discovery completely disproves preformation! The embryo doesn't even grow in the beginning, it simply divides. There's no place for a preformed seed.

EXTON: Dumas and Prévost reached this conclusion, too. A couple of years later, Mauro Rusconi of Pavia, demonstrated that the embryo in fact divides into individual spheres. Using inventive methods, he was able to harden the embryo and then break it into larger and smaller balls. The experiment, done on embryos at different stages, showed that the fertilised egg 'first split in two, then four, dividing further and further into increasingly smaller units.'

ALKIMOS: If this process continues indefinitely, then the embryo will eventually break apart into atoms. How does this become a frog?

EXTON: Cleavage doesn't go on indefinitely. To know what happens next we should travel to the 19th century and visit the greatest comparative embryologist of the time, the Prussian–Estonian Karl Ernst von Baer.

ALKIMOS: He discovered the mammalian ovum and coined the term spermatozoa.

EXTON: Ah, I see you're paying attention, Alkimos.

ALKIMOS: No, I'm just trying to delay our departure ...

EXTON: Hermes! Prepare the Deus ex Machina! Direction Saint Petersburg, 1837!

ALKIMOS: Oh no, here we go … Heeeeelp!

EXTON: Relax, we've stabilised. Now that the shaking's stopped, I'll take this opportunity to tell you about Baer. He was born in 1792 on an estate in Estonia, the son of first cousins Magnus von Baer and Juliane von Baer.

ALKIMOS: Oh, I see, when it comes to greatness, a little incest can't hurt.

EXTON: A little respect, Alkimos. Until he was six he lived in Lasilla manor, the home of his uncle, Karl Heinrich. He studied in Tallin, Dorpat, Vienna, Berlin, and Würzburg. In 1821, he became a Professor of Zoology in Königsburg. His monograph of 1828, *Über Entwickelungsgeschichte der Thiere* was a milestone in the history of comparative embryology. Although by 1837 his interests had turned to geography, anthropology, and ichthyology, I'm sure he'd love to share his discoveries about the laws of embryonic development.

ALKIMOS: Baer's laws! Let's get to it!

EXTON: We've arrived.

ALKIMOS: Thank Zeus!

EXTON: Everyone out!

ALKIMOS: If only the door wasn't so narrow …

EXTON: This really is Saint Petersburg. There's the library, to our left. Over the canal, along the avenue, and up the stairs. We should find the reading room easily.

DEMOCRITUS: What a marvellous city. It's more beautiful than Pergamon.

ALKIMOS: Oh, master, what dimensions!

EXTON: Come on, this is it. I'll open the door … there he is, Professor von Baer.

VON BAER: I've been waiting for you. Have a seat.

EXTON: We're honoured, Professor. I hope we're not bothering you at this late hour.

VON BAER: Von Echstand, I presume, from the University of Vienna?

EXTON: In the flesh.

VON BAER: I received your letter.

EXTON: These gentlemen are my pupils, Alchumbach and Demiescher.

VON BAER: I'm delighted to meet you. And I'm more than happy to discuss with you my achievements in embryology, as you requested in your letter, Professor von Echstand.

ALKIMOS: We're quite interested in the subject, especially after reviewing the debate over epigenesis versus preformation. Such a fascinating story. We were amazed to learn of Caspar Friederich Wolff's discoveries …

VON BAER: Ah, be conclusive, my friend.

DEMOCRITUS: Alkimos, let me. We've realised that to understand embryonic development, precise, comparative observations were needed.

VON BAER: Yes, that's more or less what I realised too. I started with the observations made by my friend Christian Heinrich Pandel on chicken embryos and arrived at a law that's valid for the entire vertebrate world. The parts of a vertebrate embryo all develop from three layers that form very early on.

EXTON: The outer layer is the *ectoderm*, the middle layer the *mesoderm*, and the inner layer the *endoderm*. Wolff observed these layers and called them *Keimblätter*, or germ leaves, but let's use the term germ layer.

VON BAER: Wolff was truly a brilliant observer, but he failed to understand the significance of the layers. When I made comparative studies, I discovered that the same organs of different animals all arose from the same layers. The outer layer produced the skin and the nervous system, the middle layer the skeleton and muscles, and the inner layer the gut tube. The unity of the living world is manifest in the embryo.

ALKIMOS: How? Every embryo develops from the same germ layers? It's not *omne vivum ex ovo*, all life from an egg, but *omne vivum ex Keimblätter*, all life from a germ leaf? Early embryos must then be very similar.

EXTON: That's Professor von Baer's most important law. But let him explain it.

VON BAER: Indeed, they are. The embryos of mammals, birds, lizards, snakes, and perhaps even turtles are nearly identical in their early stages, both in their appearance and in the process by which their organs develop. In fact, the only way to distinguish between the various embryos is by their size.

ALKIMOS: Incredible. A young chicken embryo could just as easily develop into a goat?

VON BAER: No, only a chicken will develop from a chicken embryo. It's just that in the early stages we can't tell the difference between the two. In my collection I have two embryos preserved in alcohol, but I forgot to label them. Now I have no idea what class they belong to. They might be lizards, birds, or very young mammals. The development of the head and torso is so similar in these animals. The extremities haven't formed yet, but even if they had, I'd still be unable to distinguish between the animals at such an early stage. A lizard's and mammal's legs, a bird's wings and legs, and even a person's arms and legs, all develop from the same rudimentary form.

DEMOCRITUS: Astounding. When do the embryos start to show differences?

VON BAER: As development proceeds, the distinguishing characteristics of the various animal groups appear. First the general, then the less general. The most specific features that are unique to an animal group or species emerge at the end. Thus, as animals develop they gradually differentiate. Young embryos resemble only other undifferentiated, general forms, but do not resemble their parents or the developed form of any other animal.

EXTON: Here are drawings from the German embryologist Ernst Haeckel of a dog and a human in the fourth week of embryonic development.

ALKIMOS: Which is the dog?

EXTON: Ah! See how similar they are. The one on the left is the dog or *Hund* in German.

ALKIMOS: The living world is uniform in the embryonic stage, but not once it has developed. I don't understand.

VON BAER: The unity of living creatures is true for everything. You'll see.

EXTON: *Naturphilosophie* or common ancestry. But we'll get to that.

VON BAER: The embryo folds into germ layers, which develop into forms according to some inner law or mechanism. If we observe the formation of the germ layers, and later the tissues and organs, the living organism seems like a mechanical structure, a machine that builds itself; life processes are nothing more than an unceasing chain of chemical reactions. We can think of the organism as a chemistry laboratory in which the organism itself is the assistant. The assistant takes from its environment the materials needed for the chemical activity in the laboratory. If something obstructs the acquisition of material, the life process ceases.

ALKIMOS: This is beyond me. Who is instructing the assistant? How does it know when and what kind of material to take and where to put it?

VON BAER: The parts of the organism are constructed according to the chemical activities of the life processes guided by the *typus* and *rhythmus*. The *typus* is the relationship between the different parts. The *rhythmus* is the order in which the parts develop. The life processes can be compared to music. The harmony is the *typus* and the melody is the *rhythmus*. It's important to note that the type and rhythm is not the product of metabolism, but its operator.

ALKIMOS: Type and rhythm. This isn't some kind of vitalist circumlocution, is it? What's the true explanation?

VON BAER: Excuse me?

EXTON: Wait a minute, Alkimos. We agreed that the idea of an unknown, guiding force driving the epigenetic process was not entirely unfounded. It's what Aristotle concluded. Let's save this question until later.

ALKIMOS: Hm ... but Exton ...

EXTON: Professor, I'm afraid we have to go.

VON BAER: What's the hurry?

EXTON: We have to meet up with Pushkin.

VON BAER: But Pushkin died a week ago in a duel.

EXTON: You don't say?

ALKIMOS: Thank you for the conversation.

VON BAER: Wait!

EXTON: Let's board. Quick!

DEMOCRITUS: But who's Pushkin?

ALKIMOS: Ohhhh, the engine is starting ...

Cell from a Cell

ALKIMOS: Master, are you awake?

DEMOCRITUS: Yes, son. Everything's okay. I was just thinking about what von Baer said about the unity of life. Although he thinks the similarities between the early embryos make it obvious, I think there is something even more intangible than the immaterialism of *vis vitalis*, but we need to find a mechanism characteristic of every living thing. Von Baer has taught us a lot, but to discover true atoms, we need to move on.

ALKIMOS: I see, master, that you were not satisfied with the concept of type and rhythm.

EXTON: We mustn't leave your inquisitiveness unsatisfied! We will now talk about the next major breakthrough in physiology: the birth of cell theory. In the middle of the 19th century, the cell theory identified a mechanism, or rather a type of organisation, that applied to all living things.

ALKIMOS: Cell theory. I suspect this is something new.

EXTON: Actually, the observations that led scientists to recognise cell organisation as a fundamental process in all living things originated in the 17th century. In 1665 Robert Hooke, an exceptionally talented English experimental scientist published an epoch-making book, *Micrographia*. It contains detailed microscopic observations and drawings of various objects and creatures. Aware of the significance of his precise observations, he wrote, 'the Science of Nature has been already too long made only a work of the Brain and the Fancy: It is now high time that it should return to the plainness and soundness of Observations on material and obvious things.' Heeding his own words, Hooke used his microscope to examine cork, and discovered it consisted of small, regular compartments. In other plants, he noticed that these 'cells' contained fluid, which he presumed was sap.

ALKIMOS: Are these cells the same as Democritus' atoms?

EXTON: Not exactly. It appeared to him that the compartments comprised the basic units of the living world. Henri Dutrochet, a French physiologist, wrote the following: 'all the organic tissues of animals are really globular cells of an extreme smallness, which are united only by cohesion. Thus all the tissues, all the organs of animals are really only a cellular tissue diversely modified. This uniformity of ultimate structure proves that organs really differ one from the other only in the nature of the substances which are contained in the vesicular cells of which they are composed.'

DEMOCRITUS: In other words, not only do various animal organs exhibit uniform cellular structures, but plant organs also have microscopic structures identical to those of animals? That means von Baer's law of similarity can be even more generally applied on the cellular level?

EXTON: That's true. But let's see what Franz Julius Ferdinand Meyen had to say in 1830: '... each cell forms an independent, isolated whole; it nourishes itself, builds itself up, and elaborates raw nutrient materials, which it takes up, into very different substances and structures.' Here is a drawing by Meyen from 1830 of a section of the leaf of the common water-plantain.

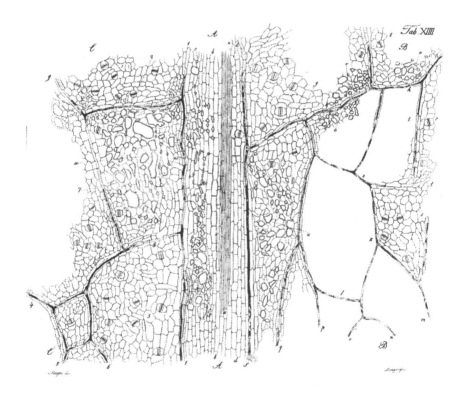

However, it is not Meyen but Czech anatomist Jan Evangelista Purkyně, the founder of histology, and his students, along with the German scientists Theodor Schwann and Matthias Schleiden, who are credited with recognizing the general significance of cells and with developing the cell theory.

ALKIMOS: Purkyně's theory of cells.

EXTON: Yes, but before we delve in, we need to talk about another important discovery. In 1791, the Austrian botanist Franz Bauer embarked on an extensive series of illustrations of European plants. While closely observing the interior of the cells of a certain kind of orchid, he noticed greenish-yellow patches. In 1831, the Scottish botanist Robert Brown also examined the cellular structure of orchids, and like Bauer, stumbled upon an unusual, opaque, seed-like formation. He wrote, 'I shall conclude my observations on Orchideae with a notice of some points of their general structure, which chiefly relate to the cellular tissue. In each cell of the epidermis of a great part of this family, especially of those with membranous leaves, a single circular areola, generally somewhat more opaque than the membrane of the cell, is observable. This areola, which is more or less distinctly granular, is slightly convex, and although it seems to be on the surface is in reality covered by the outer lamina of the cell. There is no regularity as to its place in the cell; it is not unfrequently, however, central or nearly so.' Brown managed to demonstrate the presence of this body he called *areola* or *nucleus* in almost every tissue.

ALKIMOS: Does the nucleus contain the tiny, preformed plant? Is that what Van Leeuwenhoek searched for with his damp brush in the spermatozoa?

EXTON: What a bold idea! A botanist from the University of Jena, Matthias Schleiden, did not go that far. He continued to examine the nucleus and reached the erroneous conclusion that the cell developed from it. That's why he renamed it 'cytoblast.' His conclusions, presented in his 1838 study *Beitrage zur Phytogenesis*, had a decisive impact on the work of his friend Theodor Schwann, who was investigating the origins of animal cells. Schwann was reluctant to acknowledge apparent similarities between animal and plant cells until Schleiden drew his attention to the nucleus. If the two kinds of tissue indeed have identical structures, then animal cells must also have nuclei. Very soon he found 'opaque patches' in the same regular arrangement as Schleiden had seen in plants. Occasionally, he observed that a membrane similar to what surrounded plant cells, was also present in animal cells. Shortly after Purkyněs discoveries Schwann realised that both plants and animals are composed of tiny cells crowded together, enclosed by walls or membranes and each containing a nucleus. And although Schwann incorrectly concluded, based on the discoveries of Purkyně and his students, that cells arise from the nucleus, his work *Mikroskopische Untersuchungen über die Uebereinstimmung in der Struktur und dem Wachsthum der Thiere und Pflanzen* laid the foundations for cell theory.

ALKIMOS: What a long title. I can't guarantee I'll remember it.

EXTON: Schwann stated that cells were independent organisms, and plants and animals are none other than a cluster of these organisms organised according to specific laws. In his support for the idea of the unity of life, he went further than von Baer: 'there is one universal principle of development for the elementary parts of organisms, however different, and … this principle is the formation of cells.'

DEMOCRITUS: Why, that's the foundation of growth and development! The origin of new cells.

EXTON: At first, they thought the most important part of the cell was its wall. But increasingly careful observations quickly revealed the presence of a thick moving fluid full of various granular material. According to the German botanist Hugo von Mohl, the nucleus swam in this viscous liquid he called *protoplasma*. Scientists quickly ascertained that the protoplasm of plants and animals was similar. It was increasingly apparent that the true bearer of life was not the wall of the cell, but the protoplasm.

ALKIMOS: So what is the purpose of the cell wall?

EXTON: Hooke had already discovered the rigid wall that enclosed plant cells. The presence or absence of such a wall in animal cells, however, was debated for some time. In 1884, Hugo de Vries, a botany professor from Amsterdam, plainly demonstrated that along the interior of the rigid wall of the plant cell was a flexible membrane. If the cells were placed in a salt solution, the water inside the cell flowed out, and the cell shrivelled. The membrane very visibly separated from the cell wall. Soon enough it became clear that animal cells were also surrounded by thin membranes.

ALKIMOS: But where do cells come from, if they don't arise from the nucleus?

EXTON: The discovery of the origin of cells was the culmination of cell theory. In 1832, a Belgian botanist, Barthélemy Dumortier, described how single cells in plant tissue divided to form two new cells. Most of his observations were of an alga species, *Conferva aurea*. Hugo von Mohl made similar observations of *Conferva glomerata*. This is how Mohl illustrated the division of cells:

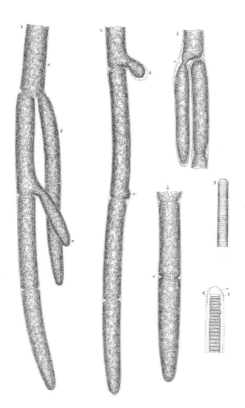

ALKIMOS: Like hydras.

EXTON: The cells grow and grow, and then only the membrane closes in and divides the cell into two daughter cells. The Polish physiologist Robert Remak demonstrated the division of animal cells. In 1841, he described the division of blood cells in chicken embryos. Two years after Schwann's work was published, Remak refuted the theory that cells form from the nucleus. Later he observed the splitting of muscle and other cells, and concluded that every cell originated from another already existing cell. With the publication of the German anatomist Rudolf Virchow's book *Cellular-Pathologie*, Remak's discovery of the process by which animal cells arise became generally known. Virchow is associated with the dictum *omnis cellula e cellula*, from every cell a cell.

ALKIMOS: But Schleiden said that too!

EXTON: Yes, but not as clearly. Catchphrases and acronyms are important in science. Later the same was found to be true of cell nuclei.

ALKIMOS: From every nucleus a nucleus.

EXTON: *Omnis nucleo e nucleo*, as Walther Flemming wrote in his book *Zellsubstanz, Kern und Zelltheilung*, published in 1882. And although Remak and others had observed, or rather inferred, the division of nuclei, Flemming's discoveries ushered in a new era in cytology, the study of cells.

DEMOCRITUS: I can barely fathom that we are just a gigantic colony of microscopic cells, each enveloped in its own membrane with its own nucleus.

EXTON: Yet, two interesting examples illustrate very well that this is indeed the case. Sea sponges have varied forms, colours, and sizes. Their bodies are composed of just a few types of cells, which makes them capable of this amazing feat: if we force a sponge through a sieve we are left with mush, but the broken-up cells will quickly find each other in water and rebuild the body of the sponge. What's astounding is that if we mix the mush of two different sponge species together, the cells separate and recreate the two original colonies.

ALKIMOS: Ah ….

EXTON: The slime mould *Dictyostelium* has an even stranger life. In times of plenty it lives as separate cells in forest soil. If the supply of nutrients begins to dwindle, huge, multicellular communities form. The clump wanders for a while, and then settles. Individual cells spurt to the top, like a fountain, forming a thin stalk with a ball at the tip. The cells on the exterior of the ball form a hard case, while the inner cells become delicate spores. In the end, the case breaks and the spores scatter to begin a new life as independent, wandering cells in moist soil. The multicellular body of *Dictyostelium* is thus composed of differentiated, self-sacrificing stalk cells and proliferating spore cells.

DEMOCRITUS: What a primitive colony, and yet how sophisticated. An animal's body, in contrast, is a repository of such complicated alliances.

EXTON: The tens of trillion human cells certainly do live in a very developed society.

ALKIMOS:

> *Cells only from cells, and seeds only from seeds*
> *Give rise to the magical concoction of life, and behold,*
> *Scores and scores, plenteous, effuse; yet without a guest*
> *Grain, tiny cells are unable to answer the joyous call,*
> *To gather at the scraggly, soft, ephemeral membrane's edge,*
> *Unable to forgo a free and selfish life, unable to*
> *Join forces, to guarantee existence to the colony's*
> *Great expanse. Flesh and blood is all I am.*

DEMOCRITUS: Your poem is a little complicated.

ALKIMOS: I had some trouble with the rhythm.

EXTON: Well, you certainly made ample use of enjambed lines. All we're missing is an acrostic.

ALKIMOS: Fine, make fun of me. It's still good, I just need a little more work.

The Feats of the Sea Urchin

DEMOCRITUS: Perhaps it's reasonable to compare multicellular organisms to well-working societies. The people who perform the various functions in society—from wagon builders to tax-collectors to the soldiers in the army—are in complicated relationships with each other, some dominant over others, much like the cells of living organisms.

EXTON: Indeed, we're extremely sophisticated examples of cohabitating cells, like the millions of other multicellular species. Several hundred types of cells create a colony in which the interdependence and division of labour assume fantastic levels of complexity. Birth and demise are in balance, suicide for the good of the colony is an everyday occurrence. It seems that every member of the colony enjoys some sort of privilege for its hard work. But this is not true. Instead a kind of tyranny prevails. The foundations of individual rights are degraded. Only very few cell groups have the opportunity of surviving into the next generation. Freedom of speech is non-existent. Every cell contains the information necessary to construct the entire organism, but they can only give voice to a fraction of it. Freedom of movement is also granted just a few. The most tightly bound are the muscle and epithelial cells. The nervous system, and the army of hormones ensure the coordinated operation of the organism. If a cell makes too much noise or proliferates uncontrollably, it will be liquidated by the immune system. Maintaining control of a crowd of hundreds of million requires exacting methods.

ALKIMOS: So this rule of terror ensures the operation of the organism. That's why so many Hellenistic city-states were short-lived. If there's no discipline, if everyone is equally free, intrigue and depravity will reign. We could learn something from our cells.

DEMOCRITUS: Let's not get too caught up in our philosophizing. Allow me to change the subject. I've come up with a clear logical sequence of thoughts. First, we know that every living thing is made up of cells. We also know *omnis cellula e cellula* and *omne vivum ex ovo*. The inevitable conclusion is that an ovum is also just a cell.

EXTON: Schwann and his immediate followers realised that in 1839. What's even stranger is that the sperms in the male seed are also just cells. In 1841, Rudolf Albert von Kölliker, a Swiss physiologist and anatomist, first demonstrated that spermatozoa arise directly from the cells in the testicles of males.

ALKIMOS: So we're not spreading parasites in the world, but our own cells.

EXTON: And not just a few. Every time it's about two hundred million.

ALKIMOS: What a waste!

EXTON: Think what a dilemma it was for Van Leeuwenhoek—all those squandered spermatozoa!

DEMOCRITUS: If every sperm contains a tiny, complete person, how could the heavens have allowed for such a squandering?

EXTON: Once scientists had moved beyond the early spermist version of preformation, such worries were forgotten. But enough problems still remained.

DEMOCRITUS: We still don't know what the relationship of the ovum is to the two hundred million sperms. What is their role? What can the ovum achieve without them?

ALKIMOS: Thanks to the lice of Charles Bonnet, we know the answer to that: a lot.

EXTON: Yes, but we know that parthenogenesis is rather rare in the living world. The eggs of most species do not develop without contact with the male seed.

ALKIMOS: Lucky for us.

EXTON: And it was our friend Spallanzani who demonstrated it. In 1768, he went to Pavia at the invitation of Empress Maria Theresa. His friend Charles Bonnet urged him to investigate the role of the male in reproduction. Spallanzani was an ovist, and therefore considered the sperm to be parasites; yet he was the one who showed that for the ovum to develop it must have contact with the sperm. Before mating, he put tight taffeta pants on some male frogs. The eggs of the females that encountered the diapered males did not develop. What's more, he also showed that exhalation from the semen—dubbed *aura seminalis*, or fertilizing air, by Swammerdam—was not sufficient to induce development of the egg; direct contact was necessary. When he carefully strained the seminal fluid, he discovered the resulting liquid was sterile. The fatty mass caught in the mesh, however, if placed in water, was still capable of stimulating development of the ovum. Finally, around 1824, Jean-Louis Prévost and Jean-Baptiste Dumas, who first described the cleavage of frog embryos, demonstrated exactly which part of the fluid was responsible for fertilisation: the sperms, once considered parasites. They also discovered that sperm originated in the testes of the male, and that of the millions of sperm, one was enough to cause the egg to develop.

DEMOCRITUS: Van Leeuwenhoek must have arrived at the same conclusion, although along a different route. For the spermists it was obvious that only one preformed person could enter the nourishing environment of the egg and grow.

EXTON: It's important to note that in Prévost and Dumas' time, no one knew about the cellular nature of the ovum and sperm or that they had a nucleus. The way in which the two made contact was also a mystery.

ALKIMOS: I know what comes next: the further development of microscopic techniques will make it possible to make crucial observations in the decades that follow.

EXTON: Indeed, Alkimos. Scientists devised new methods of dyeing and fixing the samples. By the end of the 19th century, the oil immersion technique developed by Giovanni Battista Amici in 1840 was widely used. Also common by that time were apochromatic lenses. These techniques provided much clearer images. The new microscopy and a new study animal, the sea urchin, yielded very exciting new results.

ALKIMOS: Oh, how many times have I stepped on a sea urchin walking along the shore in Abdera. My feet ache just thinking about it.

EXTON: Oskar Hertwig, student of the German embryologist Ernst Haeckel, was the first to study sea urchins. The animals were very cooperative, laying huge numbers of transparent eggs that quickly started developing, even *in vitro*, that is, in laboratory vials. In 1877, another student of Haeckel, a Frenchman named Hermann Fol, observed that in the sea star, the moment of fertilisation took place when a sperm penetrated an ovum. These illustrations from Edmund Bleecher Wilson show the moment of fertilisation in different marine animals.

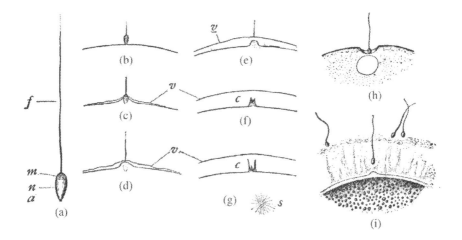

The sperm fused with the protoplasm of the ovum, and its nucleus, the *pronucleus*, negotiated the ovum until it reached the female pronucleus and merged with it. In that same year, based on his observation of the sea urchin, Hertwig also reported that the nucleus of the embryo arises from the nuclei of the two sex cells or gametes, from the nucleus of the ovum and that of the penetrating sperm. This diagram shows the approach of the male and female pronuclei after fertilisation in the sea urchin.

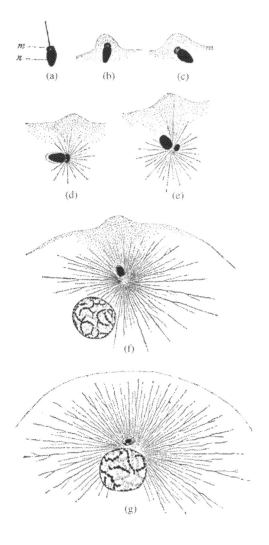

Once the pronuclei have merged, development begins. I should note that fertilisation in plants was described much earlier, in 1830, by the Italian astronomer and botanist Giovanni Battista Amici, who invented the oil immersion technique. A tube develops from the pollen that lands on the flower's stigma, and then as it grows it passes through the style until it reaches the ovary, or seed case, as you can see in this diagram

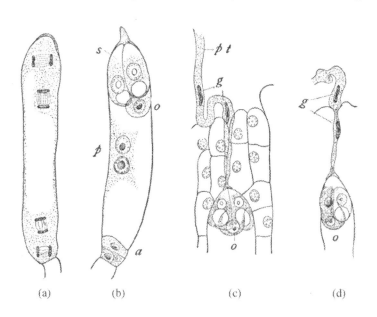

(a)   (b)   (c)   (d)

The role and origin of the pronucleus in the plant's pollen and ovule was for a long time a mystery. The most important observation, however, was that it was sufficient for just one grain of pollen to reach the stigma for fertilisation to occur.

DEMOCRITUS: Given the unity of life, we can presume that fertilisation in plants takes place in the same way that was observed in animals. First of all, the gamete cells unite, and then their nuclei. This fusion may hold the key to how the traits of the parents mix. Cell theory and the cellular nature of gametes help explain the development of an organism. The fertilised egg is in fact a separate cell with a fused nucleus. The cleavage of this cell is the only way it can grow and develop. Since every cell contains a nucleus, obviously these also divide every time the cell does. Organism development is merely the division of cells and their nuclei.

Inheritance must arise from the influence on cellular division and growth of the tiny parts that transmit the traits of the parents.

EXTON: What insight! But we should talk about the recognition of the cellular character of development, which is attributed to Robert Remak. His book *Untersuchungen über die Entwickelung der Wirbelthiere*, published in 1855, discussed the development of vertebrates and put to rest the debate inspired by Schwann by showing that organisms develop from the fertilized ovum — in other words from a cell—through a series of cell divisions. Let's have a look at this diagram of the development of the sea urchin embryo.

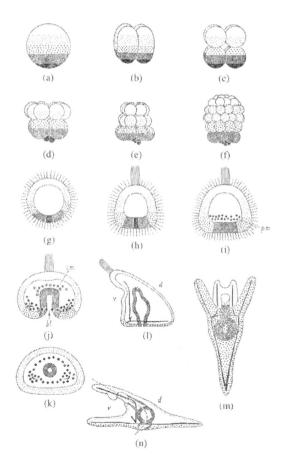

The fertilised egg, known as a zygote (A) undergoes its first cell division. The result is two daughter cells. Both divide further, in a plane perpendicular to the plane of the first division (B). The four new cells divide again perpendicular to the two earlier planes (C). The eight resulting cells are all the same size. But after that, another division takes place that results in cells of varying sizes. Eight of the sixteen cells lie in one plane and are called the mesomeres. Four others form the tiny micromeres, and the remaining four are the large macromeres between the other two groups (D). Another round of divisions produces an embryo of thirty-two cells (E). As more divisions take place, a hollow area begins to form in the interior of the embryo. As the hollow expands the embryo develops into a *blastula*, from the Greek word 'sprout' (G). Gradually cells break away from the layer of cells surrounding the cavity and wander into the open space (I). Meanwhile the part above the micromeres, the vegetal pole begins to fold into the cavity. This is the process of *gastrulation*, which results in the 'intestinal sprout' or *gastrula* (J). And then von Baer's germ layers form: the ecto-, meso-, and endoderm. After further development, a free-swimming, self-nourishing pluteus larva forms (K–M). After swimming around for a while, the pluteus transforms into a sea urchin during the process of *metamorphosis* (N).

ALKIMOS: It's a pretty elaborate way of constructing itself. There's no sign of preformation.

DEMOCRITUS: Dear Exton, this can't be all there is to the theory of development through cell divisions. After all, we can see a very important difference when macro-, meso-, and micromeres form: the cells are not uniform. The colony has only sixteen cells but it seems that it already starts assigning roles.

EXTON: This is due to unequal divisions. Finally, development leads to the process of *differentiation*. It results in the creation of different types of cells, such as muscle cells, red blood cells, photoreceptor cells, and sperm. The number of cell types in an adult animal is in the hundreds.

DEMOCRITUS: During differentiation cells thus receive their unique missions after cleavage. The development of an organism is thus a complicated interplay of cleavage and differentiation.

EXTON: I see you no longer require me. Your minds are so sharp that my thoughts on the history of embryology are superfluous.

DEMOCRITUS: Oh no, Exton. I'm afraid we're going to need you more than ever as we go deeper into the secrets of life. Now, let's proceed—the story is about to get even more interesting.

Part Two

Chromosomes, Mendelian Factors and Evolution

Roasted Capon

HERMES: How's it going?

COOK: Almost done.

HERMES: Me, too. We need to be finished in fifteen minutes. Honestly, since when does a messenger of the Gods set the table?

COOK: The big guys upstairs'll ask for anything.

HERMES: It wouldn't be such a big deal if Ganymede were here.

COOK: Oh, I know. He's not doing so well.

HERMES: But this is the age of antibiotics! If it were two hundred years ago, I could understand.

COOK: Oh, I know … hey, isn't it shocking how much they eat?

HERMES: Yes, especially the one with the crooked leg.

COOK: Well, the shaggy one certainly knows how to pack it away.

HERMES: I'll say. He eats like a horse.

COOK: The boss can't keep up with him.

HERMES: As if we need all this bother. It's too much.

COOK: Hermes, listen. I wanted to ask you earlier … what do they do in that machine?

HERMES: I have no idea. A thought strikes them and they rush over and squeeze in. I have to shut the door and pull this, push that. Sort of like what you do at the stove.

COOK: Hee, hee, hee. But what are they up to?

HERMES: I don't know, but they pack in there tight.

COOK: Maybe they can think better in a cramped space.

HERMES: Who knows? But when I pull the last lever, the thing starts to shake and they start screaming like half-stuck pigs.

COOK: Oh, I'd love to have a peek next time!

HERMES: Then everything goes silent, the shaking stops, and for an hour nothing.

COOK: I'll bet they pass out from their own stench, hee, hee, hee …

HERMES: Then the shaking starts again, and the screaming.

COOK: Blimey.

HERMES: Finally it stops, the door opens, and they wriggle out of. Then all they want to do is eat.

COOK: And the crooked-nosed one's the first one out.

HERMES: I don't even know how to describe their faces. The old, shaggy guy—Demotritus, or something like that—is usually half asleep, the other is sputtering and gasping, his hair a matted mess.

COOK: And the boss?

HERMES: Droopy as usual.

COOK: Lovely bunch. I'd just like to know where they go.

HERMES: Like I said, I have no idea. Now, let's hurry. They'll be here any minute.

COOK: Oh no, I think the capon's burning!

HERMES: Get it out of the oven fast! I'll bring the wine.

The Immortal Germplasm

EXTON: My friends, please have a seat at the table.

ALKIMOS: Ah, my favourite part of the day. Oh, what will it be? Is it …

EXTON: Yes, Alkimos, it is. The study of *chromosomes*.

ALKIMOS: Umm, I was referring to din—

EXTON: Alkimos?

ALKIMOS: Yes, of course. Chromosomes. *Kroma*, as in colour, and *soma*, as in body. So a chromosome is a coloured body. But that could describe anything.

EXTON: Let me explain to you exactly what it is. Hermes, you may serve dinner.

ALKIMOS: Oh, it's about time. I mean for you to explain things …

EXTON: We've arrived at a subject-matter that I think will excite you as no food or drink ever could—that is, assuming you're true scholars. We're approaching the greatest puzzle of genesis: Democritus' atoms. What operates living beings? How is it possible that similar creatures produce similar offspring? What ensures the immutability of a species as it grows from an egg to a fully developed organism?

ALKIMOS: It could only be the blood! But how does blood get into the sperm?

EXTON: According to Charles Darwin's theory of pangenesis, the grains, or *gemmules*, that arise in various body tissues accumulate in the sexual organs. Darwin thought of this as a kind of budding process—cells "throw off" gemmules that are alive. After fertilisation, every gemmule constructs its own appropriate tissue.

DEMOCRITUS: That's what Hippocrates argued.

ALKIMOS: Blood transports the gemmules to the gametes.

EXTON: Darwin's cousin, Francis Galton, tested the theory of pangenesis by performing blood transplants on rabbits. If the blood transported the gemmules responsible for inheritance, then the rabbits would produce offspring that resembled the blood donors and not themselves. But Galton never observed this, not even once.

DEMOCRITUS: So blood doesn't influence the hereditary material in the gametes?

EXTON: Not at all. The gametes don't contain any gemmules from the body's other cells, which are called the somatic cells.

ALKIMOS: So how do the gametes create offspring that resemble their parents?

EXTON: The gemmules or pangenes are already in them.

ALKIMOS: Just like the *animalcula seminalia*. However we look at it, the gametes are very important. New life begins with them.

EXTON: At this point we should acquaint ourselves with the theory of *Keimplasma,* or germplasm. August Weismann, a scientist active in Freiburg, proposed an alternative theory to Darwin's pangenesis. He was the greatest theoretician of the late 19th century. The scholarly world was introduced to his theoretical ideas in 1892, when he published his book *Das Keimplasma.* Before its publication, scientists believed that the embryo grows and develops before the gametes have matured. The gametes are thus created by the *soma,* the cells of the individual's body. The next generation can then construct new soma. The germs and somas alternate with each other, somewhat like this:

germ → soma → germ → soma → germ → soma → germ → soma

At the end of 1880s, Weismann developed his revolutionary idea. The roots of his theory can be traced to his research in Freiburg two decades earlier. At that time he was preoccupied with the metamorphosis of insects and the germ cells of hydras.

ALKIMOS: Hydras? I remember they were used in Trembley's experiments. Does this mean they're of greater use to the scientific world?

EXTON: Absolutely. It was Weismann's observations of the developing gametes or germ cells in hydras and dipteran insects that led him to the concept of a germline. In these animals, the germ cells do not separate from the somatic cells at the time of sexual maturation, but rather in the earliest stages of development. Weismann thus concluded that the germ cells of the hydra must contain 'something essential for the species, something which must be carefully preserved and passed on from one generation to another.'

ALKIMOS: Of course, because the gametes contain the inheritance gemmules.

EXTON: The separation of the germ and the somatic cells can also be observed in the multicellular green alga *Volvox*. The larger germ cells (kz) are distinct form the ciliated somatic cells (sz), as we can see here in Weismann's drawing:

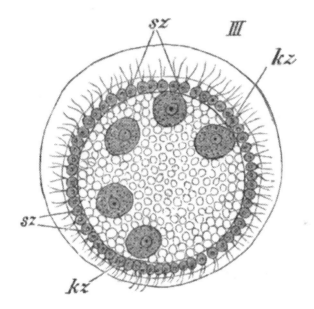

Weismann contended that the germ line was preserved intact during the process of development. Unlike the soma, the germ cell was immortal, and the germplasm it contained was passed on from generation to generation to ensure that new somas were always created. This illustration gives you an idea:

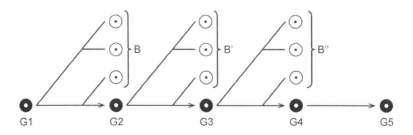

But if the germ really does separate completely, then the maturation of the soma, its differentiation, and any other kind of change does not affect germ cells. In this diagram, Weismann shows the development of the germ line of the nematode worm *Rhabditis nigrovenosa*.

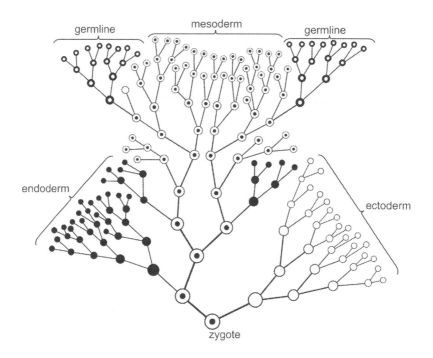

This is a cellular lineage tree that displays development from the zygote through the twelfth cell division. The germ cells, which are shown with a white nucleus in black cytoplasm, differentiate from the mesoderm at the sixth cell division.

DEMOCRITUS: This is why the characteristics an individual acquires in the course of life are not passed on to the next generation.

EXTON: That's right. But it's important to note that such an early differentiation of the germplasm is not universal. The soma of a plant, for example, can grow for years without producing germ cells. Nevertheless, Weismann's germ line theory had a profound impact on the theories of evolution and inheritance.

ALKIMOS: But how does the germplasm generate a nose or ear?

EXTON: According to Weismann, differentiation during development occurs by the gradual loss of the hereditary material during successive cell divisions. This is why specially differentiated cells came into existence: each retains only a subset of what he called *determinants.*

HERMES: Gentlemen, rooster testicles marinated in red wine.

EXTON: Excellent!

HERMES: I'm afraid the rest of dinner has mostly burned …

DEMOCRITUS: So the germ cells contain all the hereditary units necessary for the construction of the organism, but the heart cells contain only those units required by the heart cells.

EXTON: Right. Weismann's theory of development was supported by Wilhelm Roux, a German zoologist who conducted experiments on frog embryos. When the embryo was in the two-cell stage he used a hot needle to kill one of the cells. The surviving cell was only able to produce a half embryo. According to Weismann's theory the "left-hand" nuclear determinants could only enter the blastomere on the left, while the "right-hand" determinants could only enter the blastomere on the right.

DEMOCRITUS: In other words, the cells of the soma lose valuable hereditary material during development. They can't possibly be responsible for inheritance.

EXTON: At least that's what Weismann thought. But Oskar Hertwig …

ALKIMOS: … the first patron of sea urchins …

EXTON: … had a different opinion. To understand his objections, we need to examine the structure of germplasm. But where is the germplasm? Where can we find the hereditary material?

ALKIMOS: In the germ cells!

DEMOCRITUS: My intuition tells me we need to look in the nucleus. The living world is composed of cells, and within the cells are nuclei, but what lies within the nuclei? If nature is consistent, then its hierarchical arrangement requires another level.

HERMES: Gentlemen, frog heart with thyme.

ALKIMOS: Hurray! My favourite!

EXTON: Thank you, Hermes. Ernst Haeckel had similar ideas. In his book of 1866, *Allgemeine Anatomie*, he argued that the nucleus was responsible for inheritance, and therefore was the most important part of the cell. Observations made by Eduard Strasburger in 1884 confirmed Haeckel's idea. Born in Warsaw, Strasburger was later professor of botany in Jena and Bonn. During his study of the fertilisation of the orchid species *Orchis latifolia*, he discovered that when the pollen landed on the stigma of the flower, only the nucleus and not the protoplasm made it down the pollen tube and into the ovary.

DEMOCRITUS: So only the nucleus is necessary for fertilisation! We saw that in the fertilisation of the sea urchin.

EXTON: Yes, indeed. It was increasingly difficult to dispute the importance of the nucleus.

ALKIMOS: So does the nucleus contain the germplasm?

EXTON: Most likely. Now gentlemen, the time has come for us to take a look inside the nucleus.

DEMOCRITUS: Is that possible?

EXTON: At first, it wasn't easy. But as technology advanced, an enormously exciting world unfolded before the eyes of inquiring researchers. In 1848, Wilhelm Hofmeister, while examining *Tradescantia* and the cell division of other plants, noticed tiny clumps in the dividing cell nuclei as the nuclear membrane dissolved. He stained the mysterious clumps with iodine and observed that they divided into two separate clusters. The new nuclei then formed around these bundles. Have a look at Hofmesiter's depiction of what he saw:

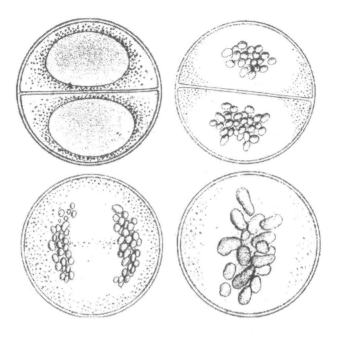

ALKIMOS: Are those the germplasms? Will we find that every sphere encloses another, which then encloses some bundles, and then more bundles? And this continues to infinity, as the preformists thought?

EXTON: Not entirely, Alkimos. In a moment you'll understand why. Similar forms were observed in animal cells during cell division. In 1873, Anton Schneider discovered thick rods in the nuclei of flatworm cells. Two years later Strasburger noticed similar thread-like structures in the nuclei of young fern cells undergoing division. The first to describe the behaviour of the strange threads was Walther Flemming, an anatomist at the University of Kiel.

ALKIMOS: So, now it's threads, not clumps.

EXTON: In 1879, Flemming tried a new type of dye in hopes of making different parts of the cell visible. The threads in the nuclei of dividing cells absorbed the most pigment. He named them *chromatin*, and the parts that absorbed less dye he called *achromatin*. The various preparations showed the dividing nucleus at different stages. His meticulous work allowed him to place the stages in order, and describe the entire process, step by step, of cell division. His book *Zellsubstanz, Kern und Zelltheilung* reveals that the threads shorten and thicken and line up in parallel rows along the dividing plane of the cell before division. At this point the chromosomes consist of a series of threads lying next to each other. The chromosomes then migrate to the two opposite poles, and the two daughter cells form around them, as you can see in Flemming's drawings.

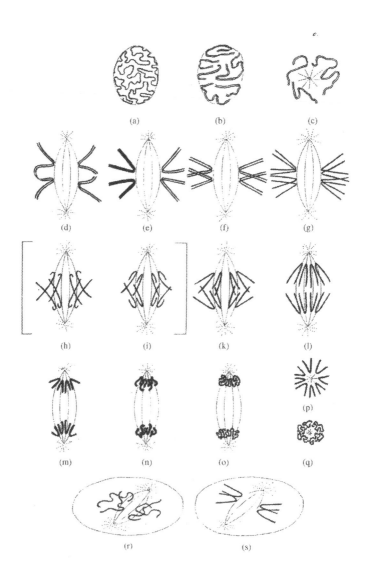

HERMES: If it pleases you, Democritus, sir.

DEMOCRITUS: Ah, paté. Thank you.

EXTON: Flemming called this process *mitosis*. He wrote, 'mitosis can be considered a kind of asexual reproduction in which the chromatin separates from the achromatin, and assembles in a characteristic formation, then divides in two. From these two new parts arise the nuclei of the daughter cells.'

DEMOCRITUS: So the crucial step in cell division is not the division of the nucleus, but that of the threads it contains. Otherwise what would justify such complicated choreography?

ALKIMOS: *Choreia* is dance, and *grapho* is …

EXTON: Thank you, Alkimos. Now, Wilhelm Roux argued the same thing. According to Darwin's theory of evolution, natural selection would not maintain a complicated mechanism if it didn't serve an important purpose.

ALKIMOS: Exton, what's natural selection?

EXTON: We'll talk about that later. First let's understand mitosis. Why is it necessary? Why don't nuclei simply split in two, as Remak suggested? According to Roux, the reason was that the chromatin is not homogenous. It contains a small number of particles with distinct qualities.

DEMOCRITUS: So a simple splitting of the nucleus is insufficient.

EXTON: That's right. The parts with their different qualities need to be carefully collected and then equally divided between the two daughter cells. This theoretical consideration led Roux to correct the diagram made by Flemming. Roux stated that during mitosis, the chromosomes lying next to each other split in two: one half becomes part of one daughter cell, and the other half part of the other. If we look carefully at Flemming's diagram, we notice that after the chromosomal threads split in two, they end up in the same daughter cell.

ALKIMOS: So, Flemming's drawing is wrong?

EXTON: Yes. Roux's recognition is one of the gems of theoretical biology. He reminds us that the best experimental–observational scientists—and

Flemming was among the best—may find themselves lost without a solid theoretical grounding. For a correct diagram of chromosome splitting we should look at Emil Heuser's drawing made in 1884:

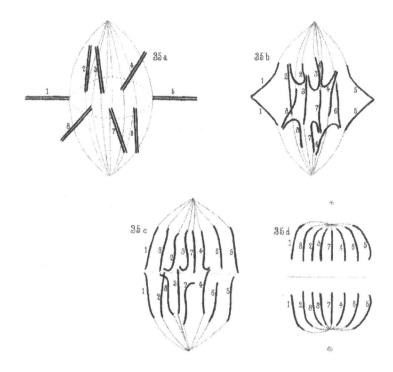

Chromosomes were found in nearly every living thing, and the complicated process of mitosis also turns out to be universal in the world of cells with a nucleus.

DEMOCRITUS: The unity of life is thus valid on the level of chromosomes.

EXTON: Because of their deep colour when dyed, Wilhelm von Waldeyer dubbed the threads chromosomes in 1888. It was Strasburger who named the first three phases of mitosis. *Prophase* is the stage at which the chromosomes become visible, *metaphase* is when they line up along the *metaphase plate*. At this point, it's possible to clearly see that the chromosomes comprise two halves, or *sister chromatids*, that lie next to each other. During *anaphase*, the chromosomes split longitudinally and the two sister chromatids migrate to the opposite poles. In the last phase—labelled *telophase* by Martin Heidenhain in 1894—two new cell nuclei are created. The chromosomes then disappear and are not seen again until the next cell division. At the next division, the chromosomes with the two sister chromatids reappeared in the same form at metaphase.

DEMOCRITUS: Allow me to summarise the process, dear Exton. In the meantime, you can finish your chicken soup. During cell division it's not the division of the nucleus that's so important, but the separation of the threads it contains. This is why the mechanism is so complicated, because the resulting daughter cells must contain the same quantity of chromatin. The coloured threads align themselves at the metaphase plate and then spilt longitudinally, resulting in two identical halves.

ALKIMOS: Please go on, master.

DEMOCRITUS: So it appears the chromosomes are the most carefully guarded elements in the nucleus. It's natural to assume that chromosomes are one and the same as germplasm. In other words we can consider them the bearers of hereditary material.

ALKIMOS: Yes, I wanted to suggest that, too.

EXTON: Weismann reached the same conclusion. In his own words: 'the complex mechanism of cell division exists practically for the sole purpose of dividing the chromatin, and that thus the latter is without doubt the most important portion of the nucleus. Since, therefore, the hereditary substance is contained within the nucleus, the chromatin must be the hereditary substance.'

DEMOCRITUS: However, this presents a serious problem. After all, Weismann posited that the hereditary material was gradually lost during the organism's development. But the exact splitting of chromosomes shows precisely the opposite: the mechanism ensures an exact division of the material; nothing is lost.

EXTON: Oskar Hertwig argued just that. Weismann's theory of unequal cell division is incompatible with cytological evidence.

DEMOCRITUS: Then we have two options. Either we reject Weismann's theory, or we reject the idea that chromosomes play a role in heredity.

EXTON: Weismann refused to flinch in the face of criticism. He asserted that although an identical quantity of chromatin thread did indeed arrive in each daughter cell, the quality was not the same. The identical cytological formations in each cell disguised an uneven distribution of the qualities of the nucleus. Hertwig, on the other hand, believed that every cell contains a matching set of chromosomes, and differentiation occurs because of the various environments surrounding the cells.

DEMOCRITUS: Do you mean Hertwig thought chromosomes weren't actually necessary during the development of the organism?

EXTON: Not exactly. He believed the entire hereditary material was present in the cells, but in each cell only the required hereditary units were active.

ALKIMOS: So the pangenes either work or don't.

EXTON: Precisely. This idea first appeared in the writing of the Dutch botanist Hugo de Vries. He postulated that animals probably didn't have even one cell in which all of the pangenes were active at the same time. Instead one or more groups of pangenes would become dominant and shape the cell in a certain way.

DEMOCRITUS: However it happens, one thing is certain: during development not only the cell and nucleus split in two, but the chromosomes too. New life springs from the fusion of the male and female gametes. After, the two pronuclei also merge, and their chromosomes come in contact with each other. Chromosome division is present throughout the entire development of the embryo. The role of the chromatin, however, is still a mystery. Does it gradually shed parts of its hereditary material, or do environmental variations causes only parts of it to be active? Another puzzle is whether pangenes exist inside or outside the chromosomes?

ALKIMOS: Or whether they exist at all.

EXTON: Unfortunately we'll have to wait to address the question of chromosome's role in inheritance. All I'll reveal is that it would have been impossible to answer without the aid of embryology. To unravel the mystery, the sciences of inheritance and embryology needed to join forces.

ALKIMOS: We saw that with the debate over preformation versus epigenesis.

DEMOCRITUS: The embryo thus inherited itself from its chromosomes. Ah, I think I'm beginning to see the answer ...

ALKIMOS: Then let's have a look at the embryos.

HERMES: Right away gentlemen. May I present to you ten-day-old partridge eggs with garlic.

ALKIMOS: And some wine ...

EXTON: As you surely remember, von Baer laboured over comparisons of various animal embryos and the morphological details of development. Although the concept of type and rhythm occurred to him, the internal cause, the mechanism, of development remained elusive.

ALKIMOS: Type and rhythm, type and rhythm.

EXTON: The influential German morphologist, Ernst Haeckel, famed for his approach to phylogeny, similarly failed to work out the mechanism. The main goal of Haeckel and comparative embryologists of the same vein was to determine the relationship and lines of ancestry based on the embryonic similarities between different species. This was all in order to support Darwin's theory of evolution from an embryological standpoint.

DEMOCRITUS: What kind of theory, dear Exton?

EXTON: You'll know in a minute. We could say that embryology was, at that time, merely the assistant science to the doctrine of evolution. Of course this period was enormously important. It was the golden age of embryology. Yet, an entirely new branch of science was needed to shed even a smidgeon of light on the details of development: experimental embryology. This new approach of investigating the reasons for development is attributed to Wilhelm Roux.

ALKIMOS: Roux, the frog pricker.

EXTON: Roux called his new approach *Entwicklungsmechanik*. His experiments with frog embryos, which supported Weismann's theories, reflect this new line of inquiry.

ALKIMOS: A half frog can't produce an entire frog!

EXTON: You both look sleepy. Perhaps it's time to retire.

ALKIMOS: But Exton, we're just getting excited about chromosomes.

EXTON: Hermes!

DEMOCRITUS: It's the gist of science, son. Hand me that cup.

ALKIMOS: Why don't you listen to my newest diptych instead!

DEMOCRITUS: All right, let's hear it.

ALKIMOS:

> *From the bulbus cordis, if ruptured, flows crimson blood*
> *Engulfing the sorrowful frog in a bath of comforting warmth.*

DEMOCRITUS: Very good. But what's a *bulbus cordis*?

ALKIMOS: The bulb of the frog's heart. I read it in Exton's book.

DEMOCRITUS: Pardon?

ALKIMOS: In this. See here: *The Handbook of Frog Dissection.*

DEMOCRITUS: My Zeus. That's something!

ALKIMOS: I'm sure the experiments with Roux's needle are in it too. Let's see … master?

DEMOCRITUS: Zzzzzzzz …

Reduction Division

EXTON: Democritus, Alkimos! Time to get up!

ALKIMOS: Hmmm …

EXTON: Come now!

ALKIMOS: Just a few more minutes …

EXTON: We have a lot to talk about. Today will be devoted to the deepest of the deep wells of the past.

ALKIMOS: I suppose I'm up for it, if we have good breakfast.

EXTON: We are going to acquaint ourselves with the astounding concept of evolution. But first we need to clarify a few important ideas concerning chromosomes.

DEMOCRITUS: Dear Exton, you've awakened my curiosity! Come on, Alkimos!

ALKIMOS: Okay, okay …

EXTON: We haven't addressed every issue surrounding chromosomal division. We saw yesterday that Weismann believed the germplasm was the bearer of heredity, but at the end of the 19th century not everyone was convinced. After all, the origin of chromosomes was unknown. Were they transient or permanent structures? Did they play a vital role in inheritance and development?

DEMOCRITUS: I'm wondering something too. We know that during development, cell divisions take place over and over again, and each time the number of chromosomes is halved. It follows that the stock of chromatin will eventually be depleted.

ALKIMOS: And then what?

DEMOCRITUS: Well, I don't believe it ever happens. I'm certain that during cell division something balances this process. The easiest solution I could come up with is that during cell division, the entire stock of chromosomes doubles, we just don't see this process because the chromosomes are not clumped.

ALKIMOS: Divided like the cells and the nuclei? But master, we don't know what happens to the threads when we can't see them. Perhaps they don't even exist, as Exton said. They disintegrate and only come into being again during cell division, when they assume their customary form. What's the point of them doubling?

DEMOCRITUS: It seems we've returned to the problem of preformation versus epigenesis. I believe that's at the heart of the debate over the origin of life and the development of organisms.

EXTON: Astounding, gentlemen! You have described precisely the great dilemma of the time. Do chromosomes exist during every phase of the cell's life, even if we don't see them? Or do they dissolve after cell division, and then regenerate in the next cycle? Strasburger favoured the latter. He believed at the time of cell division the microsomal elements of the nucleus reorganised to form a long thread. During mitosis the clustering of microsomes resulted in thin and thick sections. The thin sections could break, creating the chromosomes visible through the microscope. These chromosomes would later split in two, migrate away from each other, and after cell division, break up into microsomes. Other scientists, however, posited that between cell divisions, the chromosomes remained in the thread formation, but were invisible because they did not cluster into thicker strands. We now know that Democritus is right, and the chromosomes continuously exist, and they indeed double at every cell division. This process is called chromosome replication.

ALKIMOS: So you need to replicate, before you can split it.

DEMOCRITUS: Very elegant! But something still bothers me. If we accept that chromosomes always exist, and the daughter cells receive the same number as the mother cell had before division, then we're faced with yet another problem. What happens during fertilisation?

ALKIMOS: The ovum and the sperm fuse.

DEMOCRITUS: Yes, but if the ovum and the sperm contain chromosomes and pronuclei, and the two pronuclei unite after the sperm penetrates the ovum, then the number of chromosomes in the fertilised ovum is now double.

ALKIMOS: Then every new generation has twice as many chromosomes as their parents. But that's absurd!

DEMOCRITUS: I suspect, if such a process existed, it would be biologically unsustainable. We must assume that some kind of counteractive mechanism exists.

EXTON: Weismann made similar arguments. If we want to maintain the germplasm over the generations, then the gametes must only receive half as many chromosomes as the somatic cells, which are formed through mitosis.

DEMOCRITUS: So there must be some process by which the number of germplasm units in the gametes is halved.

ALKIMOS: And therefore half of the germplasm comes from the mother, and half from the father. That's clear.

EXTON: Earlier scientists observed that when the ovum matured, something unusual happened. The large ovum divided into one large cell and three small parts called polar bodies. Here is Oskar Hertwig's diagram of the formation of the polar bodies.

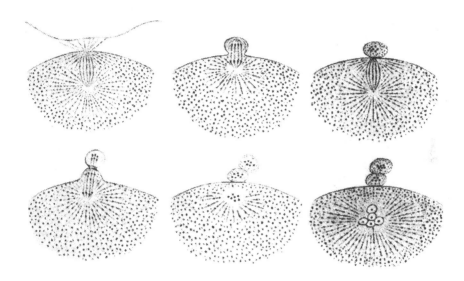

Weismann believed that the polar bodies were the receptacles for half of the germplasm of the egg. His argument was that the polar bodies remove as much hereditary material as the sperm contributes when it fuses with the egg. Weismann called this process *reduction division*, division by halving. Weismann then formulated a very insightful hypothesis. According to this it was improbable that during reduction division always the same germplasm units were removed by the polar bodies; therefore the mature ovum must always contain different combinations. Even to the extent that each and every mature ovum is slightly different. Weismann had hit upon the key to intra-species diversity.

DEMOCRITUS: That's why every individual is different, because they all arise from germ cells that contain different germplasm. If the number of sperms is in the millions, and we assume that fertilisation is random, then the number of combinations is mind-boggling. The miracle is not that two individuals are different, but that any two are the same.

EXTON: But only if chromosomes are indeed the bearers of hereditary material and the germplasm units are assorted randomly. For many scientists, it was not as obvious as it was to Weismann. But let's return to the germplasm and polar bodies. Weismann arrived at his theory of reduction division simply through logic, but his hypothesis was quickly confirmed through careful observation. This brings us to an outstanding biologist whose name we must not forget.

ALKIMOS: Exton, we've managed to remember everybody's name so far!

DEMOCRITUS: Without names and facts, the history of the study of inheritance would be nothing but jibber–jabber.

EXTON: I agree. The scientist I'm referring to is Theodor Boveri. He was born in Bamberg, Germany, in 1862. He studied at the University of Munich and then taught until he was thirty, when he was named Professor of Zoology and Comparative Anatomy at the University of Würzburg. He worked there until his death in 1915. During his years in Munich, he published some observations related to our topic. Boveri chose a truly bizarre animal to study: *Ascaris megalocephala*, the round-worm that lives in horses' intestines. His discoveries ushered in a new era in the study of inheritance.

DEMOCRITUS: Ah, how much science owes to the strangest of critters: the freshwater hydra, the aphid, the sea urchin, and the roundworm. So many huge secrets concealed in such despicable creatures! Because of the unity of the living world, basic observations of the oddest life forms can do as much to enrich science as studies of the supposedly important organisms, including ourselves.

ALKIMOS: Lay people often disparage investigations of lower forms.

DEMOCRITUS: Because they have not yet been captivated by the magic of a unified living world with all its profound interconnections.

ALKIMOS: Perhaps they never will …

EXTON: So I see you've mastered the skills of rhetoric. I think you'll likewise master *Ascaris megalocephala* and its chromosomes as easily as Boveri did. After all, it has only four chromosomes, two paternal and two maternal, as you can see in this drawing by Edouard van Beneden, a Zoology Professor from Leiden.

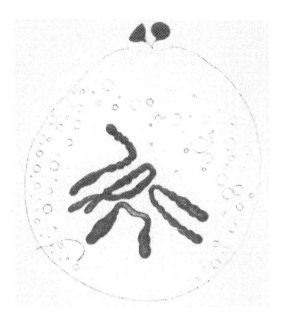

Boveri observed that during the maturation of the germ cells into gametes, the number of chromosomes dropped to two, in harmony with Weismann's theory of reduction division. In his book of 1890, entitled *Zellenstudien*, he wrote, 'there is a reduction by half of the originally existing number of chromosomes, and this numerical reduction is not therefore only a theoretical postulate but a fact.' He called this process of division by which the number of chromosomes decreased by half *meiosis*. Boveri observed that when the gametes formed, the chromosomes gathered in two groups, so that both contained four chromosomal strands. He called these groups of four *Vierergruppen*, or tetrads.

ALKIMOS: Two times four is eight. But why did four become eight?

EXTON: Wait a minute! During the first division of the ovum, both tetrads split into two dyads. One dyad from each tetrad remains in the ovum, and the other two dyads enter the first polar body. During the second division of the ovum, the two strands of chromosomes in the dyads separate. Two remain in the ovum and two enter the second polar body.

Meanwhile the first polar body divides again. Altogether there are three polar bodies and one ovum, and each one has two strands of chromosomes, as you can see in this diagram. The order of the panels is a bit confusing, but follow from A to H.

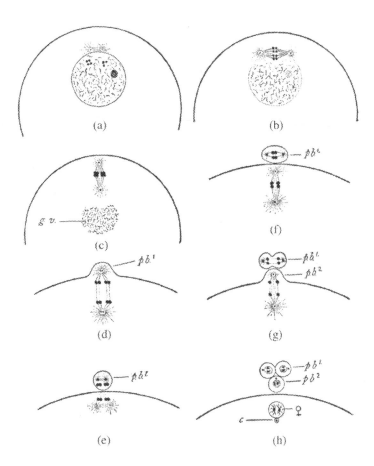

In that same year Oskar Hertwig demonstrated that sperms underwent the same process during maturation. A sperm mother cell divides, creating two spermatocytes that also divide. The end result is four sperm cells as you see here:

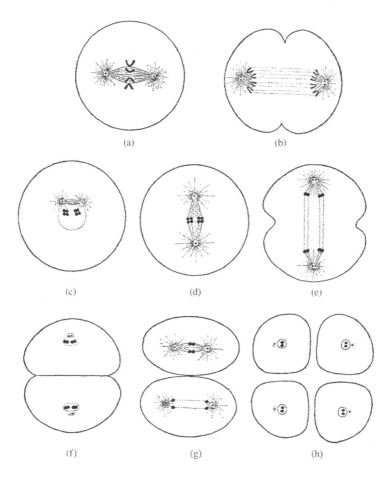

(a) (b) (c) (d) (e) (f) (g) (h)

When the ovum matures the same thing happens, but only one ovum is derived and three polar bodies.

DEMOCRITUS: So the polar bodies are just rudimentary cells.

EXTON: Exactly. They aren't the receptacles of expulsed germplasm as Weismann thought. The four sperms produced by the sperm mother cell correspond to the ovum and three polar bodies.

DEMOCRITUS: But it's strange that two rounds of cell division take place, when one would suffice to halve the number of chromosomes.

EXTON: Weismann wondered the same thing. Hertwig and Boveri soon provided an answer, which I think Alkimos knows.

ALKIMOS: Um, because ... well, four became eight. The chromosomes doubled before each cell division, so ...

EXTON: Weismann was very happy with the answer. If eight, instead of four, chromosome threads separate randomly at the time of gamete production, then more combinations of germplasm units are possible. The essence of sexual reproduction is the creation of great variety, the most thorough mixing of the hereditary units of the mother and father. I must emphasise, however, that the mixing does not increase the diversity of the hereditary units itself—for that *mutations* are necessary. But we'll talk about that later.

DEMOCRITUS: Dear Exton, did the observations by Boveri confirm my idea about the continuity of the chromosomes? If he could observe them in *Ascaris* to such incredible detail, maybe he could also have seen them in the resting cell.

EXTON: In 1888, based on his observations of *Ascaris* chromosomes, Boveri indeed concluded that the chromatin elements remained as independent structures during the resting phase of the cell, too. He was right. Despite his cogent arguments, however, most scientists refused to accept that chromosomes retained their individuality. The reason was simple: most often nothing could be seen in the nucleus except for perhaps some kind of amorphous tissue.

DEMOCRITUS: Fascinating. Let me summarise what you've told us from the perspective of preformation versus epigenesis.

ALKIMOS: Yes, master, please do!

DEMOCRITUS: We have seen that the subject of chromosomes reignited impassioned debate. Although the studies of Wolff and von Baer appear to have definitely proven epigenesis, we still might ask: is the development of complicated living organisms a process by which complexity arises from nothing, or is it the gradual transition of an already existing, invisible form into a visible one? To the school of epigenesis, any effort to tie the path of development to hereditary material attached to immortal chromosomes was suspicious.

EXTON: That's true. That's the reason why the epigenesis camp rejected the individuality and permanence of chromosomes in favour of the microsome concept.

DEMOCRITUS: In contrast, the preformists—if we can call them that—swore by the individuality of chromosomes and the immortality of germplasm.

EXTON: Strasburger, for example, believed that the units aligned on the chromosomes were identical, and the chromosomes were transient. In contrast, neo-preformists, like Weismann and Roux, stressed that development was none other than the gradual appearance of differentiation that already existed in the germplasm.

DEMOCRITUS: But if chromosomes continually disintegrate and then reassemble, as Strasburger thought, and don't contain different hereditary units, then the division of the nuclear material is not responsible for cell differentiation. The cause would have to be the interaction of the cells or the effects of their environment.

ALKIMOS: That's true epigenesis.

EXTON: Yes, in a way it is. I suggest we end this conversation with the words of the eminent scientist Oskar Hertwig. Let's try to enter his frame of thought so we can better enjoy his eloquent rebuke of Weismann's theory of preformation.

'To satisfy our craving for causality, biologists [such as Weismann] transform the visible complexity of the adult organism into a latent complexity of the germ, and try to express this by imaginary tokens.... Thus craftily, they prepare for our craving after causality a slumbrous pillow.... But their pillow of sleep is dangerous for biological research; he who builds such castles in the air easily mistakes his imaginary bricks, invented to explain the complexity, for real stones. He entangles himself in the cobwebs of his own thoughts, which seem to him so logical, that finally he trusts the labour of his mind more than nature herself.'

ALKIMOS: Hertwig could have been a great writer.

EXTON: Perhaps. You know, Boveri wanted to be a painter.

ALKIMOS: And I a poet ...

DEMOCRITUS: But you are, my friend, you are!

A London Pigeon Sale—*Thomas Henry Huxley*

DEMOCRITUS: The living world is connected everywhere by hidden threads. Von Baer's laws reveal this unity in embryos; cell theory in the tissue of plants and animals; and chromosomes, with their magical dance, in the innermost realm of living things. What we need is a new explanation of the unity of life. We must discover the origins of this unity, no matter how difficult it proves!

ALKIMOS: As we view smaller and smaller details, we find ever greater similarities. My master's atoms must lie at a level so deep that this unity applies to everything in the world. There, forms and variations cease to exist. The fundamental element rages in itself, trapped in a prison of maximum simplicity.

EXTON: I see you've both been deeply affected by the relationship between sameness and difference. May I recommend we set aside chromosomes and immerse ourselves in Charles Darwin's theory of the origins and evolution of life? Because Darwin was a rather modest, sickly and reserved person, I suggest we pay a visit to the greatest proponent and disseminator of his theory: Thomas Henry Huxley, one of the 19th century's greatest writers of biology and philosophy, also known as 'Darwin's bulldog'.

ALKIMOS: The 19th century? I have a bad feeling. I'm glad I asked for the remaining roast goose.

EXTON: Hermes. The Deus ex Machina! Direction London, 1882!

HERMES: Yes Sir Extonos.

COOK: Can I go, too?

ALKIMOS: Oh no, here we go …

EXTON: Re-laxxxx, wwwe'll beee there sooooon!

ALKIMOS: Wait, are we here already? Thank Zeus!

EXTON: Yes, everyone out. We're in London. What's this—it's raining again?

ALKIMOS: What's going on? Who are all these people, and what are they doing out in this abominable weather?

DEMOCRITUS: Quiet, son. Exton, why have you brought us here?

EXTON: To find Huxley, of course. I wrote him a letter and he said he'd be happy to meet with us today. Let's go into the abbey.

ALKIMOS: What are all those candles and that dark platform?

EXTON: This is Charles Darwin's funeral. We're in Westminster Abbey.

HUXLEY: Mr. Darwin's death is a wholly irreparable loss. And this not merely because of his wonderfully genial, simple, and generous nature; his cheerful and animated conversation, and the infinite variety and accuracy of his information; but because the more one knew of him, the more he seemed the incorporated ideal of a man of science. Acute as were his reasoning powers, vast as was his knowledge, marvellous as was his tenacious industry, under physical difficulties which would have converted nine men out of ten into aimless invalids; it was not these qualities, great as they were, which impressed those who were admitted to his intimacy with involuntary veneration, but a certain intense and almost passionate honesty by which all his thoughts and actions were irradiated, as by a central fire.

ALKIMOS: That's a catafalque!

DEMOCRITUS: Sh!

HUXLEY: One could not converse with Darwin without being reminded of Socrates. There was the same desire to find someone wiser than himself; the same belief in the sovereignty of reason; the same ready humour; the same sympathetic interest in all the ways and works of men. But instead of turning away from the problems of nature as hopelessly insoluble, our modern philosopher devoted his whole life to attacking them in the spirit of Heraclitus and of Democritus, with results which are as the substance of which their speculations were anticipatory shadows.

DEMOCRITUS: In my spirit?

ALKIMOS: Master, sh!

HUXLEY: None have fought better, and none have been more fortunate than Charles Darwin. He found a great truth, trodden under foot, reviled by bigots, and ridiculed by all the world; he lived long enough to see it, chiefly by his own efforts, irrefragably established in science, inseparably incorporated with the common thoughts of men, and only hated and feared by those who would revile, but dare not. What shall a man desire more than this? Once more the image of Socrates rises unbidden, and the noble peroration of the 'Apology' rings in our ears as if it were Charles Darwin's farewell: 'The hour of departure has arrived, and we go our ways—I to die and you to live. Which is the better, God only knows.'

EXTON: Democritus, Alkimos, come on! Let's have a walk around the abbey, and then wait for Huxley outside.

ALKIMOS: Only now do I dare to look around. What a magnificent building!

EXTON: Construction of the abbey began in the 11th century, during the reign of Edward I. It was the place of William the Conqueror's coronation on Christmas Day, 1066. Every coronation since, with the exception of Edward V's and Edward VIII's, have been held here. It's also the resting place of kings and queens.

DEMOCRITUS: And Charles Darwin. So the kings and queens of the country and the mind are placed side by side. But who is Socrates?

EXTON: The nave was built in the 14th century. At thirty-one metres high, it has the greatest interior height of any building in England. Let's go outside and have a look at the Gothic system of flying buttresses that make the architectural feats of the nave possible.

DEMOCRITUS: Nave, flying buttresses and Gothic. A land of majestic order.

EXTON: Scholastica and Abbot Suger. But let's see how the notion of common origins and its branching tree of life rocked the hierarchy of gods and mortals.

ALKIMOS: Common origins?

EXTON: According to one of Darwin's brilliant theories all living things evolved from the same ancestor. The succession of life forms can best be represented not by a ladder or a chain, but by a tree.

ALKIMOS: The tree of life?

EXTON: Exactly. This is Haeckel's phylogenetic tree of animals made in 1873.

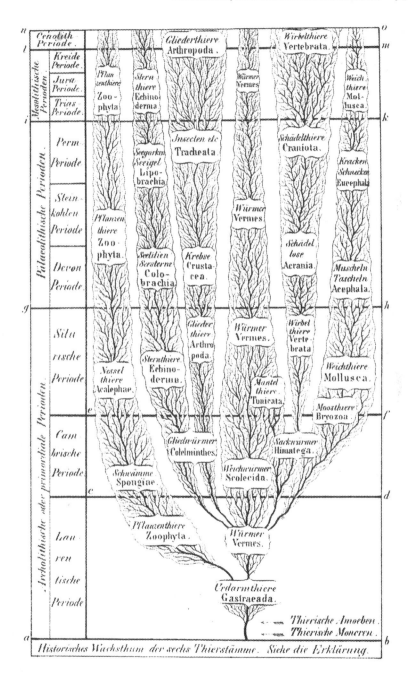

DEMOCRITUS: Why didn't we ever think of this! Common origin—that explains the similarities between living things. We're all related.

EXTON: After *On the Origin of Species* was published, the theory of common origins immediately gained widespread acceptance, unlike Darwin's other theories. But we'll hear the rest from Huxley. Let's find him!

ALKIMOS: That way! I saw him slip out the door.

EXTON: He should be expecting us. Mr. Huxley, Mr. Huxley! It's Exton here, from the University of Sussex.

HUXLEY: Ah, Mr. Exton. These gentlemen must be your students, Mr. Alsmith and Mr. Dimatthews. It's an honour to meet you. Please come this way.

DEMOCRITUS: Mr. Huxley, we heard your eulogy. Now we have a clearer picture of who Mr. Darwin was. But we still aren't sure how he contributed to our understanding of the unity of life.

HUXLEY: He shed light on the origin of species. But let's continue down the street. Around the corner is Royal College.

EXTON: I think my students would better understand if we discussed the arguments presented in *On The Origin of Species*.

HUXLEY: Of course. Please come into my office.

EXTON: The book was published in 1859 and it changed every branch of biology. I hope that after our conversation my students will appreciate how.

HUXLEY: Have a seat, gentlemen.

EXTON: Thank you.

HUXLEY: The principle tenet in Mr. Darwin's book is that species form through *natural selection*, or the survival of the fittest in the struggle for existence. To understand this we need to clarify the concepts. First of all, what does he mean by 'species'? The question seems simple but arriving at the correct answer is difficult. According to some, a species is the total number of individuals who all trace their ancestry to the same set of parents. Others contend that a species is nothing more than the smallest defined group of living things. Yet still others consider a species to be an existing group that is permanent and immutable, or just the opposite, a non-existent category arbitrarily created by people.

ALKIMOS: For my first try, I'd say a species is a group of living things that are highly similar to each other in appearance and behaviour.

HUXLEY: The main problem with that definition is how to determine the degree of similarity. The truth is the number of highly differentiated living things is greater than we could ever imagine. There are at least several hundred thousand different types of insects in collections. The total number of types of living things is more than half a million. The difficulty is that every form has different versions, and among them are many nearly indistinguishable transitional forms. The task of deciding which are species and which are mere variations is extraordinarily complicated.

DEMOCRITUS: How discouraging! Couldn't we devise a method to aid us in distinguishing between species, and then we could more clearly determine what counts as merely a variant? Why can't we formulate a clear definition of a species?

HUXLEY: Some experts have asserted that categorisation is indeed possible. Living things belong to the same species if they can mate and produce offspring. However, if they cannot produce offspring, or their offspring are sterile *hybrids*, then they are said to belong to a different species. These experts believe this inability of animals of a different species to reproduce exists to preserve the purity of the species.

EXTON: So, we can say that a species is a group that can reproduce with each other but not with other similar groups because of behaviour, physiology, or other reasons. However, we know that many living things have found ways to defy this definition.

DEMOCRITUS: If we take this all into account, it seems the breath taking variety within and between species disproves the concept of a species as an immutable product of a single act of creation.

HUXLEY: Indeed. Moreover, we might presume that every part of a living organism is perfectly adapted to its role in life. Yet, as soon as we start our examinations, we'll discover the individual properties of living things are far from perfect. We'll discover the useless teeth in a cow or whale embryo. We'll come across insects that never chew during their entire lives, yet have mandibles; or they never fly, yet have small, rudimentary wings. And then we'll encounter animals that are completely blind, yet have vestigial eyes.

DEMOCRITUS: The idea of perfect adaptation may not help us understand the existence of odd body structures, but it can perhaps explain something else: why do some creatures live only in certain parts of the world and never in others? The palm tree cannot survive in mountain climates; the oak tree will never take root in the frozen north; the bear cannot live in the lion's realm and vice versa.

HUXLEY: If we study the geographical distribution of plants and animals it seems hopeless that we should ever understand the peculiarities and capricious relationships we encounter. And although an *a priori* assumption that every country is populated by living things that are best suited to the environment is tempting, we'll run into some problems. For example, how do we explain that at the time the New World was discovered, the South American Pampa had no cattle? We can't argue that they wouldn't be able to tolerate the conditions there, considering that now millions live in the wild all over South America. Strangely enough, many plants and animals of the northern hemisphere not only do fine in the southern hemisphere, but in some cases are more ideally suited for the region and eventually outcompete indigenous species.

DEMOCRITUS: In other words, the species of a geographic region may not be best suited for the climate or other features of the place. The multiple imperfections made apparent by the peculiarities of a living thing's physical make-up or its geographic distribution cannot be the work of the gods. Hephaestus would never craft a dull spear. It seems the gods of Olympus lost their power to create life, and they let nature govern herself according to her own laws.

HUXLEY: A rather advanced notion. But let's take another example. The plants and animals on distant islands in the oceans are unlike any other known species. Yet, they most resemble those species on the nearest dry land. Wherever we look, we are overwhelmed by the perplexing mysteries of nature.

ALKIMOS: I suspect the origin of this puzzle lies in the puzzle of origins.

DEMOCRITUS: We've been struggling with these questions for some time. Now let's listen to Huxley.

HUXLEY: Our knowledge is not limited just to the living world. Beneath the surface of the Earth lie layers upon layers of rock. The accumulation of these extraordinarily thick layers occurred over huge periods of time.

Deep within these petrified cliffs we can find traces of past organic life — of long-ago plants and animals that were somehow trapped in mudflows and eventually turned to stone. Our museums have on display shells that are immeasurably old, yet in perfect condition; skeletons completely intact; petrified embryos; and even footprints.

EXTON: The petrified remains of ancient plants and animals are known as fossils. This drawing of fossilised shells is from Martin Lister's book *Historium animalium Angliae*, published in 1678.

HUXLEY: There's no question it's a marvellous book. It shows that as many distinctive species are being excavated from the depths of the Earth as can be seen on its surface. What's astounding is that these unearthed species of the past are completely different than any living species today.

ALKIMOS: How is that possible?

HUXLEY: Because the species transform. The fossils provide us an opportunity to observe the laws of diversity. Generally, the deeper we dig, the less similar the species become to present-day varieties. By the same token, the farther apart we find two extinct species, the less similar they are to each other.

ALKIMOS: That means the species of the past gradually succeeded each other over time, becoming more and more like today's species.

HUXLEY: Exactly. Some believe that this succession of species resulted from a series of catastrophic events in which all was destroyed, followed by new periods of creation. Today, however, this view seems increasingly flawed. The progression of life forms may appear to have breaks in it, but that's only because we have lacked the knowledge to interpret it properly. Earlier life forms were not eradicated all at once, as in a catastrophic event, but were gradually—step by step—squeezed out by new forms. If it were possible to observe every event of the past, the borders between the various geologic periods or the individual episodes in creation would blur, as in a rainbow, whose colours transition slowly and continuously into another separate colour.

DEMOCRITUS: If I understand, during the long history of life, newer and newer species have always appeared. Since we have rejected the concept of spontaneous generation, and we can't assume divine intervention, we must accept that the new species come into being through the transformation of the old ones. But what then defines a species? Certainly not immutability.

HUXLEY: It's difficult to reject the notion of the immutability of species. Only recently the majority of scientists stressed two major characteristics of species. First, every species is permanent and invariable. Second, new species arise through separate acts of creation. Non-scientists have accepted this as true, because they assume a wealth of evidence stands behind it. But some more thoughtful individuals have questioned the permanence of species. Among them was the famous French naturalist Jean Baptiste Lamarck. In his time, he was the greatest expert on simple life forms and was also known for his extensive knowledge of botany. Two observations appear to have exerted crucial influence on his thinking. One is that animals of the highest order are separated from lower-order animals only by degrees. The other is that living organisms may develop in a specific direction, and the impact of specific efforts and the subsequent properties acquired can be inherited by the next generation. Lamarck conjectured that if an animal is placed in a new environment, it will have a new set of needs. The animal's efforts to satisfy these needs will result in corresponding changes to the organism.

ALKIMOS: If someone becomes a blacksmith, his arm muscle will strengthen accordingly. If this new property is passed on, over the generations, his blacksmith descendants will be increasingly muscular.

HUXLEY: You haven't read Lamarck, have you, Mr. Alsmith?

ALKIMOS: No, I just inferred that from what you've said.

HUXLEY: Well, you certainly are creative. Ahem, now back to Lamarck. An example he provided was of some short-necked birds that wanted to catch fish without getting wet. Over time, their persistent efforts resulted in the lengthening of their necks, and thus the emergence of long-necked wading birds. Lamarck's hypothesis has since been rejected, and today, sadly many newcomers to the field have mocked his ideas. But I warn you, gentlemen, one should never cast aspersions on a great man because of a mistaken theory. Although his theory was based on some assumptions we now know to be false, the brilliant, logical structure of his arguments far outshine his errors. There is much we can learn from Lamarck.

DEMOCRITUS: Indeed, we've learned a lot from his theory, as we have from so many that have been proven wrong. However, I'm afraid we still don't see how to proceed. So much mystery still surrounds the origin of species, but perhaps Darwin's doctrines can help illuminate the matter.

HUXLEY: If you pay attention, gentlemen, you'll see that Mr. Darwin did indeed unlock the secret of the origin of species and the changes they undergo. Like Lamarck, he didn't believe in the permanence of species, but his theory to explain the changes went well beyond the French naturalist's. His book *On The Origin of Species* immediately grabbed the attention of naturalists and geologists and his generalisations have been widely accepted. I'd say, without question, *On The Origin of Species* is the greatest intellectual achievement of the century, and its contents have done more to promote the progress of science than anything before.

DEMOCRITUS: So, let's hear some of his brilliant arguments.

HUXLEY: Allow me to use the example of some recent London events to shed light on his reasoning. At the annual fair held at the Baker Street Bazaar, some oxen of varied appearance—some straight backed and small headed, others broad-chested, and none resembling their counterparts in the wild—and a dozen different breeds of sheep were competing for attention and prizes. Also on display were absurdly fattened pigs as unlike a wild pig as the town councillor an orangutan. After the procession of four-legged livestock came the poultry. Much can be said about those clucking and cock-a-doodle-doing wonders, but one thing is sure: they look nothing like their ancestors, the *Phasianus gallus*. Those dissatisfied with the variety exhibited on Baker Street need only take a stroll through the Seven Dials district of London. The pigeon breeds are just as varied and in no way resemble their relatives in the wild.

EXTON: My students will better understand if I show them these drawings by Mr. Darwin of various English pigeon breeds.

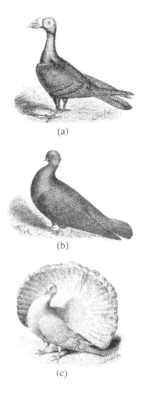

(a)

(b)

(c)

HUXLEY: Of course, very useful.

DEMOCRITUS: Our domesticated flora and fauna display a stunning variety. It's hard to believe they all descended from individuals in the wild.

HUXLEY: Yes, it is. But careful studies have demonstrated that they were all created by one method: an observant breeder notices a tiny variation in one of the individuals in the stock. To maintain this trait, the breeder selects a male and female that both possess this variation and crosses them. Then, only those offspring that best present this desired characteristic are bred. Selective breeding continues until a new group is created that is sufficiently different than the original stock. This rigorous method of selection creates new breeds. The degree of variation achieved through this method is vast. Imagine if all we knew of today's numerous breeds of dogs, pigeons and horses was from the fossil record. We'd almost certainly categorise the different breeds as independent species.

DEMOCRITUS: It's mindboggling to think that so many of the well-known plants and animals around us were completely transformed by selective breeding. If all life forms can be altered in this way, then we have no choice but to reject the immutability of species and the underlying notion of perfect form. There's no such thing as *the horse*, only *a horse*. Species have no abstract content; they are merely changeable interbreeding aggregates of individuals.

HUXLEY: One of Mr. Darwin's main arguments was that living creatures change, and their forms are variable, as we have seen in the case of the pigeon.

DEMOCRITUS: Okay, but there's a snag. For a new species of domesticated plants or animals to be created, you need the sharp eye of a breeder. But how can this explain the great diversity of species in the natural world? What force in nature corresponds to our sharp-eyed breeder? Although I suspect this is exactly what Darwin intended to show.

HUXLEY: Indeed it is. In his book, Mr. Darwin clearly stated that he had discovered the force that powered selection in nature; the *modus operandi* of what he called natural selection. The process proposed by Mr. Darwin is extremely simple, and is the result of several obvious, but often ignored, facts.

ALKIMOS: What are these ignored facts?

HUXLEY: Think about the ferocious struggle that every living thing wages with other living things day in, day out. Has any scientist thought about the outcome of this struggle? And it's not just animals that fight with animals, but plants, too, engage in fierce battles with both plants and animals. The soil is full of seeds that never develop into seedlings. Those plants that succeed in sprouting steal the sunlight, air, and water from one another. Only the most aggressive among them triumphs, obliterating the competition. If wild animals are undisturbed by humans, their number is relatively stable, but we know that the number of their offspring can span from one to nearly a million. Therefore, it's a mathematical certainty that almost as many offspring die as are born every year. Only those few offspring survive that can resist the destructive forces of nature and whose abilities surpass their siblings. The individuals of a species are in exactly the same situation as the crew on a sinking ship: only the good swimmers have a chance to reach shore.

DEMOCRITUS: So that's the foundation of natural selection?

HUXLEY: Yes. Mr. Darwin discovered the driving force of natural selection in nature's bounty and its wastefulness. Let's suppose in the never-ending struggle several individuals of a species just happen to change a little, giving them a slight advantage over their companions: they acquire nutrients more efficiently, develop more quickly, defend themselves more effectively, and produce several offspring that resemble themselves. These offspring will also have advantages over their fellow members of the species, and over time they will out compete them. It becomes clear that certain members of the group gradually go through more and more changes until a variation is arrived at that is best adapted to the given conditions. As long as the conditions don't change, no new forms will be capable of outcompeting it. It's also clear that at a certain point, the changes consolidate into what can be considered a new species.

DEMOCRITUS: Nature truly is capable of selection; this is obvious from the large number of offspring produced and the almost equally large number that are defeated. What's more, these random changes can take place before our very eyes. How unusual the workings of nature are! The essence of Darwin's theory is the survival of random variations according to the laws of natural selection. What is disconcerting about Darwin's theory, unlike Lamarck's, is that no goal drives this process. The transformations are dictated by changes in the external environment. Lamarck believed the process was driven by the efforts of internal sentiments, which he thought animals possessed, but which they most certainly do not.

EXTON: But Darwin had his problems. He didn't have a clear understanding of inheritance, and without it, it's impossible to accurately interpret the phenomenon of evolution.

HUXLEY: Very good, gentlemen. You've correctly ascertained that inheritance is one of the key elements in the theories of transmutation or transformism.

EXTON: Let's not forget that Lamarck accepted the idea of inheritance of acquired traits. He believed an individual could pass on characteristics acquired through 'internal efforts' to its offspring. Although Darwin didn't entirely reject the inheritance of acquired traits, his theory also works without it, while Lamarck's would collapse.

ALKIMOS: But Exton, Weismann strongly opposed the inheritance of acquired characters in his book *Das Keimplasma* published in 1892.

HUXLEY: I'm not familiar with this book of Weismann's. What did you say Mr. Alsmith? When was it written?

ALKIMOS: 1892.

EXTON: Alkimos!

HUXLEY: You must be mistaken. This is the year 1882. How odd ...

ALKIMOS: I wanted to say ...

HUXLEY: How is that possible? Did Weismann show you his manuscript? I don't understand.

EXTON: Mr. Huxley, we don't want to keep you. We'll be on our way.

HUXLEY: Wait a moment! The question of the inheritance of acquired traits is of key importance.

EXTON: We agree, but I'm afraid we're out of time. Alkimos, Democritus, quickly!

HUXLEY: Wait. What did Weismann tell you? What do you mean by 1892?

EXTON: Quickly, on board!

HUXLEY: Gentlemen, what sort of vehicle is this?

EXTON: I'm starting it up!

ALKIMOS: Ohh, here we go!

COOK: Did you hear that, Hermes? Is that what you were talking about? They're screaming their heads off.

HERMES: Eh, I'm used to it. Now, into the kitchen!

The Orchard of Evolution

ALKIMOS: Master, are you awake?

DEMOCRITUS: Yes, son. I was just thinking about what we've heard. It's disconcerting to think that humans are merely one of Mother Nature's special creations. Our startling resemblance to some species of apes and the correspondence of our tiniest parts, our cells and chromosomes, to theirs, points us toward only one unavoidable conclusion: Darwin's theories apply to us too.

EXTON: Huxley's drawing highlights the astonishing similarity between the skeletons of humans and apes.

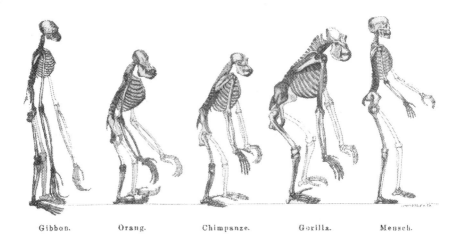

Gibbon.   Orang.   Chimpanze.   Gorilla.   Mensch.

DEMOCRITUS: We have no choice but to acknowledge our reason for placing ourselves at the centre of the universe, above all other living things and at the convergence of time, was our overestimation of our importance, our desire to dominate, and our unbounded vanity. Darwin has showed that our haphazard and transient existence is not a culmination and not even a beginning, but rather a hazy awakening.

EXTON: I think you'll understand why Darwin's theory provoked such hostility and violent attacks. The idea that humans had animal origins offended the sensibilities of the falsely modest, and consequently they employed every possible weapon to fend off their opponents. Their arsenal of arguments, however, shrunk over time, and today only a small number of extraordinarily limited and disturbingly unscientific and simplistic rebuttals remain in their armoury. Still, they fight on with such tenacity there's little hope of pacifying them. Most people have accepted that the Earth revolves around the Sun, but the idea that we have common ancestors with orangutans and also with goats is too much for some to swallow.

ALKIMOS: Luckily the questions of science are not decided by a popular vote. There'd be no progress.

DEMOCRITUS: Dear Exton, let's return our focus to Darwin. What effects did his theory have?

EXTON: I also propose that we examine each of the notions presented by Huxley and try to interpret them through the lens of Darwin's theory of transmutation, or *evolution*.

DEMOCRITUS: I already have a question. If the origin and transmutation of species give rise to variations best adapted to their environment, then why do we find so many imperfections in the living world?

EXTON: Yes, the theory of natural selection does seem to suggest the creation of the most perfect form. However, the creating force of nature has many limitations on it. Any new innovation must make use of the already existing components. This frequently results in strange solutions that work surprisingly well. Wings arose from legs, petals from leaves, nectar glands from petals, and teeth from scales.

DEMOCRITUS: Nature is like a tinker who patches the broken pitcher from whatever materials are available.

EXTON: Yes, one of the fundamental features of evolution is the ingenious tinkering from whatever materials—bone, feathers, glands—are on hand. This is precisely what separates it from the designs of an engineer or the gods. The products of evolution contain within every part of them their own history. A divine creature or a designed object, in contrast, is perfect and efficient, the relationship of all its parts carefully calculated, but nowhere does it contain evidence of its history. Moreover, it's independent; its resemblance to any other creature has no explanation. The imagination of the Creator is, *per definitionem*, endless. Unlike nature, the Creator is not limited to making variations on the huge number of basic forms.

ALKIMOS: During Creation, the different species popped from Zeus's forehead in full armour.

EXTON: There's something very important I should stress about the theory of evolution: it's a mathematical certainty that it works. The theory is in fact nothing but an algorithm, a system of certain instructions and rules. If the necessary conditions are provided, evolutionary change is inevitable.

ALKIMOS: But what are these conditions, Exton?

DEMOCRITUS: For example, the subjects of evolutionary change must be able to reproduce.

EXTON: Yes, that's the first requirement.

DEMOCRITUS: If reproduction is the emergence of many from one, then smashing an amphora to the ground would be enough. But I don't believe that's evolution. An indispensable tenet is that like beings bear like offspring.

ALKIMOS: So inheritance must be the second requirement!

EXTON: That's true. But imagine if this system of inheritance were inflexible? What would happen to evolution if every offspring were absolutely identical to its parents?

ALKIMOS: Well, that depends …

DEMOCRITUS: No, it's clear! Without variation reproduction would just mean individuals copy themselves, and evolution could not take place.

EXTON: Exactly. But it's also obvious that reproduction is not endless because the resources are finite. Perhaps we could describe the algorithm of evolution like this. We have a system by which individuals reproduce. Furthermore, the offspring resemble their parents, but with a certain amount of variation. The degree of variation—the strength of inheritance—is a crucial factor. If this system is placed in an environment with limited resources, then evolutionary change is inevitable.

DEMOCRITUS: The question is does the natural world meet these criteria?

ALKIMOS: Master, we've been talking about this for days! The mysteries of reproduction, inheritance, and variation. There's no doubt that every individual in the natural world fulfils these requirements.

EXTON: These three criteria are unique to the living world, their most internal law. The triumvirate of reproduction, inheritance, and variation can be considered the definition of life. Thus, it's no surprise that the three phenomena are among the most fundamental questions of biology.

DEMOCRITUS: Since they are an inevitable component of the living world, it's undeniable and inescapable that under their influence, living creatures will undergo evolutionary change. This change continues as long as the three conditions are met. If any one of them fails, evolution and life will cease at once. We cannot avoid this conclusion: life is the ability to evolve.

EXTON: And life can only endure and change in the presence of evolution. The end of one necessitates the end of the other.

ALKIMOS: Oh heavens! Without change, time would cease!

EXTON: That's right. Now you understand how a mere moment can last an eternity. Static eternity is Heaven itself.

DEMOCRITUS: But change can also be eternal. An eternally changing heaven, however, cannot forever remain heaven. Sooner or later it will change so much that its master will not recognise it.

ALKIMOS: Is that why the gods are constantly killing each other? Every so often the Hekatonkheires need to be released from Tartarus, otherwise it would be the end of Olympus?

EXTON: I think so, too. Darwin discovered the laws governing evolution. Examples abound in the several-hundred pages of *On The Origin of Species* that bear out the rules and algorithms of evolution. If we understand how the rules work, it will be easier to understand the main observations listed by Huxley.

DEMOCRITUS: Transformations occur, and during natural selection, these changes will either disappear or a series of selections will result in individuals of the species that are increasingly well adapted to their environment.

EXTON: This is the process of *adaptation*. Wading birds, for example, do not develop elongated necks because of inner efforts, but because of selection. Of the various water birds, those that have longer necks and can better reach the fish will produce more offspring.

ALKIMOS: Those that are better suited to their aquatic environment.

EXTON: Yes, but it's important to note that by environment, we don't just mean rocks and sunshine. Nature is so crowded that the strongest determinants of a creature's environment are often other living things. Most often, changes are adaptations to each other, minor adjustments, but they can also result from fierce competition.

DEMOCRITUS: Species only make sense if we consider the other species around them.

ALKIMOS: Now I understand how species change. But how do new species come into being?

DEMOCRITUS: I don't think it's too hard to imagine. Let's suppose the individuals of a species drift away from each other, perhaps covering such a large territory that the individuals along the perimeter never meet with one another. If these groups encounter different conditions of natural selection—which is highly possible—then they will gradually begin to differ from one another. After some time, the differences will be so great that we can call them separate species.

EXTON: This is the process of *adaptive radiation*. It's the link between adaptation and the creation of a new species. The most famous example is the origin of Darwin's finches in the Galapagos Islands. Darwin first observed that closely related finch species lived on different islands. The body structure of each—in particular the shape of the beak—was adapted to its food sources and other specifics of its environment. The original South American finch had arrived to the islands and adapted to the variety of conditions on the different islands, and in the process developed into new species.

ALKIMOS: But we hit a snag, Exton! What do we mean exactly by a new species?

EXTON: It's useful if we accept the notion that a species is a community of animals that can reproduce with each other. If a group can no longer reproduce with another, then it is a new species.

DEMOCRITUS: I suspect the moment of transition is difficult to ascertain in many cases. If two groups of species have no contact with each other, and have already begun to diversify, it's quite difficult to decide at what point a new species emerges.

EXTON: Yes, the moment of transition is blurred. We have to accept that we have no opportunity to decide, even in principle, the moment that one species becomes two. This is what makes biology so exciting!

DEMOCRITUS: But after a while it becomes clear if a new species has evolved. If no individuals in the two groups reproduce with each other under any conditions, either in the wild or in captivity, then we know.

EXTON: That is the essence of the biological species concept. Species have no birthdays. Often we can't determine if they've even been conceived. We only know once they have achieved their adult form.

DEMOCRITUS: Species thus emerge and change, fill geological layers, and distant islands. The closer they are in time and space the more closely they will resemble the species they arose from.

ALKIMOS: This explains fossils and the unusual living world found on islands. Both bear witness to the continuous transmutation and reproduction of species. But what about heredity?

EXTON: Heredity. We haven't dealt with that yet. But we cannot completely understand evolution without the laws of inheritance and transmutation. This thought should give us the strength to face the towering obstacles in our way. Gentlemen, you have struggled valiantly thus far. Now it's time to tackle this question that has confounded us for long enough: what is the essence of inheritance? It was mostly thanks to the groundwork of animal breeders and horticulturalists, that the laws of inheritance were revealed. We can consider their work the roots of our tree of understanding. However, it took a brilliant mind, Gregor Mendel, to nurture the sprout and his long series of careful experiments to help it grow into a tree. Someone was needed who would break with the traditions of plant hybridisation and ask different questions of a different sort. However, his work grew out of a rich soil, the practice and thoughts of centuries of cultivators. An early forerunner to the cultivators was the philosopher Francis Bacon. Now I'll read to you a beautiful text, an excerpt from Bacon's masterpiece, *The New Atlantis*, published in 1626. The inhabitants of his utopian island, Bensalem, performed a great feat of cultivation:

'We have also large and various orchards and gardens; wherein we do not so much respect beauty, as variety of ground and soil, proper for divers trees and herbs: some very spacious, where trees and berries are set whereof we make divers kinds of drinks, besides the vineyards. In these we practise likewise all conclusions of grafting, and inoculating as well of wild-trees as fruit-trees, which produceth many effects. And we make (by art) in the same orchards and gardens, trees and flowers to come earlier or later than their seasons; and to come up and bear more speedily than by their natural course they do. We make them also by art greater much than their nature; and their fruit greater and sweeter and of different taste, smell, colour, and figure, from their nature. And many of them we so order, as they become of medicinal use.'

Peas and Minotaur—*William Bateson*

DEMOCRITUS: Darwin's theory is simple, yet its marvellous structure has tremendous power to explain the formation and the peculiarities of the living world. The origin of forms and species, their design and adaptations did not require the intervention of an intelligent creator or a final cause. Darwin's brilliant theory puts to rest the contradictory notion of the infinitely varied work of a creator all encased within ever smaller germs.

ALKIMOS: Master, Exton said that something crucial is missing from Darwin's theories. The laws of inheritance.

EXTON: Yes, this is true. Although theories of inheritance existed in Darwin's time, all of them—including Darwin's theory of pangenes—were wrong. The only exception was the theory of Gregor Mendel, an Augustine friar from Brno. Through careful and time-consuming experiments crossing different kinds of peas, he arrived at a fundamentally new understanding. His findings were potentially revolutionary, but were ignored for thirty-five years. By 1900, however, scientific thought had developed enough that scientists understood the import of his work. In that year, three scientists independently rediscovered Mendel's 1866 essay *Versuche über Pflanzen-Hybriden*. In addition to these three scientists, the British biologist William Bateson, originator of the word *genetics*, was also working to gain quick recognition for Mendel's theories. Bateson was the most active Mendelian of the first decades of the 20th century. His efforts contributed significantly to the rapid widespread acceptance of genetics and the clarification of its basic concepts. Mendel's theories languished in obscurity until Bateson helped weave them into the fabric of science. That's why it's Bateson we'll visit, not Mendel. Let's listen to the speech he gave to the Royal Horticultural Society in 1900.

ALKIMOS: Don't you have the manuscript? I'd rather just read it.

EXTON: Hermes, prepare the Deus ex Machina! Direction London, 1900!

ALKIMOS: But we're still wet!

EXTON: On the way you can read an earlier article by Bateson. Come on, let's board!

ALKIMOS: Oh … not again ….

EXTON: Wwwwhere's the pppppaperrr? Ahhh, hhhere itttt is. Now, Bateson recognised that to understand the laws of inheritance, precise, controlled hybridisation of plants was necessary. In 1899 he wrote, 'What we first require is to know what happens when a variety is crossed with its nearest allies. If the result is to have a scientific value, it is …'

DEMOCRITUS: Son, move over.

ALKIMOS: Sorry, master.

EXTON: Pay attention! Now, where was I? Yes, '… it is almost absolutely necessary that the offspring of such crossing should then be examined statistically. It must be recorded how many of the offspring resembled each parent and how many showed characters intermediate between those of the parents. If the parents differ in several characters, the offspring must be examined statistically, and marshalled, as it is called, in respect of each of those characters separately.'

ALKIMOS: That makes sense.

EXTON: That's all I'll say. We'll hear the rest at his speech. Ah, we've arrived. Everyone out!

ALKIMOS: What is this place? The Royal Horticultural Society?

EXTON: That's it! Let's go into the auditorium.

ALKIMOS: It's already started!

BATESON: An exact determination of the laws of heredity will probably work more change in man's outlook on the world, and in his power over nature, than any other advance in natural knowledge that can be clearly foreseen. There is no doubt that these laws can be determined. In comparison with the labour that has been needed for other great discoveries it is even likely that the necessary effort will be small. It is rather remarkable that while in other branches of physiology such great progress has of late been made, our knowledge of the phenomena of heredity has increased but little; though that these phenomena constitute the basis of all evolutionary science and the very central problem of natural history has been admitted by all.

DEMOCRITUS: There's no debate. Without inheritance, there's no evolution. Lamarck accepted the inheritance of acquired characters. Darwin, however, got tangled up in the idea of pangenes mixing. I'm eager to hear what Bateson has to say and how will it affect our explanation of evolution.

BATESON: ...Nor is this due to the special difficulty of such inquiries so much as to general neglect of the subject. It is in the hope of inducing others to pursue these lines of investigation that I take the problems of heredity as the subject of this lecture to the Royal Horticultural Society.

ALKIMOS: We don't need much inducement. We understand the importance of the question. We're thirsty for answers!

BATESON: No one had better opportunities of pursuing such work than horticulturalists. They are daily witnesses of the phenomenon of heredity. Their success also depends largely on a knowledge of its laws and obviously every increase in that knowledge is of direct and special importance to them.

DEMOCRITUS: Think about how much even Francis Bacon's inhabitants of New Atlantis knew about inheritance. Given the huge numbers of cultivated plants they had, their knowledge must have been prodigious, and today's horticulturalists know even more. Why do we need to clarify the laws of inheritance for them?

BATESON: … we must distinguish what we can do from what we want to do. We want to know the whole truth of the matter; we want to know the physical basis, the inward and essential nature, "the causes," as they are sometimes called, of heredity. We want to know the laws which the outward and visible phenomena obey.

EXTON: Pay attention. This is my favourite part.

ALKIMOS: Exton, how do you know what comes next?

EXTON: I've already heard the entire speech.

BATESON: Let us recognise from the outset that as to the essential nature of these phenomena we still know absolutely nothing. We have no glimmering of an idea as to what constitutes the essential process by which the likeness of the parent is transmitted to the offspring. We can study the processes of fertilisation and development in the finest detail which the microscope manifests to us, and we may fairly say that we have now a considerable grasp of the visible phenomena; but of the nature of the physical basis of heredity we have no conception at all. No one has yet any suggestion, working hypothesis, or mental picture that has thus far helped in the slightest degree to penetrate beyond what we see. The process is as utterly mysterious to us as a flash of lightning is to a savage. We do not know what is the essential agent in the transmission of parental characters, not even whether it is a material agent or not. Not only is our ignorance complete, but no one has the remotest idea how to set to work on that part of the problem.

DEMOCRITUS: This is true. Weismann's germplasm was pure fiction. Chromosomes can be seen, but we have no basis for understanding how they are connected to inheritance. The cell nucleus and the chromosome are nothing but apparitions with no actual physical substance.

ALKIMOS: Master, I'd like to hear Bateson ...

BATESON: But apart from any conception of the essential modes of transmission of characters, we can study the outward facts of the transmission. Here, if our knowledge is still very vague, we are at least beginning to see how we ought to go to work.

ALKIMOS: Maupertuis saw it too, one hundred and fifty years ago.

BATESON: All laws of heredity so far propounded are of a statistical character and have been obtained by statistical methods. If we consider for a moment what is actually meant by a "law of heredity" we shall see at once why these investigations must follow statistical methods. For a "law" of heredity is simply an attempt to declare the course of heredity under given conditions. But if we attempt to predicate the course of heredity we have to deal with conditions and groups of causes wholly unknown to us, whose presence we cannot recognise, and whose magnitude we cannot estimate in any particular case. The course of heredity in particular cases therefore cannot be foreseen ... In dealing with phenomena of this class the study of single instances reveals no regularity. It is only by collection of facts of great number, and by statistical treatment of the mass, that any order or law can be perceived.

ALKIMOS: That doesn't sound encouraging.

BATESON: ... with heredity it is somewhat as it is in the case of the rainfall. No one can say how much rain will fall tomorrow in a given place, but we can predict with moderate accuracy how much will fall next year, and for a period of years a prediction can be made which accords very closely with the truth.

ALKIMOS: In London it will be a lot.

EXTON: Alkimos, pay attention!

BATESON: We are as far as ever from knowing why some characters are transmitted, while others are not; nor can anyone yet foretell which individual parent will transmit characters to the offspring, and which will not; ... As yet investigations of this kind have been made in only a few instances ...

ALKIMOS: It seems those investigations have been quietly forgotten, or they've been ignored.

BATESON: ... the most notable being those of Galton on human stature, and on the transmission of colours in Basset hounds. In each of these cases he has shown that the expectation of inheritance is such that a simple arithmetical rule is approximately followed. The rule thus arrived at is that of the whole heritage of the offspring the two parents together on an average contribute one half, the four grandparents one-quarter, the eight great-grandparents one-eighth, and so on, the remainder being contributed by the remoter ancestors.

EXTON: To better understand this, we should have a look at Galton's theory of inheritance as illustrated in this diagram:

In this diagram, the entire square—regardless of the trait in question—represents the genetic inheritance of the individual. The smaller squares within it, marked with numbers, each refer to one ancestor, and the size of the squares indicates the degree to which each ancestor contributes to the total inheritance. The individual is assigned the number one. Here's how we can interpret the other numbers: if we take any member of the family tree and assign them the letter n, then their father is $2n$. The mother is $2n+1$. As a result every male in the family tree is an even number and every female is an odd number.

ALKIMOS: That means, for example, that the offspring of number **2** is number **1**, and number **2**'s partner was number **3**? The offspring of **6** and **7** is the individual marked **3**.

EXTON: That's correct. To make it easier to understand, Galton indicated the females with a black background. The numbering could go on to infinity; the diagram stops, however, at the fourth generation. As you can see Galton thought the contribution of the great–great grandparents was just 1/16.

ALKIMOS: I don't understand, Exton. How can the two parents be responsible for only half of the inherited traits? I thought distant ancestors didn't make any contributions to the characteristics of the individual. But Galton's diagram suggests that the parents carry on their backs the four grandparents, who carry the eight great-grandparents, and so on, all of them industriously working to create one offspring. I wonder if Mr. Galton understood how difficult that would be.

EXTON: Obviously that's not what Galton meant when he constructed his theory of inheritance. He wanted to express the mixing of traits and the contribution of earlier ancestors according to rules of mathematics. But you're right, Alkimos, that the mixing doesn't exactly work that way. The diagram may be rather misleading, but before Galton, no one had ever tried to express the law of inheritance in mathematical terms. Except, of course, for Mendel. Now, let's resume listening to Bateson.

DEMOCRITUS: He's still talking about Galton …

BATESON: … Galton points out that it takes no account of individual prepotencies. There are, besides, numerous cases in which on crossing two varieties the character of one variety almost always appears in each member of the first cross-breed generation. Examples of these will be familiar to those who have experience in such matters. The offspring of the Polled Angus cow and the Shorthorn bull is almost invariably polled or with very small loose "scurs."

ALKIMOS: Polled?

DEMOCRITUS: Meaning it has no horns.

ALKIMOS: Ah. As you see, I have little expertise in this area. What do you suppose Minotaur would say?

EXTON: Sh!

BATESON: ... seedlings raised by crossing *Atropa belladonna* with the yellow-fruited variety have without exception the blackish-purple fruits of the type. In several hairy species when a cross with a glabrous variety is made, the first cross-breed generation is altogether hairy.

DEMOCRITUS: These examples clearly show that the mixing of traits is not general. In instances like these, alternative traits not only don't mix, but one squeezes out the other.

EXTON: Sh! He's about to discuss Mendel!

BATESON: Until lately the work which Galton accomplished stood almost alone in this field, but quite recently remarkable additions to our knowledge of these questions have been made. In the year 1900, Professor de Vries published a brief account of experiments which he has for several years been carrying on ...

EXTON: The Dutch botanist Hugo de Vries was one of the three scientists who independently rediscovered Mendel's treatise. The other two were Carl Correns, a German geneticist, and Erich Tschermak, an Austrian botanist. Now let's hear what he has to say about de Vries' description.

BATESON: The description is very short, and there are several points as to which more precise information is necessary both as to details of procedure and as to statement of results. Nevertheless, it is impossible to doubt that the work as a whole constitutes a marked step forward, and the full publication which is promised will be awaited with great interest. The work relates to the course of heredity in cases where definite varieties differing from each other in some one definite character are crossed together. The cases are all examples of discontinuous variation: that is to say, cases in which actual intermediates between the parent forms are not usually produced on crossing. It is shown that the subsequent posterity obtained by self-fertilising these cross-breeds or hybrids, or by breeding them with each other, break up into the original parent forms according to fixed numerical rule.

ALKIMOS: Wait a minute! The progeny of Minotaur and his polled girlfriend would all be polled as well; the next generation, however, would again contain horned individuals?

EXTON: Yes. Moreover the proportion of horned and hornless would be fixed.

BATESON: Professor de Vries begins by reference to a remarkable memoir by Gregor Mendel, giving the results of his experiments in crossing varieties of *Pisum sativum*. These experiments of Mendel's were carried out on a large scale, his account of them is excellent and complete, and the principles which he was able to deduce from them will certainly play a conspicuous part in all future discussions of evolutionary problems. It is not a little remarkable that Mendel's work should have escaped notice, and been so long forgotten.

ALKIMOS: Why didn't we leave a copy on Huxley's desk?

BATESON: For the purposes of his experiments Mendel selected seven pairs of characters as follows:

1. Shape of ripe seed, whether round, or angular and wrinkled.
2. Colour of "endosperm" (cotyledons), whether some shade of yellow, or a more or less intense green.
3. Colour of the seed-skin, whether various shades of grey and grey-brown, or white.
4. Shape of seed-pod, whether simply inflated, or deeply constricted between the seeds.
5. Colour of unripe pod, whether a shade of green, or bright yellow.
6. Shape of inflorescence, whether the flowers are arranged along the axis of the plant; or are terminal and more or less umbellate.
7. Length of stem, whether about 6 or 7 inches long, or about ¾ to 1½ inches.

EXTON: We should note that Mendel chose these seven from twenty-two pairs of traits based on previous experiments. The seven selected pairs of traits behaved in a similar fashion to the bulls' horns. There were no intermediary forms; for example the seed skin was always grey–brown or white, but never greyish white.

DEMOCRITUS: The rest is easy to deduce. If Mendel crossed peas with opposing traits, the progeny showed the trait of just one parent.

EXTON: Exactly. The trait that overpowers the other is called *dominant*. The trait that is suppressed is called *recessive*.

ALKIMOS: So Minotaur's recessive horns were defeated by the cow's dominant hornlessness?

DEMOCRITUS: We can also infer that if we take those first-generation hybrids, which all exhibit the same characteristic inherited from just one of the original parents, and breed them with each other, their offspring will exhibit the characteristics of both the original parents.

EXTON: Mendel didn't even cross the individuals in the first generation of hybrids with each other; instead he pollinated each plant with its own pollen—in other words artificial self-pollination. To avoid unwanted cross-pollination, he covered the pea flowers in little sacs.

ALKIMOS: Like Spallanzani's taffeta pants covering the male frogs' behinds?

BATESON: Gentlemen in the back, quiet please.

EXTON: Excuse us.

BATESON: By self-fertilising the cross-breeds Mendel next raised another generation. In this generation were individuals that showed the dominant character, but also individuals that preserved the recessive characters. This fact is also known in a good many instances. But Mendel discovered that in this generation the numerical proportion of dominants to recessives is approximately constant, being in fact *as three to one*. With very considerable regularity these numbers were approached in the case of each pair of his characters ... These plants were again self-fertilised, and the offspring of each plant separately sown. It next appeared that the offspring of the recessives *remained pure recessive*, and in subsequent generations never reverted to the dominant again.

DEMOCRITUS: Those individuals bearing recessive traits thus irrevocably lost something—the ability to exhibit dominant traits.

ALKIMOS: And what happened to the individuals with dominant traits when they reproduced? Did their progeny show only dominant traits or did they show both dominant and recessive traits like their parents?

EXTON: Surprisingly, one portion produced only progeny with dominant traits, and the other a mix. In other words, one group remained irreversibly dominant, meaning every consequent generation achieved through self-pollination exhibited only dominant traits. The other group produced progeny with dominant and recessive traits in the same 3:1 ratio. The overall ratio between these two groups was 1:2.

DEMOCRITUS: It's starting to get complicated. Let's have another look at the proportions. One quarter of the second generation is pure recessive, and if they further reproduce with each other, their progeny will all be recessive. Another quarter is pure dominant and will only produce dominant progeny. And half consists of individuals who appear dominant but will produce offspring that bear both dominant and recessive traits in a 3:1 ratio. I suspect that these mixed offspring will produce a mix of purebreds and hybrids in the same proportions: ¼, ½, ¼.

EXTON: Exactly! But let's listen to what Bateson has to say.

BATESON: … of one dominant to two cross-breeds. The process of breaking up into the parent forms is thus continued in each successive generation, and the same numerical law is followed. Mendel made further experiments with *Pisum sativum*, crossing pairs of varieties which differed from each other in two characters, and the results, though necessarily much more complex, showed that the law exhibited in the simpler case of pairs differing in respect of one character operated here also … Professor de Vries has worked at the same problem in some dozen species belonging to several genera, using pairs of varieties characterised by a great number of characters: for instance, colour of flowers, stems, or fruits, hairiness, length of style, and so forth. He states that in all these cases Mendel's principles are followed … When we consider, besides, that Tschermak and Correns announce definite confirmation in the case of *Pisum*, and de Vries adds the evidence of his long series of observations on other species and orders, there can be no doubt that Mendel's law is a substantial reality… That we are in the presence of a new principle of the highest importance is, I think, manifest. To what further conclusions it might lead us cannot yet be foretold. Thank you for your attention. If you have any question, please don't hesitate to ask.

DEMOCRITUS: Professor Bateson, allow me …

BATESON: Yes, in the back?

DEMOCRITUS: Your speech was very impressive. I'd just like to make one comment. Knowing the laws of fertility discovered by Hermann Fol, Oskar Hertwig, and Eduard Strasburger, I believe we need to think about something. If we suppose that fertilisation is random, and also that the gametes contain only one half of every pair of attributes, will our final results show the same Mendelian proportions?

BATESON: Excellent observation. Indeed Mendel calls our attention to the same obvious conclusion. The results of hybridisation are just what we would calculate, if we presumed that the pollen and ovum produced by each plant only carried one of two opposing characters. If this were the case, and the characters were divided evenly between the pollen grains and the ovum, then obviously during random pollination, the progeny would show traits that correspond to Mendel's laws. After all—according to the laws of randomness—one quarter of the pollen grains carrying the dominant character would fertilise the one quarter of the ova also carrying the dominant character. The same would happen for the one quarter of the pollen grains and ova carrying the recessive character. And of course the remaining half carrying the opposite characters would also meet. Based on this reasoning, in accordance with Mendel's laws, the conclusion that the gametes can only pass on one or the other of the opposing characters, but not both, is sound.

DEMOCRITUS: What is found separately in the gametes, is reunited at the moment of fertilisation. The traits thus also halve and then double during the germ-line cycles, just like the chromosomes!

BATESON: Possibly, but I seriously doubt the two phenomena can be related. I don't believe that formations in the simple nucleus could possibly be identified as the complicated forces that govern heredity. By the way, have we met? Are you members of the Royal Society?

DEMOCRITUS: Yes, or rather, no. We're just passing through.

BATESON: Where are you from?

ALKIMOS: My master has come from Abdera, and I'm from Thrace.

EXTON: Alkimos!

ALKIMOS: And Exton is from the Institute of the Deus ex Machina in Alexandria!

BATESON: Deus ex Machina? Is this something related to the occult?

EXTON: I'm afraid we must be off. We're on our way to Cornwall.

BATESON: Ah, a bicycle tour?

EXTON: A what?

DEMOCRITUS: I'm shocked. How could Professor Bateson not find my brilliant comparison more compelling?

EXTON: Come on!

BATESON: Wait a minute! We can debate the matter!

EXTON: Quickly, let's start the motor!

DEMOCRITUS: The germplasm therefore manifests itself in the attributes inherited according to Mendel's laws! An incredible coincidence!

EXTON: Get in, get in!

BATESON: Thrace? Whoever heard of such a thing?

ALKIMOS: Ah … here we go!

Galton and Mendel

ALKIMOS: Master, are you awake?

DEMOCRITUS: Yes, I'm just dizzy from the shaking. Travelling is difficult at my age.

ALKIMOS: Yes, but as long as your mind is sharp, you have no reason to complain.

DEMOCRITUS: Ah, that's true, Alkimos. Quick feet are not what make a philosopher great.

EXTON: If you're tired, let's have a rest in the dining room. A quick morning meal will give us the mental acuity we need to further explore peas and hybrids and the theories of inheritance.

ALKIMOS: Excellent idea!

EXTON: We've seen that before Mendel's work was rediscovered many theories of inheritance had been formulated. Darwin, Weismann, and de Vries all developed their own—erroneous— theories. According to these theories, the gametes contain numerous copies of every unit of inheritance. Mendel's ingenious proposal was that only one copy of each hereditary unit, which he called factors, existed in a germ cell. This is the only way to explain the fixed 3:1 ratio of opposing traits. And it was this theoretical innovation, not his laws, that was so radical. In fact his laws are not really laws at all. Take the law of independent assortment: it's not true in the case of linkage. What's more, his laws are not new. After all, horticulturalists have been aware of the uniformity of the F1 generation for some time. Mendel's greatness lies in the conscious connection he made between theory and practice; through carefully planned and conducted experiments he succeeded in quantifying inheritance patterns and interpreted his results according to his radically new theory.

DEMOCRITUS: Dear Exton, I'm quite taken by your enthusiasm, but I believe you've forgotten something.

ALKIMOS: Indeed you have. What do you mean for example by *efwon*?

EXTON: Gentlemen, please excuse me. Mendel is one of my favourite subjects. Let's slow down. Bateson was the first to use the terms F1 and F2 to refer to the first and second *filial generations*, in other words the offspring resulting from hybridisation. The characteristics of these generations can be understood if we suppose that every germ cell contains only one element, or factor, of every pair of opposing traits. To understand this better, let's break down the question. In his experiments, Mendel chose peas that differed from each other in one characteristic. He used plant lines that were pure bred, in other words the plants had been bred only with like plants. The germ cells of these plants were uniform with respect to this characteristic. He then labelled one of the opposing characteristics, for example a smooth seed, with a capital *A*, and the other, for example a wrinkled seed, with a lowercase *a*. The germ cells of the pure line of smooth seed peas contained only the *A* factors, while those of the wrinkled seed had only *a* factors. When the two plant lines are crossed, the gametes fuse and the offspring will carry one capital *A* and one lowercase *a* factor.

ALKIMOS: But the offspring all look alike since they all have one *A* and one *a* factor and *A* is dominant!

EXTON: That's right, Alkimos. If Mendel crossed plants with a difference in just one characteristic, the next generation of plants looked the same in that characteristic. This is known as the *principle of uniformity*.

DEMOCRITUS: When germ cells are created the capital *A* and the lowercase *a* separate, and half of the cells contain the capital *A* and half the lowercase *a*.

EXTON: Exactly. The gametes are *haploid*, because they contain only one hereditary factor, but when two gametes fuse during fertilisation, they return to a *diploid* state.

ALKIMOS: Now I understand. These plants are the F1 generation. If these plants are bred with themselves, as they were in Mendel's experiment, then their progeny will produce the capital *A* and lowercase *a* characteristics in a 3:1 ratio.

EXTON: Assuming the meeting of the gametes is random. Based on Mendel's experiments, Thomas Hunt Morgan diagrammed the hybridisation of short-stemmed peas, labelled with a lowercase *k* (klein), and long-stemmed peas, labelled with a capital *K* (Groß).

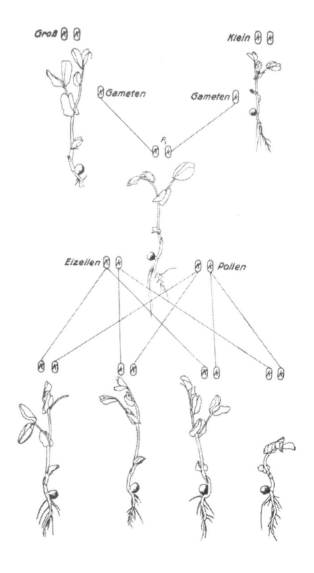

The English geneticist Reginald Punnet created a table, known as a Punnet Square, to demonstrate the cross-breeding process.

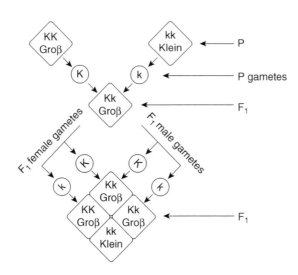

What the square shows is that the 3:1 ratio is in fact 1:2:1. One quarter of the offspring contain only the *A* characteristic, and if self-pollinated, or "selfed," would behave like a pure line. One half contain one *A* and one *a*, and so will behave like the F1 generation when selfed. The remaining quarter carry two lowercase *a* factors and thus will also act as a pure line.

ALKIMOS: That's exactly what Mendel found!

DEMOCRITUS: It's hardly a coincidence.

EXTON: We should note that in the F2 generation, the 3:1 ratio reflects only the probability of achieving this numeric breakdown. Actual experiments won't necessarily produce precisely this result. Here look at this table. It shows Mendel's actual results.

Seed shape (smooth–wrinkled)	5474	1850	2.96:1
Seed colour (yellow–green)	6022	2001	3.01:1
Flower colour (purple–white)	705	224	3.15:1
Pod shape (inflated–constricted)	882	299	2.95:1
Pod colour (green–yellow)	428	152	2.82:1
Flower position (axial–terminal)	651	207	3.14:1
Plant height (tall–short)	787	277	2.84:1
TOTAL	14,949	5010	2.98:1

ALKIMOS: So that's what you mean by probability. The results are a little more or a little less, but not necessarily exactly what we expect.

DEMOCRITUS: That's the beauty of mathematics.

EXTON: Before we continue our study of cross-breeding, we need to expand our vocabulary of genetic terms. Let's discuss some fundamental concepts. First of all, we have the elements that determine heredity. Weismann used the term *germplasm units* to describe these elements, whose nature was not understood until the early 20th century. Wilhelm Johannsen, a Botany Professor from Copenhagen, was the first to call them *genes*. He also introduced the notions of the *genotype* and the *phenotype*. The first refers to the collection of genes found in a living thing, also known as an organism's genetic make-up. The second refers to the collection of observable traits.

DEMOCRITUS: That means that in Mendel's experiments the genotype of the F1 generation is *Aa*, while the phenotype is *A*. So a plant's seeds appeared smooth, but the plant carried the genes for both smooth and wrinkled seeds.

EXTON: Correct.

DEMOCRITUS: Mendel's work thus revealed that genes come in pairs. One member of the pair comes from the father and the other from the mother. When the gametes form, the pair separates again, and each gamete contains only one member of the pair. Mendel called the single members of each pair a factor. In pure lines both factors of the pair are the same, but in hybrids the factors are different. One can only produce wrinkled seeds, and the other only smooth.

EXTON: I should mention that Mendel thought that when germ cells containing the same factors fused, then the factors also merged. That's why he labelled those individuals that were, for example, pure dominant *A* instead of *AA*.

DEMOCRITUS: That makes sense. After all, Mendel wasn't familiar with the behaviour of chromosomes.

EXTON: Right. When Mendel's work was rediscovered, the single *A* was immediately corrected to a double *AA*, thanks to the progress made in cytology during the thirty-five year interim. But let's return to our discussion of terminology. It will now be easier to understand what is meant by *allelomorph, homozygote,* and *heterozygote.* All three terms were coined by Bateson in 1902. Let's start with allelomorph, or *allele.*

ALKIMOS: *Alello* means other, and *morph* means form. So an allelomorph means "other form."

EXTON: But another form of what? I'll give you a hint. The key lies in what Democritus said earlier.

ALKIMOS: Perhaps an alternative form of the gene, a variant.

EXTON: Excellent! The variants of a gene are known as its alleles. Genes can have a huge variety of alleles; later we'll understand why. In Mendel's study, for each of the seven genes he studies he used just two alleles. For example, one version of the gene that determines the surface of the pea results in a smooth skin (*A*), while the other version causes wrinkled skin (*a*), although only in the homozygote form. These are the two alleles of the *A* gene.

DEMOCRITUS: Let me guess the rest! The expressions *homo-* and *heterozygote* refer to the genotype of the individual that develops from the fertilised ovum, or the zygote. *Homo,* or same, means the zygote received an identical allele from each parent. *Hetero,* or different, means the zygote received a different allele from each parent.

EXTON: That's right.

ALKIMOS: This means the smooth-skinned pea with *AA* genotype and the wrinkle-skinned pea with the *aa* genotype are both homozygotes. The smooth-skinned peas with *Aa* genotypes are heterozygotes.

EXTON: Yes, but only with respect to the gene that determines the surface appearance of the pea. But at the moment the other genes are not of interest to us. You only need to note that the terms homo- and heterozygote refer to just one allele pair of a single gene of the individual, but an individual has many genes.

ALKIMOS: These terms aren't useful when we're talking about phenotypes. I think I understand why.

DEMOCRITUS: Of course, son. The smooth-skinned pea can be either the *AA* or *Aa* genotype, so whether it's a homozygote or a heterozygote has no impact on the phenotype, since the dominant allele prevails. Although the seed skins look identical, when the germline continues in the next generation the hidden *a* alleles can join together to produce wrinkled skin.

EXTON: Mendel experienced this with his F2 generation. He called it the *principle of segregation*, or separation. But many before Mendel had noticed that the alleles separate when the gametes are formed.

DEMOCRITUS: According to the principle of segregation, the properties of an allele are not affected by its being paired with a different allele in a heterozygote. Genes are independent units that don't fuse together. This is why there is no such thing as blending inheritance.

ALKIMOS: Then what exactly did Galton calculate? Don't tell me his law of inheritance was based on nothing but a gut feeling?

EXTON: Strangely enough, Galton was also to some extent correct. The theory of blending inheritance cannot be rejected in its entirety. But before we attempt to resolve this intriguing contradiction, let's have a quick look at Mendel's experiments with dihybrids.

ALKIMOS: *Di* means two, so two hybrids. Does that mean he crossed plants that differed in two characteristics?

EXTON: Yes. Smooth or wrinkled skin and yellow or green pods. We can take any of the seven traits and pair them, which is what Mendel did.

ALKIMOS: I suppose we have to guess the result.

EXTON: That's right.

ALKIMOS: I give up.

DEMOCRITUS: Now, let' see. When Mendel examined two alleles of one gene, in the first generation he got identical peas. We have no reason to think this principle of uniformity doesn't hold for more than one pair of alleles. In the F2 generation of a regular hybrid, however, we have a 3:1

ratio of dominant to recessive. The big question is how this ratio changes when we have two pairs of alleles.

ALKIMOS: I'm listening, master …

DEMOCRITUS: In the F2 generation, for every pair of alleles we have two groups. Clearly, if we have two pairs, then we'll have four groups altogether.

ALKIMOS: We could have smooth-yellow, smooth-green, wrinkled-yellow, and wrinkled-green.

DEMOCRITUS: Furthermore, if we have the 3:1 ratio for one pair of alleles, then in the case of two pairs, the proportions break down further. The 3 and the 1 are each in a 3:1 ratio, too.

EXTON: So in the end we have a ratio of 9:3:3:1. Every dihybrid cross done by Mendel gave the same approximate results. For this kind of distribution to occur, the alleles of the two genes must separate and regroup independently during gamete formation. This is known as the *principle of independent assortment.* Look, here. While Democritus was brilliantly explaining the principle, I found this diagram that shows the process of dihybrid crossing.

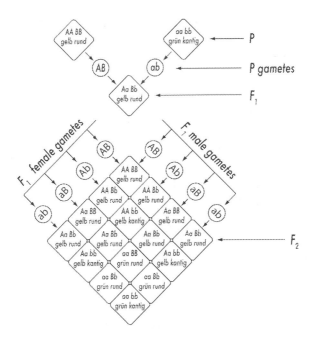

We can use the same method to calculate the distribution of the F2 generation in the case of numerous independently assorting alleles. For example, we can figure out how many individuals contain only homozygotes for all the given genes.

DEMOCRITUS: I'm beginning to understand what Bateson was referring to when he said his research would be of immediate use to horticulturalists. After all, their goal is not only to retain the desired characteristics, but to gather them together in one individual, producing the most ideal plant or animal for a given purpose. If they understand Mendel's laws, they can predict in what proportions and groupings they'll find the desired traits in future generations and more effectively improve and reproduce their stock. In fact, with all the time and effort they save, they will no doubt fatten their wallets.

ALKIMOS: That's what Francis Bacon wrote about 240 years before Mendel! But what about Galton? You've piqued my curiosity!

EXTON: Let's remember what Galton asserted. He believed that every single ancestor of an individual contributed to his characteristics in increasingly smaller degrees as we go back through the generations. But what did Mendel's laws tell us?

ALKIMOS: Well, if we consider Mendel's laws, we can see that homozygotes for one allele no longer carry the other allele, which came from one of the grandparents.

EXTON: Right. In the **1:2:1** ratio, the two **1**s have no connection to one of the grandparents.

DEMOCRITUS: At least with respect to that particular gene.

EXTON: Ah, but let's consider this further. What happens if we study two pairs of alleles? If we look at dihybrid crossing using a Punnet Square, we'll notice some interesting things. First, let's see how many offspring on average inherit alleles from only one grandparent.

ALKIMOS: Two of sixteen: the genotypes *AABB* and *aabb*.

EXTON: Similarly we can figure out that on average four offspring receive three alleles from their grandmother and one from their grandfather, six offspring receive two alleles each from their two grandparents, and four receive one from their grandmother and three from their grandfather.

DEMOCRITUS: So the distribution is 1:4:6:4:1.

EXTON: For three pairs of alleles the distribution is 1:6:15:20:15:6:1, for four it's 1:8:28:56:70:56:28:8:1. We can figure the rest out using Pascal's triangle.

```
                    1
                  1   1
                1   2   1
              1   3   3   1
            1   4   6   4   1
          1   5  10  10   5   1
        1   6  15  20  15   6   1
      1   7  21  35  35  21   7   1
    1   8  28  56  70  56  28   8   1
  1   9  36  84 126 126  84  36   9   1
1  10  45 120 210 252 210 120  45  10   1
```

For both Galton's and Mendel's theories we need to progress downwards through the triangle if we want to know the distribution for an increasing number of allele pairs. But look here! If we illustrate the distribution of individual groups, we get very interesting configurations that appear more and more like a bell. Galton invented the Galton box, a mechanical device shown in this picture to illustrate such a distribution.

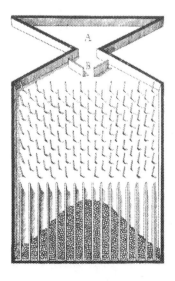

If you throw beans into the box, it will show a distribution that matches Pascal's triangle. Such a bell curve is called the *normal* or *Gaussian distribution*.

ALKIMOS: I see. But what does this tell us about Galton's theory?

DEMOCRITUS: Why, Alkimos, it's clear! When we have a large number of genes, which are represented in Galton's box by the beans, we find ourselves back at Galton's theory of blending inheritance; we just need to interpret it slightly differently. A grandchild does not resemble his ancestors in accordance with Galton's theory if we consider just one characteristic, but if we consider *all* of them, then the grandchild does. Most likely organisms are determined by an enormous number of genes, and the distribution of ancient gene forms is described by such a bell shape. In fact if there are many beans then almost no progeny exist that don't contain a little something of the allele set of each and every ancient ancestor.

ALKIMOS: I think I understand now. It's not the characteristics that blend, but the genes. The huge number of genes explains why we resemble just a little this and that ancestor.

EXTON: Now I'll tell you about a seemingly simple case of blending inheritance, known as *intermediate inheritance* or *partial dominance*. In this case, one of the alleles of a given gene exhibits an in-between phenotype. Mendel observed that pea plant offspring flowered at almost exactly the same time as their parents, but sometimes the offspring of white- and

(a)   (b)   (c)

purple-flowered plants produced a pale purple flower. Another example is willow trees with either wide or narrow leaves, which if crossed could produce offspring with leaves of intermediate width.

DEMOCRITUS: In this case, in the F2 generation, we'd have a ratio of 1:2:1, if I'm not mistaken. Only the homozygotes have a pure colour, but the heterozygotes—which comprise half of the offspring—will exhibit this mixed colour. Thus, when we have partial dominance, we know the genotype of the offspring.

EXTON: If we're acquainted with these phenomena, then we can understand to a certain extent why the belief in blending inheritance was so tenacious. Imagine how bewildering the results were when early horticulturalists crossed plant lines with differing characteristics and heterozygous genotypes. And of course no one paid attention to the ratios. No wonder even Darwin was stumped when he tried to understand the true nature of inheritance.

DEMOCRITUS: It seems what was needed was monastic solitude and the quiet of the Brno monastery. And although offspring seem to show an enormous amount of blending—their characteristics indeed appearing as a mélange—deep within them, we can still see the Mendelian factors and the rigid choreography of the inscrutable genes, set to the rhythmic alternation of the germ and the soma.

EXTON: Ah, see, von Baer's concept of a type and rhythm doesn't sound so foreign.

ALKIMOS: Yes, it does! I still have no idea what he meant. Especially considering he knew nothing about genes.

EXTON: Perhaps he had an idea …

ALKIMOS: Nah. His thoughts were so vague, I think all they really reveal is his belief in a mystical life force.

EXTON: Possibly. In any case, we're about to witness the attack on another theory. It's the flip side of von Baer's: mechanistic-reductionism.

ALKIMOS: Hm … I can hardly wait.

Two Sperm, One Ovum—*Theodor Boveri*

ALKIMOS: Can I come in?

COOK: Is that you, lad? Come in. Had enough of science?

ALKIMOS: My master and Exton have started talking about things that are beyond me. I was able to grasp Mendel, but I can't follow polygenic inheritance or epistasis.

COOK: Poly-what?

ALKIMOS: Hey, are there any leftovers from yesterday's dinner? My stomach's growling.

COOK: There sure are. How about a turkey leg?

ALKIMOS: Perfect.

COOK: Some wine with it?

ALKIMOS: Ah … delicious. My friend, if only our grandparents could have counted better. But they didn't. They just blindly mixed up their genes and distributed them to their offspring.

COOK: Whatever you say…

ALKIMOS: If Mendel had told them that every gene in their body has two copies, and one might be shoddy—you know, curved or crooked or maybe six-fingered—they could have paid closer attention to their spry little sperms. 'That one's frail, that one's hearty. Oh Look! That one's the most energetic. Come on, dear wife, let's make our little tyke from that!'

COOK: I like where this conversation is going, lad!

ALKIMOS: But they didn't. They just went at it with no plan. If I think about the number of haphazard events that went into my making, I feel sick. I'm a catchment basin for the streams of my ancestors' seed; I am my own randomness, my own imperfection.

COOK: Um, well, sorry to hear that.

ALKIMOS: I'm the winner of my ancestors' genetic game of chance. If my mother had passed on to me my great-grandfather's allele and not my great grandmother's, I'd no longer be me; I'd be my brother.

COOK: Oh, I see. Like, I'd be my older brother if I'd been born before him. Yeah, I get it. It doesn't take a lot of science.

ALKIMOS: How is it possible to climb out of this philosophical labyrinth?

COOK: Don't take this the wrong way, but don't you think there are better things to think about?

ALKIMOS: There's no solution.

EXTON: Alkimos, is that you? We've been looking for you everywhere!

ALKIMOS: Exton?

EXTON: Come on, we've reached a crucial point in our discussion. We must shed light on the connection between chromosomes and genes! Democritus and I have thought it over. He can tell you what we need to do.

ALKIMOS: Okay, I'm coming.

DEMOCRITUS: Alkimos, I'm glad you've rejoined us. Here, I'll outline our task.

ALKIMOS: I'm listening, master.

DEMOCRITUS: Mendel's research and cytological observations have helped us to develop the concept of the gene, a hereditary unit that appears in pairs, but separates when the gametes form, and affects one characteristic in an individual.

ALKIMOS: Right, I understand that.

DEMOCRITUS: But we should notice that Mendel's laws reveal very little about the physical nature of genes. We know that they never blend, they can be independently assorted, and are almost certainly indivisible.

ALKIMOS: Like your atoms, master!

DEMOCRITUS: But there's so much more to know! Where are they located in the cell? What materials comprise them? What happens to them during cell division and the formation of gametes? What relationship do they have to chromosomes?

ALKIMOS: Master, after Bateson's lecture, you asked whether genes are attached to chromosomes when they're passed on.

DEMOCRITUS: Yes, that does seem to be the most obvious interpretation, since the chromosomes also double and halve, like genes.

ALKIMOS: Why didn't Bateson like that answer?

DEMOCRITUS: The central problem is we have no evidence of any connection between genes and chromosomes. We need to prove somehow that the development of the phenotype from the genotype ...

ALKIMOS: ... the development of the organism ...

DEMOCRITUS: ... is impossible without a proper distribution of the chromosomes, because every chromosome is different; they carry the genes that regulate different stages of development.

ALKIMOS: Master, that sounds like a crazy preformist argument! No wonder so many scientists didn't like it. What happens to the self-organizing power of the protoplasm, the inner force driving development? It's reduced to a dyed chromatin rod that gives out orders like a zealous overseer. It makes a mockery of epigenesis.

EXTON: Let's not forget that some scientists think the chromosomes are all alike. But if they are, then there are at least as many copies of each gene as there are chromosomes. In that case we would certainly not have outcomes that correspond to Mendel's simple ratios.

ALKIMOS: Or if the chromosomes are all alike, then they have no connection to genes.

EXTON: There are several contradictions we need to resolve.

DEMOCRITUS: To know the truth, we need to investigate the exact distribution of chromosomes. We have to devise an experiment that ensures that certain cells don't receive certain chromosomes, while other cells receive more than one copy. If the inheritance of traits, in other words the way in which organisms develop, is indeed linked to the chromosomes, then the lack or the surplus of some chromosomes will result in developmental abnormalities, defective cells and organs, a confused developmental path.

ALKIMOS: A brilliant idea befitting such a great mind as yours. But how do you propose to carry this out?

DEMOCRITUS: I must confess, I have no idea.

ALKIMOS: Aha. Truly brilliant. Well, lots of luck mixing up chromosomes.

EXTON: You'd be surprised. The task is not impossible.

ALKIMOS: Really? I suppose all we need is some fabulous machine no one has ever heard of.

EXTON: A chromosome separator?

ALKIMOS: Yes, that's it.

EXTON: Alkimos, you may find it hard to believe, but what we need is not elaborate technology, but resourcefulness. That's the heart of true science. We're about to travel to the early years of the 20th century and meet a scientist who had an idea for an experiment very similar to Democritus', and he didn't use any fancy technology. Theodor Boveri devised and conducted the most ingenious set of experiments in the history of chromosomes.

ALKIMOS: The great expert on intestinal roundworms. But we're not …

EXTON: Hermes, the Deus ex Machina! Direction Naples, 1905!

ALKIMOS: Ah … here we go!

EXTON: Wherrrrre arrre my nnnnnotesssss … ah, here they are. We have some things to discuss on route. At the beginning of the 20th century it had become obvious that chromosomes retain their individuality during every stage of the cell's life. They never disintegrate and then reassemble as Strasburger thought. All that happens is that during mitosis, they clump together, or condense, which makes them visible, but after cell division is over, they relax. At this point it was also certain that during fertilisation the sperm and the ovum contributed the same number of chromosomes to the nuclear stock of the new individual.

ALKIMOS: Just like genes. One from the mother and one from the father.

EXTON: Yes, but some organisms are capable of developing with just one set of either maternal or paternal chromosomes. In other words one set of chromosomes can single-handedly regulate an organism's development. For this to be true, the set must contain all the information necessary to construct a complete individual. Unfortunately, this observation was both illuminating and misleading.

ALKIMOS: I think we've arrived.

EXTON: Beautiful. The Bay of Naples, 1905. Everyone out!

DEMOCRITUS: Dear Exton, again you've brought us to a marvellous place. Boveri certainly knew how to choose his workplace.

EXTON: It is lovely. And just think, with such proximity to the sea, naturalists in this part of the world begin observing marine life before they can walk. There it is! The famous Stazione Zoologica, founded by Anton Dohrn. We'll find Boveri inside, but before we enter, let's finish our discussion of the groundwork for Boveri's experiments. A few years earlier, Jacques Loeb, an American biologist, discovered that cell division could be induced in the unfertilised ova of the sea urchin by changing the salinity of seawater. In the right solution, completely intact pluteus larvae can develop without the presence of sperm.

ALKIMOS: Artificial parthenogenesis …

EXTON: This is a picture of a pluteus larva as depicted by Karl Belar in 1928.

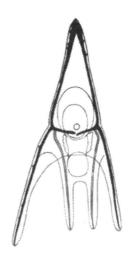

DEMOCRITUS: This observation confirmed what ovists already believed when, to their delight, the asexual reproduction of aphids was observed: the ovum alone is sufficient to create an offspring.

EXTON: It's easy to understand why Loeb's findings were welcomed by women suffering from infertility. Saltwater baths became a popular treatment.

ALKIMOS: So researchers once again failed to understand the role of the male.

EXTON: That's true. Loeb believed adding the right amount of salt would trigger development. To him it was obvious that the sperm's contribution to fertilisation was the salty fluid, and the nucleus and its chromosomes were unimportant. Development was regulated by a complicated epigenetic interplay of salts, solutions, membranes, and other chemical substances. He determined that preformed genes didn't exist.

ALKIMOS: Incredible! So what did Loeb think Mendel had discovered if not genes?

EXTON: The laws of probability of life's complex biochemical pathways; the inner dynamic of epigenesis; the peculiar working of a chemical kaleidoscope of salts, sugars, and proteins.

DEMOCRITUS: But the laws of genetics can't possibly be the result of just some chemical reactions. My Zeus! Exton, you must show us a way out of this!

EXTON: Not I. Boveri will. Now, let's go in.

ALKIMOS: Wait, the door's opening.

EXTON: Ah, Professor, so good to see you.

BOVERI: Please come in!

EXTON: These are my students, from the University of Sussex ...

BOVERI: Yes, yes, I know. Now gentlemen, come into my lab. You must have a look at this. The miracle of fertilisation. It's the most crucial of life's many, incredible and varied processes. Two cells that until this point were wandering independently come in contact to create something qualitatively different: a developing organism. Believe me, gentlemen, this is the most astounding process that mortal eyes have ever seen.

EXTON: Professor, this is precisely why we've come. We'd like to understand this mysterious process and know the role chromosomes play in it. Could you give us an account of your recent observations?

BOVERI: Of course. But first you need to understand the experimental process which helped me to discover the vital role of chromosomes.

ALKIMOS: Let's get started!

BOVERI: Ever since Hermann Fol and Oskar Hertwig conducted their research on sea urchins, which I'm sure you gentlemen are familiar with, we've known that if two sperm simultaneously fertilise the sea urchin ovum, then four spindle poles form in the zygote. As a consequence, the first cleavage results in four cells rather than two. You see, the two sperm have contributed two centrosomes to the ovum. Because both divide, we're left with four spindle poles, as you can see in this illustration:

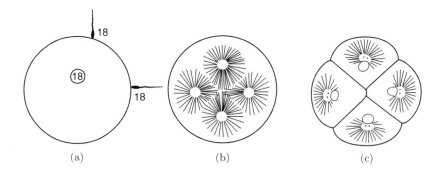

(a)   (b)   (c)

The whole thing is very simple; yet it reveals so much! I've always said, one experiment is far more valuable than one hundred pages of Gegenbaur.

EXTON: I agree. But give me a moment. I need to explain to my students what centrosomes and spindles are.

BOVERI: If you think it's necessary …

EXTON: During fertilisation, the sperm contributes not only its nucleus to the zygote, but also its *centrosome*, which regulates the formation of *spindle* poles during cell division. The centrosome is an independent organelle that, like chromatin, first duplicates and then one copy is transferred to each daughter cell. The spindle threads form around the centrosomes and pull the chromosomes toward the two poles. During fertilisation, the sperm's centrosome divides, creating the two mitotic poles in the zygote. The ovum's centrosome doesn't generally participate in cell division, and instead disintegrates.

ALKIMOS: But who is Gegenbaur?

EXTON: Karl Gegenbaur was a prominent comparative embryologist of the 19th century.

BOVERI: Yes, that's true… Now, back to sea urchins. Hans Driesch examined numerous quadripolar zygotes and discovered that most of them die very soon. Only very rarely does a defective blastula develop.

ALKIMOS: Ah yes, Dumas and Prévost had already shown that one sperm is enough for fertilisation. The fragile ovum apparently can't cope with two sperm.

BOVERI: But the question is why can't it? Why is development quickly derailed by the formation of four poles? To find out I began to study eggs fertilised by two sperm. After a series of lengthy experiments I was able to reject a host of potential reasons and arrive at the one and only possible explanation, which a recent discovery has helped to confirm. Curt Herbst's studies showed that the blastomeres of sea urchins voluntarily separate from each other in seawater that lacks calcium ions. That allows us to follow the individual fate of each of an ovum's four daughter cells.

ALKIMOS: What else could develop from them but four quarter embryos?

BOVERI: Ah, but that's what's so surprising! Driesch devised his own clever methods to show that the separated blastomeres of normally fertilised ova, in other words ova penetrated by just one sperm, developed perfectly fine, although the pluteus larvae they developed into were dwarfed.

DEMOCRITUS: Incredible! If only Weismann and Roux had heard that! What else could they conclude from the undisturbed development of quarter embryos than that their theories of development were wrong!

ALKIMOS: Master, I don't understand.

DEMOCRITUS: Weismann and Roux asserted that differentiation during development could be explained by an unequal distribution of hereditary material between the daughter cells. If this were true, then the four separated blastomeres, which according to their theories would have contained unequal amounts of hereditary material, could not have produced four intact larvae. It would be impossible! The only conclusion is that the four blastomeres of a normally developing embryo must be completely equal in terms of hereditary content, and also equal to the ovum that produced them.

ALKIMOS: My word.

DEMOCRITUS: But what happens when we separate the four blastomeres of a quadripolar ovum, one fertilised by two sperm?

BOVERI: My observations show that only very rarely do these kind of blastomeres develop into pluteus larvae; however, they frequently form more or less intact gastrulae. Moreover, if we compare the fate of separated blastomeres to that of blastomeres that develop together, we find that the independent quarters frequently advance further in their development than those that remain attached. My most important result, however, is that the four separated blastomeres develop differently.

DEMOCRITUS: So, a naturally dividing ovum produces four equal blastomeres. A quadripolar ovum, on the other hand, produces blastomeres that differ from each other significantly.

BOVERI: Yes, that's correct. This difference is especially pronounced if we follow the fate of the individual blastomeres as they develop into quarter embryos and form body parts. For example, only in one of the quarter embryos will a gut develop and in another a primordial skeleton. But to understand the reason for these astounding facts, we need to look at all of the possible causes of differentiation.

ALKIMOS: All? I can't even name one.

DEMOCRITUS: Wait a minute, Alkimos. Let's think about it—what *are* the possibilities? Do we believe that the protoplasm is unequally divided? Or that the reason for developmental defects is that the protoplasm is distributed between four instead of two cells? Is it possible that without a certain amount of protoplasm, development cannot proceed normally?

BOVERI: Yes, this might seem obvious, but let me explain why we can reject this idea immediately. The first division of a dispermal cell seems to result in four daughter cells with identical protoplasm, just as the division of a normally fertilised ovum would. But if four healthy blastomeres are capable of developing independently, then why shouldn't we expect the same from the blastomeres produced by a quadripolar ovum, since they also contain identical protoplasm? Moreover, if only the protoplasm was responsible for the defects, how could the development of each quarter be so markedly different?

DEMOCRITUS: Using a similar train of thought, we can understand why centrosomes can't possibly explain defects either.

BOVERI: Indeed. From observing normal development we know that the properties of the two centrosomes at the two poles are identical. Furthermore the centrosomes of quadripolar ova do not differ in any way from the centrosomes of normal ova. But if the centrosomes of differing origins possessed different properties, then they still couldn't be responsible for the endless developmental variations of quadripolar ova. We can only explain our observations by calling upon the aid of a process which itself is as varied as the outcomes it produces.

DEMOCRITUS: I suspect that such variety can only be discovered in the irregular distribution of chromosomes.

ALKIMOS: The irregular distribution of chromosomes?

BOVERI: There is simply no other explanation. Let me show you. The four daughter cells of a quadripolar ovum, which is caused by the penetration of two sperm, must carry different numbers and an unequal distribution of chromosomes.

DEMOCRITUS: When a normally fertilised ovum divides, every pair of chromosomes separates, and one migrates to one pole, and the other to the other pole. But when there are four poles, things get a little confused.

BOVERI: To say the least! Let's consider it. There are eighteen chromosomes in the ovum of the sea urchin species I studied.

ALKIMOS: Then the sperm also must contain the same number!

BOVERI: A healthy cleavage nucleus contains eighteen pairs of chromosomes, so thirty-six chromosomes altogether. Before cell division, all of them replicate and divide, so that each daughter cell receives thirty-six chromosomes.

DEMOCRITUS: It's easy to count from here. The ova penetrated by two sperm contain three times eighteen, or a total of fifty-four chromosomes.

If they all divide then the cell will have one hundred and eight chromosome strands in all. These are distributed between the four poles. The end result is …

ALKIMOS: Twenty-seven chromosomes in each of the four daughter cells. The two daughter cells of a normal ovum, however, have thirty-six.

BOVERI: And in reality the one hundred and eight chromosome strands are very rarely dispersed evenly among the four cells. About twenty years ago I managed to demonstrate that the distribution of chromosomes in the four daughter cells of a quadripolar ovum is random. This diagram will help explain why:

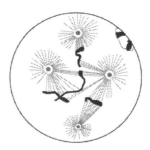

ALKIMOS: The spindle pole fibres pull the chromosomes in different directions.

BOVERI: Random distribution is the reason the daughter cells inherit differing numbers of chromosomes—more or less than what would be expected mathematically. This diagram shows one of the possible distributions.

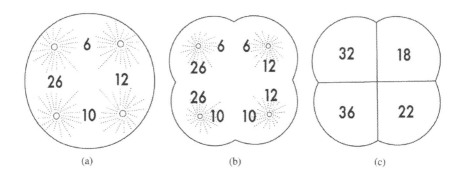

(a) (b) (c)

But the distribution is not random in only this respect. The daughter cells always inherit chromosomes in different combinations.

DEMOCRITUS: A question immediately arises: do the daughter cells develop differently because the number of chromosomes, or chromatin, is different or because specific chromosomes in the cell are present or missing? If the latter is the reason, that means the chromosomes possess different capacities to guide development.

ALKIMOS: You mean one generates the sea urchins dreadful spines, another its gut, and yet another its …

BOVERI: That's what all our evidence suggests. At the same time, we can certainly reject the notion that the ovum must have the full number of chromosomes typical of its species for it to develop normally. Loeb showed that half the chromosome stock is enough. My experiments with an enucleated ovum showed the same thing.

ALKIMOS: Enucleated ovum? An ovum without a nucleus? But how can you remove one minuscule thing from another?

BOVERI: You need only to be resourceful. If we thoroughly shake up the ova, they fall apart, and the various parts ball up into smaller ova. Since the nuclei don't fall apart, some of the newly formed tiny ova don't have any. If you fertilise one, it will develop into a pluteus larva entirely under the guidance of the sperm nucleus.

ALKIMOS: So the male sperm directs development! It's the formative force, the vital spark! That's what Aristotle said, too. The formation of the foetus is regulated by the male seed. I feel like a man again!

DEMOCRITUS: Very well, son, but you seem to have lost your common sense in the process. How quickly the careless philosopher forgets the facts that disprove his most cherished beliefs.

ALKIMOS: Aphids and Loeb's artificial parthenogenesis … I know, master. Please excuse my moment of unbridled joy.

DEMOCRITUS: Both the enucleated ovum and Loeb's experiments clearly demonstrate that half the amount of chromatin is enough for normal development.

ALKIMOS: But Professor Loeb thought sperm cells were nothing but sacks of salty solution. It was an insult to our manliness. Then Boveri came along and with his modest experiments showed that a sperm nucleus is capable of the same thing as an ovum! Why wouldn't I rejoice? There's a reason why Aphrodite's symbol isn't a saltcellar.

DEMOCRITUS: Now we have proof that a sperm and an ovum can do separately what they can also do together. They both possess the power to form and perpetuate life.

BOVERI: Yes, it's undeniable. And with this in mind, we should continue our discussion of chromosome distribution. We could argue that the entire quantity of chromatin is unnecessary for development; in fact half is enough, but no less. In other words a certain amount of chromatin allows for the development of a normal offspring, but any less results in different sorts of defects.

DEMOCRITUS: Yes, but we can see this is not the case. The separated blastomeres of quadripolar ova inherit on average twenty-seven chromosomes, which is just slightly more than what the single sperm contributed to the enucleated ovum. If this ovum could develop normally with eighteen chromosomes, then we would expect a certain percentage of the separated blastomeres to develop normally too.

EXTON: If all that's needed is a minimum quantity of chromosomes.

BOVERI: But that's not all that's needed. It matters which chromosomes the cell inherits. The only answer that remains is that for a cell to develop properly it requires a certain combination of chromosomes. This combination is none other than the stock of chromosomes present in the male and female gametes. They carry within them the potential for the complete development of the individual. Normally the chromosomes are evenly distributed between the two daughter cells. In other words during mitosis each daughter cell gets the full chromosome stock. The law of randomness helps us to understand why the daughter cells of an ovum fertilised by two sperm only rarely received the full stock of chromosomes when they migrate to four instead of two poles. This, and only this, is the cause of their abnormal development.

DEMOCRITUS: But this means only one thing: every single chromosome possesses different properties.

BOVERI: No other conclusion is supported by the observations.

DEMOCRITUS: So, does this mean that genes are linked to the chromosomes? If a cell is missing a set of genes, then it dies or is malformed? No wonder Mendel's laws about the transmission of genetic traits, which are revealed in abstract ratios, are compatible with the behaviour of chromosomes observed under the microscope. They're in perfect correspondence. And although one study involved plant cultivation and the other aimed to understand the secret of cells, they both arrived at the same place. Not only does it inspire confidence in our results, but demonstrates that the study of living things should not proceed along one single path. Only through a joint effort, relying on the widest variety of approaches, will scientists begin to make sense of nature's infinite complexity.

ALKIMOS: Master ...

DEMOCRITUS: Now that we recognise the importance of chromosomes we need to investigate further to discover the true nature of genes.

ALKIMOS: Master, look!

DEMOCRITUS: ... we've seen cells and chromosomes ...

BOVERI: Ohhhh...

DEMOCRITUS: ... but we haven't seen the atoms that create life.

EXTON: Democritus, we need to go!

DEMOCRITUS: We need to delve deeper into the laws of the tiniest beings ...

ALKIMOS: What are those spikes on Professor Boveri's back?

EXTON: Oh heavens! The chronic turbulence caused by our time travel is having a strange effect on the Professor! We need to leave before the situation gets worse!

BOVERI: Grrr ... agggghhhhh ....

DEMOCRITUS: We must revolutionise biology at the atomic level! This is the next step!

ALKIMOS: Master, please ... Boveri seems to be turning into a... a... a sea urchin!

EXTON: Quickly!

ALKIMOS: But wait! Shouldn't we put him in the aquarium?

BOVERI: Agghh …

ALKIMOS: Ow! He pricked me!

DEMOCRITUS: What's happened? Where's Professor Boveri?

EXTON: Let's start the motor!

ALKIMOS: Oh no, here we go again ….

Part Three

The Triumph of Genes

Trickster Mendelians—*Thomas Hunt Morgan*

ALKIMOS: Master …

DEMOCRITUS: I'm here, son.

ALKIMOS: I don't understand anything. What are chromosomes?

DEMOCRITUS: I understand them less and less, too.

ALKIMOS: Boveri's experiments are clear, but I can't wrap my head around how a gloppy mass can transform into an animal that croaks or has spikes all under the direction of a pair of threads in the nucleus. How can chromosomes cause jelly to develop legs and scamper around?

DEMOCRITUS: I'm afraid we're still a long way from a complete answer. We can infer the various components of life, but we haven't managed to apprehend even one. We can engrave the mathematical laws of inheritance in clay tablets, but we've still failed to understand its fundamental nature.

ALKIMOS: Master, you sound like a crazy atomist! Why do you think life has real, tangible components? I no longer believe it. The deeper we delve into the mysteries of life, the less we understand. Maybe some life spark actually does exist.

EXTON: Gentlemen, don't tell me you've given up?

DEMOCRITUS: Not at all. It's just very difficult to digest all this new knowledge. It's so far from our experience that we're not quite sure what to make of it.

EXTON: You need to get used to this strange feeling. Knowledge of phenomena at very small scales is indeed nothing but abstraction. So many of the phenomena of this world cannot be directly experienced. All we can do is develop an image of it in our minds. As our knowledge expands, we merely sharpen and elaborate on this image. We constantly reassess and modify it based on our observations of reality. The world, as such, is reflected in theoretical constructions, but these constructions are not the world itself.

DEMOCRITUS: This is the great tragedy of learning. It's never real and never complete and so it haunts us endlessly, like an unfulfilled love.

EXTON: That's why we mustn't give up.

DEMOCRITUS: We haven't.

EXTON: Good. Then we're ready for a ruthless confrontation. Let's take our theoretical constructions about chromosomes and genes and set them against the facts. Perhaps we were too quick to judge. Hermes, prepare the Deus ex Machina! Direction Columbia University, 1910! We're going to visit Thomas Hunt Morgan, one of the best embryologists of his time. Gentlemen, let's board!

ALKIMOS: If only we didn't need to …

DEMOCRITUS: Son, that's my seat.

ALKIMOS: Oh … here we ggggo!

EXTON: Your fffirst trippp accccross the Ooooocean! Tttto the shores of North America, nnnnear the fffffabulous island of Bbbensalem. Ah, the shaking has stopped, much better. Now, about Thomas Morgan: he worked in Woods Hole, Massachusetts, on the eastern edge of the continent. At the time he was a committed developmental biologist, passionate about embryos, dedicated to uncovering their secrets. The theory of chromosomes and Mendel's laws were, in his mind, unacceptable. He considered them unfounded, preformist hogwash. And he had reason. After rigorously reviewing earlier research, he concluded there was a lack of evidence to support the role of chromosomes in inheritance. His arguments posed a considerable challenge to the Mendelian concept of hereditary units, or factors.

ALKIMOS: My Zeus! If chromosomes have nothing to do with inheritance, why did we have to listen to Boveri's long-winded explanation? Why didn't we go right to Morgan?

EXTON: I wanted you to understand that science never progresses in a straight line. Sometimes it takes major detours; other times it moves ahead so rapidly it leaves a great many scientists in the dust. And of course there are always those scientists who are able to race ahead of it, and then its science that needs to catch up. Only in hindsight do we clearly see the progression, one discovery building upon another. Our present knowledge makes it easy for us to say who was correct one hundred years ago. Our confidence overflowing, we might smirk at or even ridicule scientists who take a wrong turn. We seem to forget that if we possessed only the knowledge of the century before, we might have made the same mistakes. Oh good, we're almost there. Prepare to disembark.

DEMOCRITUS: This is the island of America?

EXTON: It's a little large to be called an island. We've arrived in the New World.

ALKIMOS: This is what Huxley told us about?

EXTON: Come on. Columbia University is over there, on the other side of the street. Morgan is expecting us.

DEMOCRITUS: My Zeus! What's that?

EXTON: Careful. That's an automobile. A Ford Model T.

ALKIMOS: It's so much more spacious than our contraption!

EXTON: And much more modern. Now, let's proceed. The entrance is over there.

ALKIMOS: The Ins-ins-ins-tutite for Developmental Biology.

EXTON: In-sti-tute. You'd better practice your English. At the beginning of the 20th century, most scientific publications were in German, but English quickly emerged as the dominant language. The reason is that an increasing proportion of new discoveries were made in the New World.

ALKIMOS: Exton, we're here.

MORGAN: Welcome, gentlemen.

EXTON: Ah, Professor Morgan.

MORGAN: Please come into my laboratory.

ALKIMOS: Look, a microscope!

MORGAN: No there. Please, have a seat here.

EXTON: We've come because we'd like to ask you some questions about Mendel's laws and chromosomes.

MORGAN: I hope you're not Mendelians!

DEMOCRITUS: I myself am an atomist, but I am open to new ideas. I must confess, however, that I find nothing objectionable in Mendel's laws. In fact I think it's astounding how well the behaviour of chromosomes conforms to Mendel's theory on the segregation of characteristics.

MORGAN: I agree. Abstract Mendelian ratios shed light on many things. However, today's practice of Mendelism has caused us to be too quick to use Mendelian factors to explain certain facts, and if one factor will not explain an observation, then try two, and if not two, then three. This kind of, let's call it superior jugglery, is needed to account for the results, and it blinds us to the truth: the results are so well "explained" because the explanation was invented to explain them. I appreciate the advantages that this method has in handling the facts, but I'm afraid we're rapidly developing a sort of Mendelian ritual by which to explain the extraordinary facts of alternative inheritance. Isn't it absurd, for example, to assume that the height of a pea plant is determined by one particular Mendelian factor? Although we can quite easily explain the 3:1 outcome of tall to dwarf plants if we presume each parent carries an alternative factor, but consider what an astonishing simplification this is. If we accept this factor hypothesis—and many have done so without a second thought—then we are forced to accept a huge number of highly improbable conclusions. What must take place in the germ cells of these hybrid pea plants for the factors to work as many believe? We must assume that alternative factors, such as the tall factor and the dwarf factor in the pea plant, retire into separate cells without further ado, after having lived together, and competed with each other, through countless generations of cells. Can they really separate, without ever having produced any influence on each other?

DEMOCRITUS: The notion that characteristics are immortal and never blend clearly has its roots in preformationism.

MORGAN: That's right. Mendelism is nothing but preformationism in fancy new clothes. We can shuffle the factors like a pack of cards, but they won't mix. The deck contains all of the individual's characteristics, and development means activating those preformed factors. No, gentlemen! I believe the simplicity and the temporary success of the preformationist-Mendelian explanation don't justify its use! Preformation gives us no answer. It merely points its finger at some already-made unit—a tiny organism, or organ, or factor. In contrast, the epigenetic conception, although laborious and uncertain, keeps the door open for further examination and re-examination. This is the path to scientific advancement!

ALKIMOS: I don't understand. Mendel's theory is so simple and elegant. An epigenetic approach to inheritance would be much more complicated.

MORGAN: See, gentlemen. That's what I'm talking about. It's very tempting to explain everything with Mendelian factors, but its brilliance blinds us to the facts of science and leads to intellectual stagnation. If we can't describe something using magical Mendelian calculations, then we don't dare to dig anywhere near it. Let it finally be clear: embryos do not understand the language of mathematics!

ALKIMOS: But mathematics understands embryos …

DEMOCRITUS: I agree that there is no difference in principle between genes and the preformists' *animalcula seminalia*. But I can't imagine how the exact ratios that result from hybridisation can be explained by the theory of epigenesis.

MORGAN: Allow me to enlighten you with a very simple example. Let's say that two alternative characteristics correspond not to two factors but to two states of protoplasm. These two states may alternate without any intermediary stage, similar to many well-known chemical compounds that are alternative and separate, yet each has its own properties. Then it's easy to see that a tall or dwarf plant are the result of alternate states of protoplasm.

DEMOCRITUS: If I understand correctly, the germ cells don't need to contain the characteristics of the offspring, but only protoplasmic components which in some way or another, by unknown processes or interactions during development, produce the particular characteristics of the offspring.

MORGAN: Exactly! So you see how this is not the same thing as the notion of characteristics produced in the developing organism by the operation of factors in the germ cells.

ALKIMOS: But if the factors are the same thing as stable protoplasm states, then don't Mendel's laws apply?

MORGAN: In describing the probability of a phenomenon we can of course use the concept of factors and Mendel's ratios, but it will not bring us any closer to understanding the true organizing power of protoplasm.

ALKIMOS: And what about Boveri's chromosome experiments?

MORGAN: Gentlemen, do not blindly trust in Boveri's results! I repeated his experiment with enucleated ovum fragments and got completely different results.

EXTON: If I recall, in his article Boveri stated that if the enucleated ovum of a sea urchin is fertilised with the sperm of another species of sea urchin, the larva that develops will bear only the characteristics of the father.

MORGAN: This is true. The larval skeletons of the two different species are very different. This makes it easy to decide if the mother or the father was responsible for inheritance. Boveri's experiments resulted in larvae with the skeletal structure of the father. Mine however resulted in a wide array of skeletal structures.

DEMOCRITUS: So it's uncertain if the nucleus guides development.

MORGAN: Very. In fact I have absolute proof of the opposite. I can show that the cytoplasm is responsible for guiding development.

DEMOCRITUS: What's your proof?

MORGAN: Comb jellies.

ALKIMOS: Comb jellies?

EXTON: Comb jellies, or ctenophores, are transparent marine creatures that swim by beating with hair-like cellular protrusions, called cilia.

MORGAN: Indeed. If I remove a small portion of the ovum just as it begins to develop, but without disturbing the nucleus, I end up with abnormal embryos.

DEMOCRITUS: The nucleus remains intact, and still development goes awry. Incredible.

MORGAN: The embryo's shape is determined by the protoplasm, or, after first cleavage, by the conditions of the blastomeres. The conclusion is unavoidable: during early development the regulative powers reside in the cytoplasm, not the nucleus.

ALKIMOS: Master, let's go home. This is too confusing. I don't know who to believe.

DEMOCRITUS: Son, I think what we've discovered is that the question is far more complicated than we thought. If the regulative power resides in the cytoplasm, then clearly it must govern the chromosomes too.

ALKIMOS: But not too long ago we concluded the opposite was true.

DEMOCRITUS: Which means the question has not yet been decided.

ALKIMOS: And what if both sides are a little bit correct?

DEMOCRITUS: Perhaps sometimes the cause becomes the effect, and the effect the cause.

ALKIMOS: You mean the chromosomes regulate the cytoplasm and the cytoplasm the chromosomes?

MORGAN: Excellent idea.

DEMOCRITUS: But who gives orders first …?

ALKIMOS: And what happens before that? There's no end to this.

DEMOCRITUS: The command gives rise to another command.

ALKIMOS: Over and over, for generations.

DEMOCRITUS: The mother produces an ovum, and the ovum a mother.

ALKIMOS: And the ovum …

EXTON: I think it's time to go.

ALKIMOS: Exton, let's go in a Model T.

MORGAN: Gentlemen, would you like to ride in my car? It's brand new.

EXTON: Thank you, but ours is waiting for us outside the gate.

MORGAN: Hm … what an interesting vehicle.

EXTON: Yes, it's a factory prototype.

MORGAN: Could I give it a try?

ALKIMOS: Of course. Come along with us!

MORGAN: Excellent.

EXTON: I don't think that …

ALKIMOS: Please, Professor Morgan. After you.

MORGAN: What an ingenious design.

ALKIMOS: I suppose you could say so.

MORGAN: Hm, is this the ignition lever?

EXTON: Professor, no!

ALKIMOS: Aaagghh ... here we go!

EXTON: No wait! Democritus!

Sex Chromosomes

ALKIMOS: Master? Where are you?

EXTON: Oh no!

MORGAN: We've left our friend Democritus behind!

EXTON: That's the lesser of our problems.

MORGAN: And what's the greater problem?

EXTON: That our dear Professor Morgan has been taken unwittingly on a trip through time.

MORGAN: You don't say …

EXTON: Have no fear. I'll do everything I can to rectify the problem.

MORGAN: Time travel?

EXTON: Exactly.

MORGAN: Now just a moment …

ALKIMOS: What do we do, Exton?

MORGAN: Forward or backward?

EXTON: Forward. Ninety years.

MORGAN: Oh my!

EXTON: Don't worry, Professor, I'll get you back.

MORGAN: Can you?

EXTON: Now, let's see …

ALKIMOS: Exton, do something!

EXTON: I'm trying. Let's see, I can calibrate this so we return to … hmm, I can't get us back to 1910, but I can get us to 1911.

MORGAN: And what shall we do until then?

ALKIMOS: Never mind that! What will my master do for a whole year in the New World?

EXTON: Democritus will find some way to get along, I'm sure.

ALKIMOS: He'll probably try out that beautiful Model T …

EXTON: It will take a while to reprogram the Deus ex Machina. We'll be able to leave in about two hours.

MORGAN: Thank goodness.

EXTON: While the machine is calculating let's have lunch and chat a bit.

MORGAN: Ah, very well. But do tell me, what year are we in now?

ALKIMOS: Three thousand years after Troy.

EXTON: Two thousand fifteen years after Christ.

MORGAN: The year two thousand fifteen. Amazing.

ALKIMOS: I suppose …

MORGAN: Gentlemen, you are also scientists, and if I'm correct, embryologists. Tell me, now that we're ninety years into the future, how far has our science advanced? What have we learned about epigenetic mechanisms?

EXTON: I'm afraid we can't do that.

MORGAN: Why not?

EXTON: Professor, just think how it would upset the orderly progress of science if someone from the future arrived with advanced knowledge.

MORGAN: Why, it would help scientific progress! It's in the best interest of us all!

EXTON: I still can't tell you. I'm not allowed.

MORGAN: All right, just tell me this: were the theories about Mendelian factors proven wrong?

EXTON: No. They were proven to be absolutely true, to the last letter.

ALKIMOS: Exton, you shouldn't have said that!

EXTON: In fact, we could call the 20th century the century of genetics.

MORGAN: Incredible. You must be joking!

EXTON: Not at all. What you may find even more surprising is that it was your work in the 1910s that sparked the ascent of Mendelian genetics.

MORGAN: But I wasn't even there …

EXTON: That's why it's better if we examine how the concept of the gene, as the Mendelian factor came to be known, developed up to the 1910s.

MORGAN: Excellent. I'd like to hear what the role of the preformed factor is in the development of an organism.

EXTON: Let's start at the beginning. We agree that the question requires elucidating the role of the nucleus in the life of the cell.

MORGAN: Without question. The nucleus and the chromosomes it contains are the bearers of inheritance according to the Mendelians. But the more level-headed scientists who represent the epigenetic school of thought believe the events that take place in the nucleus—like every episode in the life of the cell and its development—are regulated by chemical changes in the cytoplasm.

EXTON: I'm sure you're familiar with Nussbaum's experiments with the one-celled *Oxytricha*, and Gruber's and Lillie's with *Stentor*, depicted here.

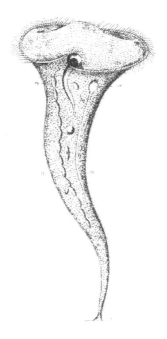

When they cut up these unicellular organisms, they found that only those fragments that contained the nucleus regenerated.

ALKIMOS: Ever since Abraham Trembley first minced an organism, it seems that chopping things up has become an important part of the scientific repertoire.

EXTON: The results of these experiments support the hypothesis that the nucleus controls cellular regeneration. One of the great embryologists of the period, Edmund Bleecher Wilson—I believe you know him well—asserted that these 'beautiful experiments prove that destructive metabolism ... may go on for some time undisturbed in a mass of cytoplasm deprived of a nucleus. On the other hand, the formation of new chemical or morphological products by the cytoplasm only takes place in the presence of a nucleus.'

MORGAN: Yes, I've read Wilson, but I don't agree with him. After all, we mustn't forget my experiments with comb jellies. Those clearly showed that parts of the cytoplasm had a crucial role in development. The cytoplasm, therefore, can't be uniform. Imagine, for example, that the essential elements of the cytoplasm are located near the nucleus in those strange one-celled creatures. Then, understandably, those fragments that contain this essential bit of the cytoplasm would also contain the nucleus.

ALKIMOS: It seems to me these so-called scientific explanations can be twisted and turned to support any theory and its converse.

EXTON: Quite true, Alkimos. But after a while these twists and turns become so complicated that only the most dedicated defenders of a theory continue to rely on them. The rest give up on the old theory and devise a new one that better explains the observations.

ALKIMOS: I definitely prefer the simplest theories. They require a lot less thinking.

EXTON: And assumptions. This principle is called Occam's razor.

MORGAN: Yes, yes, but simplifications only suit the scientist and do not in any way reflect the true complicated nature of the living world.

EXTON: Ah, but you must acknowledge how useful they can be. Wilson also said, 'This fact (of the necessity of the nucleus for protozoan regeneration) establishes the presumption that the nucleus is, if not the seat of the formative energy, at least the controlling factor in that energy, and hence the controlling factor in inheritance. This presumption becomes a practical certainty when we turn to the facts of maturation, fertilisation and cell-division. All these converge to the conclusion that the chromatin is the most essential element in development.'

MORGAN: I don't believe it's that straightforward. After all, what is the proof that the chromatin of a nucleus and not the material around the chromatin is the controlling factor?

ALKIMOS: Boveri's experiments.

MORGAN: Why is everyone so sure of the exclusive role of the nucleus?

ALKIMOS: Because an ocean of experiments demonstrates the central role of the nucleus.

MORGAN: And another ocean disproves it. The main weakness of the structuralist-Mendelian interpretations is that within all of them you can detect a kind of preformist speculation! It drives me mad!

EXTON: Nevertheless, Wilson's work was certainly remarkable. He was the first to try to reconcile the seemingly contradictory embryological facts.

Wilson believed there was no essential difference between mosaic and regulative egg development.

ALKIMOS: I agree, although I have no idea what either one is.

EXTON: Mosaic eggs are those that, after some kind of disturbance, aren't capable of developing into a complete embryo; their parts, you see, are pre-determined.

ALKIMOS: Like Morgan's comb jellies.

EXTON: Exactly.

ALKIMOS: That means the eggs of a sea urchin are regulative. If we break them into four parts they produce four intact larvae.

EXTON: The difference between mosaic and regulative is not clear-cut. This diagram shows the development of *Cerebratulus*, a nemertean worm (A to C).

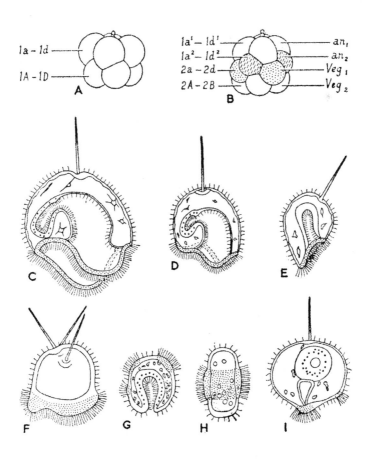

Blastomeres isolated at the two- or four-cell stages develop into normal, although smaller-sized larvae (D and E). In its eight-cell form (A), however, the four separated cells of the animal pole form without a gut (F), while the four cells of the vegetal pole form without a head (G). In its sixteen-cell form (B), the isolated cells also develop abnormally (H and I). According to Wilson—who assumes the pre-eminence of the nucleus—the only difference between the two is that in a mosaic egg the nucleus places the regulatory components in the cytoplasm before cleavage; in a regulative egg this step comes later. Consequently, the regulative embryos are able to reorder themselves more easily if disturbed. As Wilson put it: 'The primary determining cause of development lies in the nucleus, which operates by setting up a continuous series of specific metabolic changes in the cytoplasm.'

MORGAN: Believe me, gentlemen, if we start with the opposing view that the cytoplasm is the primary determining cause, we can postulate a theory of equal force. The only difference is that we imagine true chemical reactions rather than particles of preformed characteristics.

HERMES: Excuse me, Sir Extonos, but shall I serve lunch?

EXTON: Yes, of course.

HERMES: Right away, sir.

EXTON: Hermes, wait!

HERMES: Yes?

EXTON: My name is Exton. E-X-T-O-N.

HERMES: Yes, sir.

EXTON: Ah, those Greeks …

MORGAN: Now, where was I?

ALKIMOS: Preformists.

MORGAN: Yes! The most shocking manifestation of Mendelian preformism is the theory of sex chromosomes. How could they be so naïve as to postulate ideas that contradict everything we know about physiology? How can anyone think the tiny bit of chromatin visible in the male germ cell could determine the sex of the individual?

EXTON: Indeed, disagreement over the nature and role of the sex chromosome represents the greatest chasm between epigenetic and Mendelian interpretations. Hermann Henking first observed this unusual chromatin formation in 1890 while examining sperm production in the firebug *Pyrrhocoris apterus*. In half of the sperm he observed eleven chromosomes. In the other half, he also observed eleven chromosomes, plus a darkly stained chromatin particle. Having no idea what its function was, he called it X. In 1902, Walter Sutton, who worked in the laboratory of Clarence McClung, discovered a similar accessory chromatin element in the cells of the testes of the grasshopper species *Brachystola magna* during meiosis. The structure clearly had the shape of a chromosome, but after the first meiotic division—unlike its companions—it migrated to only one pole. Consequently it appeared only in half of the developing sperm. To highlight the importance of the sex chromosomes, we should examine the theory of Carl Correns, one of the rediscoverers of Mendel's work. At the end of the 19ᵗʰ century, most people believed that sex was determined by environmental factors. Mendel's theory offered a radically new explanation. Correns postulated that a Mendelian factor might have a role. Perhaps the male is hybrid and the male trait is dominant over the female.

ALKIMOS: A female can only be a recessive homozygote. In other words the male is *Aa* and the female is *aa*.

MORGAN: How simple! Um, I mean err … simplistic.

EXTON: Yes, the notion seems very naïve. However, Sutton had found the nuclear material to support the theory. Inspired by Boveri's work, he was the first to publish an article on the mysterious similarities between segregation of meiotic chromosomes and Mendelian factors. Soon after, Nettie Stevens demonstrated the sex differences in chromosomes in more than fifty species, mainly insects. Her most thorough investigations were of *Tenebrio molitor*, or the mealworm beetle, in which she discovered twenty large chromosomes in the somatic cells of females, but only nineteen of various sizes in the male. The ovum, however, always contained ten large chromosomes and the sperm either ten large or nine large and one small. Wilson posited that a male offspring was produced if a sperm with nine large and one small chromosome fertilised the ovum, restoring the 19 + 1 chromosome count characteristic of males.

ALKIMOS: Clear. And females' 10 + 10 chromosome count is restored if the other type of sperm penetrates the ovum. Who'd think sex is determined by such a little ...

EXTON: Moreover, if identical numbers of each sperm type form during meiosis, then the laws of randomness can be used to explain the ratio of males to females.

ALKIMOS: The perfect Mendelian pair of alternative characteristics!

EXTON: In addition to the 9 + 9 chromosomes, the large X chromosome and the little Y chromosome also form a pair.

ALKIMOS: But Henking called the little male chromosome X ...

EXTON: In some species the little Y shrunk so much that it disappeared. In those cases, male sex is determined by X0.

MORGAN: Ah yes, I see you're familiar with the principle of Occam's razor. Shave away as much as you can from a hypothesis—simplify, simplify! Well, Gentlemen, have you finished shaving? Because I'd like to offer another interpretation of those same observations, *contra Occam*. First of all, why do they think the two kinds of chromatin are the determinants of the two sexes? Why couldn't they be a consequence of the two sexes? It's very possible the primary cause in this case is chemical and not morphological. However, I'm ready to concede that these studies have indeed directed our attention to something of extraordinary importance: the connection between the formation of the germ cells and early development.

ALKIMOS: I still don't understand the problem with Mendel's theory. It explains everything! During meiosis the small and big chromosomes separate from each other—like every Mendelian pair—and then in their own way create an offspring of a certain sex.

MORGAN: The time has come for me to deliver the *coup-de-grace* to Sutton's and Boveri's Mendelian theory. It is the observations of my colleague, Lucien Cuénot, on the fur colour of mice. His results clearly refute the notion that Mendelian pairs segregate in the germ cells.

EXTON: Are you referring to the agouti mice?

MORGAN: Yes, indeed. Not long ago, Cuénot noticed yellow-furred mice among the population he was breeding.

EXTON: Hundred ten years ago.

MORGAN: When he crossed these with the original black-haired mice, black and yellow were born.

ALKIMOS: Well, that's easy to explain. The first yellow mice were heterozygotes and the gene for yellow fur is dominant over the gene for black fur. Let's say *SF*. When crossed with the *FF* mice the offspring were *SF* heterozygotes or *FF* homozygotes. But where did the yellow come from?

MORGAN: That's the lesser of our problems. After several generations of inbreeding, the mice still produced yellow- and black-furred babies. Cuénot found it impossible to breed only yellow-furred mice.

ALKIMOS: Impossible. If SF heterozygotes are further bred, they result in SS, SF and FF genotypes in a 1:2:1 ratio. A pure line of SS homozygote can easily be established. Cuénot must have botched something.

MORGAN: Not at all. What is means is that the cellular components that determine yellow or black fur never segregate as they should according to a Mendelian interpretation. His entire theory falls apart.

EXTON: I think it's time to end this debate. We need to go. Get ready, professor.

MORGAN: I agree, there' no point in further discussing this, since the agouti mice have so perfectly refuted the chromosome theory.

ALKIMOS: Exton, wait! I didn't understand a word. Where did Morgan hide my homozygote yellow mice? It's some kind of trick!

MORGAN: One must learn to give up the most cherished theories. It's the way of science.

ALKIMOS: My master!

MORGAN: Science cannot advance if we don't demolish idols and rigid dogma.

EXTON: Let's go, Professor!

ALKIMOS: What about me?

EXTON: Wait for me to return with Democritus.

MORGAN: One moment, my dear colleague, if this is time travel … What I mean to say is, will Democritus be the same Democritus?

EXTON: Of course not. Only their names will be the same.

ALKIMOS: And what about the agouti mice?!

EXTON: Professor, please climb in.

ALKIMOS: Exton, before you shut the door, tell me, where can I find the truth?

MORGAN: In the triumph of epigenesis.

EXTON: The homozygote mice were never born.

MORGAN: Exton, I see you don't …

EXTON: They perished in the womb.

MORGAN: Aaagghhh … here we go!

The Telltale White Eye

ALKIMOS: They perished in the womb? What does he mean? If they perished, then they had to have lived. If they lived, well, then I have my homozygote yellow mice. They were just never born and therefore couldn't produce progeny. That's why all the yellow-furred mice are heterozygotes. And that's why no pure line could be established. Professor Morgan's proof is not worth a hill of beans. Mendel and Boveri were right!

COOK: They left you here?

ALKIMOS: Yes, but I understand why. Hm, and they also left behind this marvellous stew, but I'm afraid I can't eat another bite.

COOK: Have some of this fruit punch. It'll go down easier.

ALKIMOS: Ah, heavenly!

COOK: Do you know how many fruit flies I squashed while making it?

ALKIMOS: Pesky fruit flies!

COOK: I'd love to know where they come from. If I leave out a bowl of fruit, they swarm it within an hour.

ALKIMOS: Nasty buggers.

COOK: But look at this!

ALKIMOS: What is it?

COOK: A fruit fly trap!

ALKIMOS: A what?

COOK: It's a slender jar full of wine and soapy water.

ALKIMOS: Okay, so?

COOK: Those wee-beasties think there's wine in it, they fly in and, whoops! They slip on the soap, hee, hee! Then they drown.

ALKIMOS: Excellent!

COOK: I've caught more than a hundred. Here they are—swimming.

ALKIMOS: Be careful, you might drink it by accident!

COOK: That's why I wrote on it: Fruit Fly Trap. Don't Drink!

ALKIMOS: Listen. I think they're back.

COOK: The whole house is shaking.

ALKIMOS: Master!

DEMOCRITUS: Alkimos, it's so good to see you!

ALKIMOS: Where were you?

DEMOCRITUS: Now, where should I start?

ALKIMOS: The automobile. Did you ride in it?

DEMOCRITUS: Ah, it's amazing. It goes with such speed. My beard was flying.

ALKIMOS: I would have loved to try it.

DEMOCRITUS: Alkimos, I have so much to tell you. I spend an entire year in America, in Morgan's lab.

ALKIMOS: Meanwhile Morgan was here having a meal.

DEMOCRITUS: Research didn't stop without him. In fact it accelerated.

ALKIMOS: That's what I thought. Having a boss around slows things down.

DEMOCRITUS: Only if the boss is no longer a true master. But Morgan was. As soon as he returned, he entered into an animated debate with his students. But even Morgan couldn't deny the implications of our new discoveries. By the time Exton and I left, he'd become the greatest proponent of the Mendelian theory of genetics. He knew immediately what to do.

ALKIMOS: What happened? How were you able to persuade him? Exton tried in vain.

DEMOCRITUS: His associates and I made a fantastic discovery! The key to everything was an amazing little animal. Morgan had begun studying it around 1909. It reproduces quickly, a new generation can be produced every ten days, and it fits in tiny vials; in other words, it's perfect!

ALKIMOS: So, an animal that's easier to work with than peas and mice? Hmm, it can't be a sea urchin, can it?

DEMOCRITUS: A fruit fly.

ALKIMOS: What?

DEMOCRITUS: *Drosophila melanogaster.*

ALKIMOS: Wow, if the cook only knew …

DEMOCRITUS: It has splendid red compound eyes, six legs, antennae, a proboscis, and wings. Quite a complicated body structure, and four pairs of chromosomes, as you can see here.

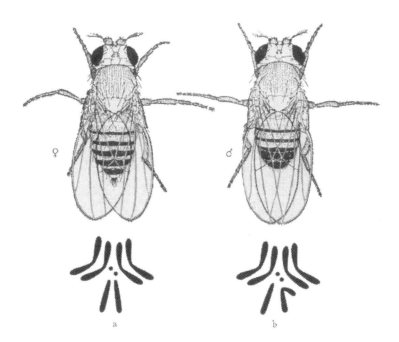

ALKIMOS: I assume you started studying its development.

DEMOCRITUS: Not at all. Only its genes.

ALKIMOS: But …

DEMOCRITUS: It's incredibly exciting. One day a white-eyed male appeared among all its red-eyed companions. A *mutant*. No one knows why. We crossed it with a red-eyed female. The first generation contained only red-eyed individuals.

ALKIMOS: That means the white-eye trait must be recessive.

DEMOCRITUS: Then we bred the first generation again. The second generation produced red-eyed males and females and white-eyed males, but no white-eyed females.

ALKIMOS: That doesn't make sense.

DEMOCRITUS: At first we didn't understand it either. But then we thought about it. What if the XY combination determines the sex and not the X0 as we thought. The only way to interpret our results is by supposing that the gene for white eyes or *white* is linked to the X chromosome, and that's how it's passed on.

ALKIMOS: So *white* is attached to the X chromosome and never exists independent of it? I still don't understand.

DEMOCRITUS: We labelled the allele of the *white* gene that causes white eyes with a lowercase *w*. The dominant red-eye allele of *white* we marked with a capital *W*. The male is XY and the *w* is linked to the X, the Y chromosome has no *white* gene attached to it. The female is XX and both X chromosomes carry a *W*.

EXTON: Alkimos, look here; this diagram will help you follow the combination of alleles.

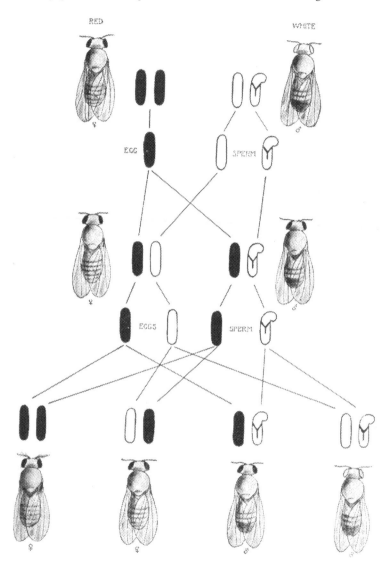

DEMOCRITUS: Thank you, Exton. So, the sperms carry either an X with a *w* attached to it, or a Y without a *white*. The eggs, however, all have one X chromosome with a *W* attached to it. Flies in the first generation all have red eyes: either *w*W–XX females or *W*–XY males in a 1:1 ratio. The eggs produced by the females will all be *w*–X or *W*–X, while the sperm will all be *W*–X or Y genotypes. In the second generation, we get every possible type except for the white-eyed female (*ww*–XX). This is because the *w*–X eggs are never fertilised by a *w*–X sperm since no sperm containing this combination emerge from this hybridisation. But all this wasn't enough to persuade Morgan.

ALKIMOS: Honestly, what does it take? Boveri was enough to persuade me.

DEMOCRITUS: He claimed a physical relationship between the X chromosome and the *white* gene was not proven by these results. It's possible that both are under the effect of the same cytoplasm, which means the connection is not direct but is regulated by indirect external causes.

ALKIMOS: Morgan will never be satisfied.

DEMOCRITUS: We thought so, too. But in the summer of 1911 we discovered other genes that affect body colour and wing shape. Amazingly these are also linked to the X chromosome. Morgan had no choice but to concede. The theory of hereditary cytoplasm crumbled. It was impossible that one state of cytoplasm could determine the inheritance of so many traits using the same mechanism. Only a very elaborate epigenetic theory could explain it, and thus it would be no match for the extraordinarily simple Mendelian interpretation. Morgan was converted. I'll never forget his words: 'Fantastic! During segregation certain factors are more likely to remain together than to separate, not because of any attraction between them, but because they lie near together on the same chromosome. [It's now clear that] sex-limited inheritance is explicable on the assumption that one of the material factors of a sex-limited character is carried by the same chromosome that carries the material factor of femaleness. I never would have thought!' That was the moment that genetics truly triumphed. I'm so happy I was present.

EXTON: That was truly a revolutionary moment. It marked the beginning of extensive research into the role of genes, their location on the chromosomes, and their physical substance. Morgan's words represent the culmination of Mendelian genetics and laid the foundations for genomics. Mendel studied only a few genes, and luckily those all happened to be independent—by that I mean they were all located on different chromosomes. If he had studied linked genes, he might have been so lost in the tangled web of his findings that he could have never established his laws of heredity. It should be clear to us, however, that the study of genetics did not arise exclusively from Mendel's work, but also from the considerable research done by others in embryology.

ALKIMOS: Like Boveri's experiments.

DEMOCRITUS: Or Morgan's. They were both embryologists.

EXTON: That's right. Sadly, however, shortly after its birth, genetics disowned its own birth mother, embryology, and arrogantly declared its superiority. Geneticists wrote with an almost religious fervour about the development of their field, endowing it with a mythical quality as they jumped directly from Mendel to Morgan. One of the most flagrant examples was Morgan himself. Nevertheless, Morgan's conversion opened the doors to dizzying possibilities. The activities of the Drosophila laboratory marked the beginning of one of the most daring paths of discovery in genetics: the mapping of the entire genetic stock of living things, also known as the *genome*. But before we embark on this fascinating journey, let's take a break. In the meantime I'll read to you the closing thoughts of Edmund Wilson in his monumental book *The Cell in Development and Inheritance*, published in 1896.

ALKIMOS: We're all ears!

EXTON: 'What lies beyond our reach at present, as Driesch has very ably urged, is to explain the orderly rhythm of development, the coordinating power that guides development to its predestined end. We are logically compelled to refer this power to the inherent organisation of the germ, but we neither know nor can we even conceive what this organisation is. The theory of Roux and Weismann demands for the orderly distribution of the elements of the germ-plasma a prearranged system of forces of absolutely inconceivable complexity ... How are we to conceive the power which guides the countless hosts of migrating pangens throughout all the long and complex events of development?'

ALKIMOS: If the book was published in 1896, then Mendel can't be mentioned in it.

DEMOCRITUS: The question is to what degree does Mendel's work bring us closer to answering the question posed by Wilson. I suspect, not at all.

EXTON: Wilson says, 'The truth is that an explanation of development is at present beyond our reach. The controversy between preformation and epigenesis has now arrived at a stage where it has little meaning apart from the general problem of physical causality.'

ALKIMOS: Master, did we become preformists like Mendel?

DEMOCRITUS: Mendel's work does not give an answer to the questions of development. Of course, if we look at it from the perspective of an embryologist, there's no doubt that the concept of the immutable gene stands closest to preformationism. Hertwig's theory, however, has shown us that some kind of in-between solution is possible. Perhaps genes don't determine the characteristics of an individual, but only those of the cells or the cell fluids. The complete individual is the result of the epigenetic organisation of the preformed parts.

EXTON: Wilson asserts that these questions about the 'historical origin of the idioplasm'—by that he means the hereditary units—'brings us to the side of the evolutionists. The idioplasm of every species has been derived, as we must believe, by the modification of a pre-existing idioplasm through variation, and the survival of the fittest. Whether these variations first arise in the idioplasm of the germ-cells, as Weismann maintains, or whether they may arise in the body-cells and then be reflected back upon the idioplasm, is a question on which, as far as I can see, the study of the cell has not thus far thrown a ray of light.'

DEMOCRITUS: If we think about it, Mendelism didn't throw a ray of light on the question either...

EXTON: Wilson continues: 'It is true that we may trace in organic nature long and finely graduated series leading upward from the lower to the higher forms, and we must believe that the wonderful adaptive manifestations of the more complex forms have been derived from simpler conditions through the progressive operation of natural causes. When all these admissions are made, and when the conserving action of natural selection is in the fullest degree recognised, we cannot close our eyes to two facts: first, that we are utterly ignorant of the manner in which the idioplasm of the germ-cell can so respond to the play of physical forces upon it as to call forth an adaptive variation; and second, that the study of the cell has on the whole seemed to widen rather than to narrow the enormous gap that separates even the lowest forms of life from the inorganic world. I am well aware that to many such a conclusion may appear reactionary or even to involve a renunciation of what has been regarded as the ultimate aim of biology. In reply to such a criticism I can only express my conviction that the magnitude of the problem of development, whether ontogenetic or phylogenetic, has been underestimated; and that the progress of science is retarded rather than advanced by a premature attack upon its ultimate problems. Yet the splendid achievements of cell-research in the past twenty years stand as the promise of its possibilities for the future, and we need set no limit to its advance. To Schleiden and Schwann the present standpoint of the cell-theory might well have seemed unattainable. We cannot foretell its future triumphs, nor can we repress the hope that step by step the way may yet be opened to an understanding of inheritance and development.'

DEMOCRITUS: Yes, let's hope!

Genetic Mapping

EXTON: She sunk into the foam again?

ALKIMOS: Yes.

DEMOCRITUS: Incredible.

EXTON: And then?

ALKIMOS: Nothing.

EXTON: What do you mean?

ALKIMOS: I never saw her again.

EXTON: You're very lucky, Alkimos.

ALKIMOS: You think so?

EXTON: Hm.

ALKIMOS: Master, tell us about your past loves! Whenever the topic comes up you're quiet. You have secrets!

DEMOCRITUS: It's not easy to talk about after so many years.

ALKIMOS: You can tell me!

DEMOCRITUS: I suppose. She was beautiful. Thracian. I would have given my life for her. I wasn't interested in philosophy when I met her. I was young and happy. We travelled the world, from Egypt to Italy, and sometimes we even dared to journey further. We traded with Phoenicians, Punics, Jews, and Moors. Packed full with goods, my galley crossed the sea. We didn't have to think. All we did was enjoy life. Our home was in the belly of my ship, a vessel made from the wood of towering cypresses. Waves rocked us to sleep. Time and space ceased to exist. One day near the shores of Attica, a terrible storm struck. Poseidon shook his trident in uncontrollable rage. The ship was tossed against the rocks and destroyed. The gods spared me, but she was drawn into Persephone's lair. Since then on only the games of the mind give me any peace. If for a second I stop thinking, I am thrust into a bottomless well of despair.

ALKIMOS: Master, you never told me this story.

DEMOCRITUS: I'm telling you now.

ALKIMOS: Exton, what's that fellow doing here again?

EXTON: Ganymede, you don't need to take notes yet!

GANYMEDE: The folks upstairs told me I can start.

EXTON: When we talk about science.

GANYMEDE: It makes no difference to me. I'll just take notes on everything.

EXTON: Fine. If you insist on writing down everything we say, then I guess we'd better continue yesterday's conversation.

ALKIMOS: Let's. Then we won't have to retrieve Democritus from any bottomless wells. Master, I'm sure your grey matter is eager.

DEMOCRITUS: Well, I must say, I don't need much encouragement. I spent last night reviewing everything we heard yesterday. At first, Wilson's parting words left me discouraged, but then slowly I began to see the light. If we aren't ready to answer the greater, more complicated questions about inheritance and development, then we should address only the simpler ones. If we can resolve them, then we'll gain the confidence to pursue those larger questions that at present loom hopelessly far in the distance. That doesn't mean, of course, that we lose sight of these larger questions; we just shouldn't deal with them quite yet. First we need to discover as much as we can about the location and physical nature of genes.

ALKIMOS: I don't understand, master. We know that genes are linked to chromosomes, and we know that chromosomes are made of chromatin. What more do we need to discover?

EXTON: Much, much more. Morgan's train of thought was much like Democritus'. In 1919, he published his book *The Physical Basis of Heredity*. His opening words were: 'That the fundamental aspects of heredity should have turned out to be so extraordinarily simple supports us in the hope that nature may, after all, be entirely approachable. Her much-advertised inscrutability has once more been found to be an illusion due to our ignorance. This is encouraging, for, if the world in which we live were as complicated as some of our friends would have us believe, we might well despair that biology could ever become an exact science.'

ALKIMOS: Is Morgan picking at Wilson again?

EXTON: I don't know. But his words are heartening.

DEMOCRITUS: Indeed they are. Let's get started!

EXTON: When Mendel crossed flies that differed for two or more genes, it was discovered that each allele pair segregates independently from all the other allele pairs. This is what we would expect if all the genes are located on their own separate chromosome, and if the maternal and paternal parts of any one chromosome pair independently migrate to one or the other of the daughter cells during meiosis. However, it is also obvious that there are far more genes than chromosomes.

ALKIMOS: So that means that one chromosome carries more than one gene?

EXTON: Yes. Because of this, a lot of dihybrid crosses produce results that contradict Mendel's principle of independent assortment. Only under certain circumstances does this principle work.

DEMOCRITUS: Such as when the genes studied are all on separate chromosomes. This was the case in Mendel's experiments.

ALKIMOS: So the pea plant has at least seven chromosomes!

EXTON: The pea plant has exactly seven chromosomes. The genes Mendel studied, however, are located on just three chromosomes. The contradiction can only be resolved by a rather long explanation.

ALKIMOS: Oh. I thought I understood.

EXTON: It's easy to imagine that those genes located on one chromosome stay together during meiosis. Consequently the alleles that are together in the first generation also appear together in the second generation; in other words, they're not independent. This is called *linkage*. And of course not just two but many genes can be linked. By 1919, Morgan and his team of scientists had identified more than one hundred fruit fly genes based on the analysis of mutants. These formed four *linkage groups*. The first group, located on the sex chromosome, contained one hundred, the second group contained seventy-five, the third sixty, and the fourth just two genes.

DEMOCRITUS: Each linkage group corresponds to one chromosome.

EXTON: And that's not all. But before we proceed, let's have a closer look at what happens to two linked pairs of alleles during hybridisation: the *black* (*b*) and *vestigial* (*v*) recessive mutations. In homozygotes, the first mutant produces a black coloured individual and the second stunted wings. The *b, v* phenotype is crossed with grey coloured, long-winged (*B, V*) so called *wild types*, the F1 generation has only *B, V* offspring. If the F1 males are backcrossed with double recessive homozygous females of the parental generation with a *b, v* phenotype, then half of the F2 generation will be *b, v* and the other half *B, V*.

ALKIMOS: If we had independent assortment, then a backcrossing should have resulted in a 1:1:1:1 ratio. Here, however, two groups—*b, V* and *B, v* — are not represented.

EXTON: Because *b*, *v* and *B*, *V* are linked.

DEMOCRITUS: There's another explanation. It's possible that the *b* and *v* phenotypes indeed originate from the same gene. In that case they would never separate from each other.

EXTON: Yes, that is possible. It's known as *pleiotropy* when one gene affects more than one characteristic. But that's definitely not the case with the *black* and *vestigial* traits. The two traits can appear separately. Therefore the above example can only be explained by linkage.

DEMOCRITUS: We saw this with sex-linked inheritance. That's when we first discovered linkage.

EXTON: To be more precise, that's when it was first interpreted correctly. In 1905, Bateson and Punnett discovered that in the F2 generation of a dihybrid cross the ratio of 9:3:3:1 was not achieved. If we follow two traits of the sweet pea, the colour of the flower and shape of the pollen, the results cannot be explained by the known *epistastic* interaction effect, or the effect of one gene on another. The double dominant and double recessive individuals appear in much larger numbers than those with mixed traits.

ALKIMOS: Which means these two traits are also linked.

EXTON: Yes, but a group with mixed traits appeared too.

ALKIMOS: So linkage is not complete. Wait a minute! How is that possible? Either the two genes are on one chromosome or not. So the genes are either linked, or they're not.

EXTON: It's possible, Alkimos, because it's not that simple. You see, it results from the rather peculiar behaviour of chromosomes. If this phenomenon didn't exist, genetics would be nowhere. We'd be unable to decipher the order and location of genes on chromosomes. What's more the evolution of the living world would have happened in a different way. The point of sex is for the genes to mingle, for a diverse generation of offspring to come into being, and for mutations to be disposed of. Populations that reproduce without sex are sentenced to slow death. The mixing of genes is one of the most crucial events of life. The magical gene-mixing embrace of chromosomes is called *crossing over*. Now real genetics begins!

DEMOCRITUS: If the mixing of genes or, to be more exact, alleles is so important then why do chromosomes exist? If there were no linkage, genes could mix freely.

EXTON: Yes, but it's very difficult to imagine a mechanism that would ensure that one copy of every gene makes it into the daughter cell during cleavage. The loss of genes has serious consequences. The chromosomes help to hold genes together and ensure that they replicate simultaneously. But still there is a great need for mixing.

DEMOCRITUS: Mixing and staying together all at the same time. That's not any easy to achieve.

EXTON: No, it isn't. That's why crossing over is one of the great accomplishments of a cell. But let's return to hybridisation. The genes for *black* and *vestigial*, for example, demonstrate that linkage can be complete. Bateson and Punnet's sweet pea hybrids, however, are good examples of weak linkage. Between the two extremes, independence and complete linkage, every gradation is possible. It's as if the genes stay together and don't stay together on the chromosomes.

ALKIMOS: So when we cross dihybrids—regardless of what Mendel's laws prescribe—pretty much any result is possible. If I were a scientist, this is the point at which I'd give up my research in genetics and become a fisherman or a shoemaker.

EXTON: Bateson was confounded by the sweet pea too. It wasn't until Morgan and his team came along that those seemingly incomprehensible results were properly interpreted. Thanks to their genius, true genetics emerged.

ALKIMOS:

> *Ah, the foul dihybrid vetch, how it causes such grief,*
> *Only in his last does the poor shoemaker obtain relief.*

EXTON: Although the explanation is rather simple. We need to presume that a small piece of the father and mother chromosomes occasionally change place in the germline. When the *homologue chromosomes* line up side by side during meiosis, there is a perfect opportunity for crossing over.

DEMOCRITUS: The alleles of the father chromosome might move to the mother chromosome and vice versa.

ALKIMOS: So genes that were originally linked can separate from each other?

EXTON: That's all there is to it. Cytological descriptions of chromosomes' crossing over were produced that reflected the actual results. One of these was Frans Alfons Janssens' theory of *chiasmatypes*, which posited that duplicated maternal and paternal chromosome threads twisted around each other and at their points of contact broke off and changed places, as illustrated by this drawing in Morgan's book:

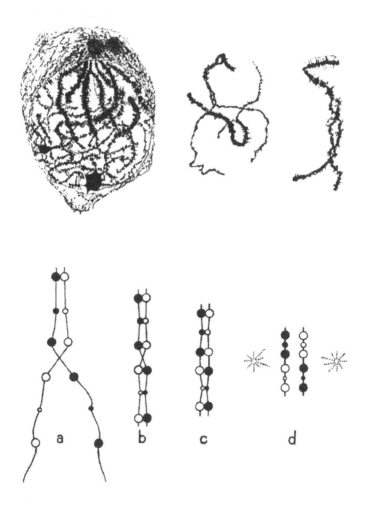

Therefore, the tetrads of paired chromosomes and the daughter cells contain chromosomal crossovers. Unfortunately, microscopes lacked the resolution necessary to produce images that could corroborate the results of genetic methods. Morgan noted that cytological evidence lagged so far behind genetic evidence that it was impossible to make an exact statement about the mechanism of crossing over. Like Janssens, the German biologist Richard Goldschmidt also developed a theory to explain crossing over. He believed that it was sufficient to assume that genes are freed during the cell's resting phase, a view shared by Oskar Hertwig, and that homologous pairs combined at random. When the chromosomes crystallised again, new allele combinations arose.

DEMOCRITUS: Goldschmidt clearly didn't accept the idea that chromosomes were permanent structures.

EXTON: Because until then no one had been able to prove it beyond a doubt. Thus, several contradictory theories about the mechanism of crossing over existed. Nevertheless, genetic methods were able to demonstrate that genes appeared in a particular order, one alongside the other, on the chromosomes.

ALKIMOS: Boveri's experiments make it clear that genes are located on the chromosomes, and the chromosomes are long threads; thus it follows that genes line up on the threads.

EXTON: That's right. But there's a big difference between thinking something is true and proving it with experiments. Morgan and his associates managed to measure what many at the time didn't even dare to think. But let's return to black coloured flies with vestigial wings. Remember we backcrossed males form the F1 generation with B, V (grey, long-winged) females. It quickly became clear that crossing over didn't take place in Drosophila males, but it did in females. This is why if we did the reverse experiment—crossing F1 females with parental b, v males—we get all four possible types.

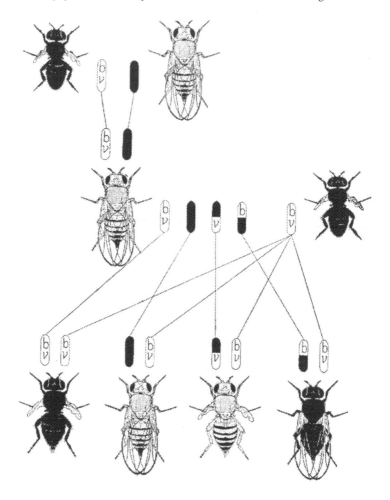

The various groups always appear in similar proportions. The individuals that carry a different combination of alleles than their parents are called *crossovers* or *recombinants*.

Non-Recombinants		Recombinants	
black, vestigial (*b*, *v*)	grey, long (*B*, *V*)	black, long (*b*, *V*)	grey, vestigial (*B*, *v*)
41.5%	41.5%	8.5%	8.5%
Total: 83%		Total: 17%	

Morgan and associates performed numerous dihybrid crosses, and received fixed ratios for every linked pair of alleles. Independent genes always produced a ratio of 1:1:1:1; in other words each group comprised 25% of the total. For *black* and *vestigial* the ratio was always 83% to 17%. For the *yellow* (*y*) allele, which results in a yellow body colour, and *white* (*w*) allele, which results in white eyes, the frequency of recombinants was much smaller: 1.2%. Morgan and associates called the values expressed as a percentage of the total number of offspring *crossover value*. This shows how often linkage between two genes is broken.

ALKIMOS: But how does this suggest that genes line up in a row?

EXTON: It doesn't. But the relationship of crossover values to each other certainly does. The crossover value between *yellow* and *white* is 1.2%; between *white* and the *bifid*, a gene that determines the wing vein pattern, is 3.5%; and between *yellow* and *bifid* is 4.7%. The crossover values are additive. See, the value between yellow and white and white and bifid equals the total value between yellow and bifid.

DEMOCRITUS: Yes, I see. Goldschmidt's model couldn't explain this. If the chromosomes disintegrate and the free-moving genes reassemble, we would not receive definite, additive crossover values. Our strongest argument in favour of the permanence of the chromosome is therefore not cytological, but genetic. What all this shows is that if two genes are in close proximity to each other, crossing over is less likely to take place along the short section of chromosome between them.

EXTON: But if the genes are located at opposite ends of the chromosome, then crossing over is much more common. The frequency of crossing over depends on the distance genes are from each other on the chromosome. Crossing over allows for not only the mixing of genes, but something else, too: *genetic mapping*. This is of lesser importance to the chromosomes, but of great importance to geneticists. One of the fundamental goals of genetics is to determine the location of the genes on the chromosomes, in other words to map the genes. The first map was made by one of Morgan's students, Alfred H. Sturtevant, in 1913.

DEMOCRITUS: Incredible! The secrets of nature reveal themselves to the educated eye! In Hertwig's time we thought chromosomes were just mysteriously wandering nuclear threads that absorbed large amounts of dye. In just a couple of decades we discovered how to measure the genes' distance from each other. In fact, if we know the distance, we can determine the order. This is what crossing over tells us! Whoever thought we'd have an opportunity to glimpse in such detail the organisation of the units that regulate life!

ALKIMOS: Master, I don't understand a word. How is that possible?

DEMOCRITUS: You don't understand, son? The likelihood of crossing over, by this I mean the number of recombinants produced in one crossing, depends on the distance of two genes from each other on the chromosome. In other words we can use the number of recombinants to calculate the distance. Of course actually measuring the distance is impossible at present—and perhaps never will be—but genetics can help us to infer it.

ALKIMOS: Lots of recombinants mean a great distance. Few recombinants mean a short distance.

DEMOCRITUS: Exactly! Morgan and associates knew of more than one hundred genes. By performing enough crossings, in theory we should be able to determine the order and also the distance.

ALKIMOS: If we have to cross all the possible combinations then we would need several thousand vials.

DEMOCRITUS: If the distances can be calculated, then there's no need to carry out every crossing. It's enough if every gene is mapped in relationship to just a few others. Sooner or later a complete picture will form.

ALKIMOS: Master, I'm in awe of your brilliant mind!

EXTON: But we still have a few challenges ahead of us. Morgan and associates showed that among genes that experience frequent crossover, the frequency of crossovers, or *recombinant frequency*, is lower than the sum of all the crossover values in the entire area.

DEMOCRITUS: You mean the values are not always additive?

EXTON: Although this seems to contradict what we've discussed thus far, if we understand the reason, we'll see that this result only confirms our theory about the ordering of genes on the chromosomes. Here's an example. The genes for yellow body colour and bar eyes have a recombinant frequency of 43.6%. According to the principle of additive numbers, however, this value, which tells us the distance between the two genes, should be 57%. Something must be happening in the section between the two genes. To understand, we need to pay attention not only to the frequency of recombination of the two genes in question but also to the behaviour of the genes that lie between them on the chromosome. What we'll discover is that a number of double crossovers take place, as depicted here.

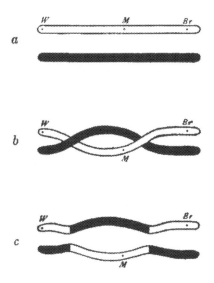

If we take these into consideration when determining the frequencies, the discrepancy between the number of the actually happening crossovers and the observed recombinant frequency disappears.

DEMOCRITUS: If two crossovers take place between the *yellow* and *Bar*, we won't notice it in the offspring, since the alleles, after changing place once, change place again, returning to their original position. Therefore, the recombinant frequencies are not proportional to the number of crossovers that actually occur between the two genes. Over large distances the frequency of crossovers is not equal to the frequency of recombination.

EXTON: The highest observed recombinant frequency is around 50%. This occurs in genes whose distance based on the sum of partial distances can be more than one hundred. Of course these partial distances need to be small enough that we can safely eliminate the possibility of double crossovers. Another student of Morgan's, Calvin Bridges produced the following example. At one end of the second chromosome is the so-called *star* gene, and at the other end the *speck* gene. If we examine the two genes, we find a recombinant frequency of 48.7%. But if we add up the crossover values of the seven genes between them then we get 104.4.

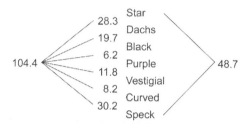

DEMOCRITUS: If a sufficient number of crossovers take place on a chromosome, then what we find is that linked genes, ones that are attached to the same chromosome, behave as independent genes in genetic crosses. If Mendel's allele pairs assorted all independently, but are located on just three chromosomes, then some of these genes must have been located on the same chromosomes. The explanation for the independent assortment then must be that these linked gene pairs had a high recombination frequency. Mendel was very lucky in his choice of allele pairs.

EXTON: He certainly was.

ALKIMOS: Using recombinant frequency values, we can determine the order of the genes and their distance from each other. Amazing!

EXTON: The the crossover frequency depends on the distance. But I must emphasise that the distances we determine are only genetic distances, not actual physical distances measured on the chromosomes. After all we don't know how the crossovers are distributed on the chromosome.

DEMOCRITUS: The actual chromosomal distances will only correspond to genetic distances if the crossovers are equally distributed on the chromosomes.

EXTON: But we have no idea at all about the actual distribution or the actual distances.

ALKIMOS: Well, we can't know everything.

DEMOCRITUS: Even if the true distances are still a mystery, we've still taken huge strides.

ALKIMOS: I see the future. Genetics will focus on creating more and more precise maps. We'll use up enormous numbers of vials, until we've lined up every single gene. The split wing next to the purple eyes, and between them the crooked nose and the twisted leg genes. I dare say it sounds rather monotonous to me. I'd rather have dinner.

EXTON: You're right, Alkimos. Gene mapping is one of the most repetitive chapters in science. The challenge grows increasingly technical and some of the beauty of research is lost.

ALKIMOS: Ugh, then let's skip it. Let others do the dirty work.

EXTON: I agree Alkimos. Let's stick with the exciting questions.

ALKIMOS: But when the mapping is done, I'd like to have a look.

EXTON: This diagram shows the fruit fly chromosome map as it was in the 1920s.

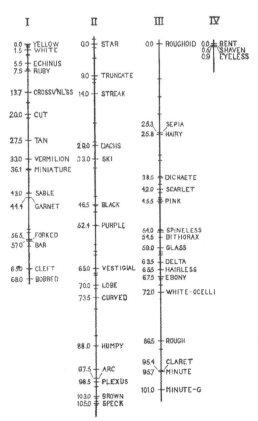

ALKIMOS: While the map is being completed, can we have a rest?

EXTON: Absolutely not! We're going to venture in another direction. Let's see how the embryos are doing.

ALKIMOS: Now that sounds better! Embryos are a lot more intriguing than chromosomes.

EXTON: Yes, but we'll eventually return to chromosomes. There's more for us to chew on.

ALKIMOS: And there's more of that leg of mutton to chew on ...

Part Four

Forces and Reactions

The Mathematics of Life—*D'Arcy Wentworth Thompson*

ALKIMOS:

> *Genes arose from dust, and while they tussled, for rights*
> *To nose, to ear, to venous wing, whooping in battle,*
> *Zeus, god of Heaven and Earth, in a fit of divine zeal*
> *Pierced their throats with a skewer of fiery chromosomes.*

DEMOCRITUS: Their groans can still be heard. Only crossing over allows them some freedom.

EXTON: Alkimos, I see your enthusiasm has returned.

ALKIMOS: I enjoy a little sarcasm when I can't understand something.

EXTON: When our thinking has reached a dead end, sometimes a dose of sarcasm might help to approach the problem from a different angle.

ALKIMOS: Or a few hexameters.

EXTON: Or the purest of all philosophies: mathematics.

ALKIMOS: Mathematics can make genes easier to understand?

EXTON: Some say it helps us understand embryos better, too.

ALKIMOS: The development of organisms in the language of numbers?

EXTON: You could say that. But for us to better understand how mathematics can be used to describe life processes, we should visit the greatest mathematician-biologist of Morgan's time, D'Arcy Wentworth Thompson. I believe the most concise way to describe him is a classical genius, an astonishing intellectual giant. He was a physicist, classical philologist, and biologist all in one. Some say biology was his weakest field; yet he managed to unravel the complicated effects of physical forces, chemical reactions, and evolutionary history on the living world. His work has had a profound and enduring impact on biological thought. He was not an experimental scientists but a theorist who revolutionised our approach to science. After all, science is not just about experiments. Of equal importance is the development of theories and hypotheses. Without it, experimentation would eventually atrophy. Hermes, prepare the Deus ex Machina! Direction St. Andrews, Scotland, 1920!

ALKIMOS: Scotland?

EXTON: You'll see. It'll be worth the effort. Now, let's go!

ALKIMOS: Aaaggh, here we go!

DEMOCRITUS: Do stop yammering, Alkimos. Be happy that we're finally going to meet a real mathematician. Mathematics is the queen of science; it's the language of Pallas Athena.

ALKIMOS: I understand the language of Dionysius better.

EXTON: Without mathematics the natural sciences would not exist. Copernicus, Galileo, Newton, Mendel—they were all, among other things, mathematicians, although mathematics is not a natural science.

ALKIMOS: Then what is it?

EXTON: Philosophy. The most perfect abstraction.

ALKIMOS: You mean all philosophers are mathematicians?

EXTON: No, but all mathematicians are philosophers.

ALKIMOS: Then what are you, master?

DEMOCRITUS: Me? Naturally, I am both.

EXTON: We're here. Everyone out!

ALKIMOS: It's raining again?

DEMOCRITUS: At least we'll be able to concentrate on science.

ALKIMOS: You mean mathematics.

DEMOCRITUS: Yes, of course.

ALKIMOS: Master, I don't understand what maths can tell us about life? How can you replace a bug with numbers?

DEMOCRITUS: Perhaps what we need is a lot of numbers.

ALKIMOS: You could fill three books with numbers, and it still won't show me a dragonfly.

DEMOCRITUS: What if the numbers are very small? Let's say as small as atoms?

ALKIMOS: We'd be old and grey before the books were ever finished.

DEMOCRITUS: I'm already old and grey.

EXTON: Follow me! D'Arcy Thompson isn't far from here. We'll find him in the Department of Natural History.

DEMOCRITUS: Ah, I've longed for this meeting.

EXTON: Wait! There he is—in the coffee shop across the street.

DEMOCRITUS: That rather large, wise-looking man?

EXTON: That's him.

ALKIMOS: I knew he'd be old and grey.

EXTON: Professor, do excuse me for bothering you.

THOMPSON: How may I help you, gentlemen?

EXTON: Allow me to introduce myself. I'm Exton—

THOMPSON: Of course. Mr. Exton and his students. Please have a seat. Would you like anything?

ALKIMOS: Yes, I'd like—

EXTON: Oh please, don't trouble yourself.

THOMPSON: Why, it's no trouble at all. Waiter, three more coffees, please. And a cognac.

EXTON: The reason we've come, Professor is we'd like to ask you about the mathematics of life. It seems such an unlikely combination, but I must say it's captured our imagination. We'd love to delve deeper into the subject.

ALKIMOS: It hasn't captured my imagination. Excuse me for being so honest, but I can't understand why we'd want to reduce nature's perfection to a bunch of numbers.

THOMPSON: Young man, numbers themselves are perfect. Perfect things can only be measured against other perfections.

ALKIMOS: Okay, but I'm not interested in swapping one for the other.

THOMPSON: Swap? We have no intention of swapping anything. All we're saying is that the language of numbers can shed light on the myriad phenomena of life. Often mathematics is the most effective tool in understanding reality.

ALKIMOS: I don't get it. How can we talk about life in the language of mathematics?

THOMPSON: Cells and tissue, shells and bones, leaves and flowers; these are all the magical manifestations of life. Yet, there's not one particle of them whose formation did not accord with the laws of physics and mathematics.

ALKIMOS: What does that mean?

THOMPSON: Only that our questions about their form are in fact mathematical questions; our questions about their growth are physical. A morphologist is *ipso facto* a practitioner of physical science.

DEMOCRITUS: Von Baer said something similar.

ALKIMOS: He said type and rhythm govern the construction of living things, and that chemical reactions are of great importance. But I still don't understand.

EXTON: Without chemical reactions, or metabolism, life would stop. Without regulation by type and rhythm, metabolism and organ development would dissolve into chaos.

DEMOCRITUS: I think we can confidently use the term *chromosomes* in place of *type and rhythm*. Would that swap be acceptable to you, Alkimos?

ALKIMOS: Chromosomes are like command headquarters?

THOMPSON: Let's recall a more general assertion of von Baer's: the forces that shape the body of the animal, *die bildenden Kräfte des thierischen Körpers*, can be traced back to the universal forces that affect the world, *die allgemeinen Kräfte des Weltganzes*.

DEMOCRITUS: That's like saying the atoms of life can be traced to the atoms of the inanimate world. One thing is certain: we have less and less need to call upon *vis vitalis* to explain the creation of life.

THOMPSON: That's true. As science advances, mathematics, physics, and chemistry have gradually replaced *vis vitalis*. The great physicist Hermann von Helmholtz can perhaps illuminate this principle for us. He stated that while the forces that act upon the living body may be different than those that act upon the inorganic world, they must be very similar when their effect is chemical and mechanical—at least in the respect that these effects must be governed by necessity. In other words, in the same conditions, the effects must always be the same.

ALKIMOS: Necessity?

THOMPSON: If we want to understand living systems, then we must subject the forces that act in living bodies to mathematical analysis, just as we have the other forces at work in the world. This does not mean that I believe that we will replace living systems with mathematical formulas, but I am convinced there is a great deal we can learn from a mathematical approach.

ALKIMOS: We are here to learn. I guess as long as we don't drown everything in numbers.

THOMPSON: I'm so glad you're persuaded.

ALKIMOS: Let's dive in.

THOMPSON: I had only one goal. To correlate the simpler outward phenomena of growth and the structure of organisms with the laws of physics and mathematics. All the while I was regarding the fabric of the organism, *ex hypothesi*, as a material and mechanical configuration.

ALKIMOS: He's an atomist, like my master.

THOMPSON: First we need to recognise that the form of living matter and the changes it undergoes as a result of movement and growth can be described as the result of forces acting upon the material. The form of living things is none other than a diagram of forces; this includes the total forces that have ever acted upon it.

ALKIMOS: What do you mean by force?

THOMPSON: In the language of Newtonian physics, force is an action capable of producing or changing motion.

DEMOCRITUS: Because growth entails the movement of material, then the material must be moved by forces. We need first to understand forces before we can understand the materials set in motion!

THOMPSON: Exactly! Yet, most books on inheritance place too much emphasis on the material components involved and not enough on the causes of development, variety and inheritance. But matter all on its own cannot create, change, or do anything. How much easier it would be to explain phenomena if it did, but this is not the case. What we must first establish is that spermatozoa and cell nuclei, and chromosomes and germplasm have no effect by virtue of their material make-up. Their role is as seats of energy and the centres of force.

DEMOCRITUS: So what is important about the material is its ability to carry and exert force. But I can't imagine how an atom can exert force.

THOMPSON: Don't worry, my friend, neither could anyone else. James Clerk Maxwell and other physicists all reached the same conclusion: to explain life's systems, we must assume the existence of far more complicated atoms and molecules than those required by the theories of chemists.

ALKIMOS: Master, your theory of atoms is getting ever more intricate. Molecules wrestling, stabbing one another. Heracles and the lion of Nemea, the Titans and the immortal gods of Olympus, their struggle played out on the level of atoms. Incredible!

THOMPSON: Isn't it interesting? But let's examine a few examples more closely so you understand what I mean by the forces on living things. Let's have a look at cell division. The changes undergone by the parts of the cleaving nucleus and the distinctive behaviour of the chromosomes reveal forces that bear strong similarities to well-known physical phenomena. Hermann Fol recognised that the lines of force between opposite magnetic poles bore startling similarity to the configuration of the spindle threads during cleavage. But we can induce artificial cell cleavage in other ways. If we drip ink into salt water, and then put a more dilute solution on either side of the drop, the forces that arise between the solutions of differing density will pull apart the ink in a pattern similar to that of cleavage spindles. I'm not saying that cell cleavage happens in the same way, but this experiment can certainly be considered a useful model of the forces at work in the cell. These three illustrations will help you understand. The first shows a cell undergoing division. The second shows artificial division using ink. The third is a diagram of forces, and it matches both, even though the materials that contain the forces are different. We also know that in cells, the forces are centred on the centrosomes.

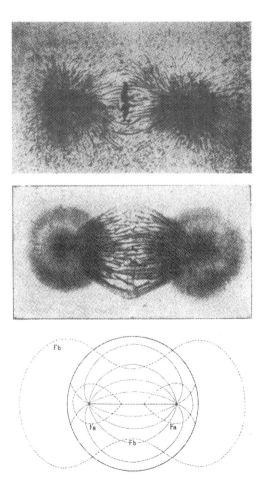

DEMOCRITUS: The centrosomes must exert force on the chromosomes through the spindle threads. The distinctive picture of cell cleavage and the migration of chromosomes are a result of this orientation and grouping of the forces.

THOMPSON: There's no question about it. But we have many examples on the level of organisms, too. The shape of jellyfish, if you consider its symmetry and ribbed pattern, bears a spooky resemblance to gelatine when it is poured into a liquid containing coagulants. A raindrop when it strikes water or a whirlpool can also take the vorticular form of a hydra. Look at these pictures of liquid jets and hydras.

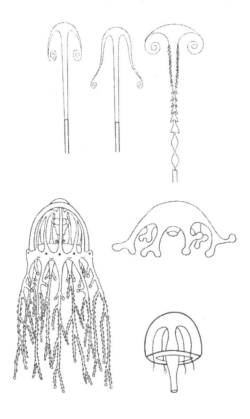

The similarity of these forms suggest the same arrangement of forces. If we look at jellyfish in the same way, the limitations placed on it by physics are immediately apparent.

DEMOCRITUS: The simpler a living organism, the greater the degree to which it is governed by the laws of physics, and the more it reflects the outer world. But I suspect more complex organisms are governed more by internal regulation and less by the laws of physics. One-celled organisms assume a simple drop shape, while large plants and animals come in far more varied forms.

ALKIMOS: Because they're able to overcome the limitations of physics and maths.

DEMOCRITUS: Hmmm, except that it's not possible. I think rather, these laws have their impact on a more minute level, assisting in the construction of more elaborate forms.

THOMPSON: That is absolutely correct. Of course this makes the mathematician's job more difficult. But let me show you how complex forms can also be analysed, even though they cannot be directly interpreted using physical forces. Mathematics, however, can illuminate the pattern by which these similar forms transform into one another. The Cartesian coordinate system can help us. The organism we're studying can easily be drawn on a rectangular grid. We can even create a table of numbers to describe our organism's shape. But this is not all that important to a morphologist or researcher of evolution. What matters to them is that they can distort the axis of the coordinate system as they wish. Naturally, this distorts the grid and the organism drawn on it. See, we can curve it or tilt it. The body parts of related species, or the entire creature, can be transformed into another, with the help of simple mathematical transformations, as you can see in these drawings:

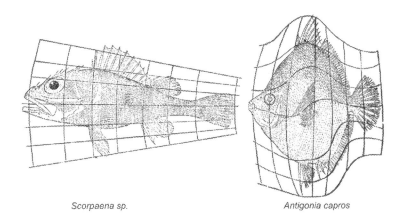

Scorpaena sp.   Antigonia capros

ALKIMOS: This is the mathematical interpretation of evolution?

THOMPSON: Not exactly. This merely helps us to see the relationship between forms. It shows us in what direction and proportions various body parts grew or shrunk as species evolved and assumed different shapes.

EXTON: That's why we must be careful when we examine transformations. Here we're transforming one adult form into another. Forms, however, didn't evolve in this manner. Instead it was the modifications in the developmental path of embryos that led to these differences.

ALKIMOS: Yes, I see. Just think how much we'd have to twist those axes to turn a fish into a fruit fly.

THOMPSON: It would be naïve to think that one method could help us explain the entire history of the living world. That's not even our goal. But isn't there something fascinating about our ignorance? In every era, when we discover our limits, we must recognise that '*la science est plutôt destiné à étudier qu'à connaître, à chercher qu'à trouver la verité.*' Yes, the goal of science is not to know the truth but to study it; not to discover the truth, but to search for it.

ALKIMOS: And philosophy?

THOMPSON: Philosophy and science stand side by side and one upholds the other. Without the strength of science, philosophy would be weak, and without philosophy's wealth, physical science would be weak. '*Rien ne retirera du tissue de la science les fils d'or que la main du philosophe y a introduits.*'

EXTON: Ah yes. Nothing can remove the golden threads woven by the philosopher into the fabric of science.

ALKIMOS: Master, the quote is about you!

DEMOCRITUS: No, it's about wisdom.

ALKIMOS: Yes, but what a poetic way to put it! Hey, what's that shaking?

DEMOCRITUS: I didn't feel a thing.

EXTON: Oh, blast it! We're out of time. It's chronoturbulence!

ALKIMOS: Master, what's happening to us? What's happened to the coffee shop?

DEMOCRITUS: We seem to be floating out of it. Farewell, Professor!

THOMPSON: Gentlemen, what's the hurry?

ALKIMOS: Hey, my biscuits!

THOMPSON: Wait! Please don't tell my wife about the cognac.

EXTON: Let's try to direct ourselves toward the Deus ex Machina!

DEMOCRITUS: Hmmm, cell cleavage, development, and evolution can all be described in the language of numbers. Now all we need to do is uncover the secret of the atoms of life that carry these forces!

ALKIMOS: What will happen to D'Arcy Thompson?

EXTON: Don't worry about him. He'll forget the whole thing this afternoon, when he has to sit through a meeting of the Classical Philological Society. Come on, we're almost there!

DEMOCRITUS: Living things are composed of matter that contains both forces and the ability to regulate. The workings and organisation of this matter is the ultimate riddle.

EXTON: Squeeze in. We have to go immediately!

DEMOCRITUS: Wait a minute!

ALKIMOS: Aaaagghhh … here we go!

The Two-Headed Newt

COOK: I don't get it. The boss ordered a six-course meal, but they just had lunch. What is it about that contraption that makes them so hungry?

HERMES: You're asking me?

COOK: Who should I ask?

HERMES: I don't know.

COOK: Come on, Hermes. Let's take a look inside. It might help me work up an appetite. You know, working in the kitchen all day makes food so unappealing. But the truth is I'd love to sample the six-course meal. I just need to feel hungry.

HERMES: I don't see how poking around the Deus ex Machina is going to make you hungry.

COOK: Come on, Hermes.

HERMES: No. It's out of the question.

COOK: Listen here. I'll make you the finest pâté. My granny's recipe.

HERMES: I'll pass.

COOK: Come on, I'll cook you anything you want!

HERMES: Anything?

COOK: Anything.

HERMES: All right. Ambrosia pudding and I'm in.

COOK: Ambrosia pudding? That's it?

HERMES: Tonight, after midnight. When they're asleep.

COOK: Thank you, thank you! I knew I could count on you!

HERMES: Quiet. They're here.

ALKIMOS: Exton, something's not quite clear to me.

EXTON: What's that?

ALKIMOS: What did D'Arcy Thompson discover?

EXTON: Strictly speaking, nothing. I told you he wasn't an experimental scientist. He was more like Democritus.

ALKIMOS: I noticed that.

EXTON: We could say that Thompson provided a synthesis of self-organising phenomena that occur in a living organism based on physical–chemical processes. An example of this is the hydra and a whirling stream of water. He taught us to examine living things from a different angle. From the perspective of physics and mathematics.

ALKIMOS: I couldn't follow everything he said.

EXTON: Let me explain how this pertains to embryos. If simple physical factors govern the development of an organism, then embryos don't develop purely by their own devises. In other words internal regulation, or some kind of *vis vitalis* cannot be exclusively responsible for it. Physical factors impact the process.

DEMOCRITUS: Wait a minute. If we think that physical forces determine the shape of living things, then we have no choice but to conclude that the similarities between certain creatures is not necessarily a sign of an evolutionary relationship, but rather a result of identical physical environments or constraints.

EXTON: Yes, and that was exactly what bothered Thompson about an evolutionary system based on morphological similarities. Recognising the importance of the laws of physics, he argued that selection could not arbitrarily sculpt the living world, because physical laws would not allow it.

ALKIMOS: Ah, the coercive power of physics.

EXTON: But living things are, after all, living things. Perhaps the best way to express it is that internal regulation and natural selection determine shape within the parameters allowed by the laws of physics. You see, Thompson attributed too much to physics and too little to selection and genes. In a word, he rejected the context in which these physical forces operate. Evolution, however, cannot take place in response to physics alone; it requires certain historical factors and inheritance.

ALKIMOS: Exton?

EXTON: Yes.

ALKIMOS: You said we were going talk about embryos.

EXTON: Yes, yes, I'm getting there.

DEMOCRITUS: Alkimos, what's the hurry? This has been one of our most exciting trips. What D'Arcy Thompson showed us is absolutely critical. It seems that through the immeasurable complex integration of the physical world, living things are capable of marvellous biological feats. Like the complex mixing of simple atoms to create the atoms of life, as Maxwell suspected.

ALKIMOS: Then D'Arcy appeared on stage and exposed this complex integration.

EXTON: And over dinner, we'll expose how embryos grow increasingly complicated.

ALKIMOS: Finally!

EXTON: I'm glad you're so interested.

ALKIMOS: There's nothing that interests me more! How many courses?

EXTON: Let's start with Hans Spemann.

COOK: Excuse me, shall I serve dinner?

EXTON: Yes, it's about time.

COOK: Sir, I only just cleared the table!

EXTON: Excuse me?

COOK: I, umm, I said I'll set the table.

ALKIMOS: Please be quick.

EXTON: Hans Spemann was a German experimental embryologist. A master of microscopic surgical techniques. He was a doctoral student of Boveri's in Würtzburg. He wrote his dissertation on the development of the sea urchin. But he quickly tired of spiky creatures and turned his attention to frogs and newts.

ALKIMOS: We haven't heard about frogs since our old friend Roux.

EXTON: His first studies were of the development of the eye in amphibians. However, his experiments on embryos during the early stages of development are what interest us most. To understand his work, we need to take another look at the early development of amphibians.

ALKIMOS: The development of the ovum, as Dumas and Prévost described.

EXTON: After fertilisation, the zygote begins to undergo cleavage, but it takes place in a slightly different way than in the sea urchin. The reason is that the cytoplasm of an amphibian's egg is much richer in yolk. However, the yolk generally inhibits cleavage. And interestingly, the yolk is not evenly distributed in the ovum, but is concentrated at the vegetal pole. The first cleavage furrow begins at the animal pole, which has less of the yolk, and then slowly progresses toward the vegetal pole, dividing the zygote into two equal cells. Before the first cleavage finishes, the second begins in a plane perpendicular to the first. The result is four uniform cells. The third cleavage is perpendicular to the first two, but does not take place exactly along the equatorial plane, because of the concentration of yolk at the vegetal pole, the cleavage furrow is closer to the animal pole. The end result is four smaller micromeres at the animal pole and four large macromeres at the vegetal pole, as we can see in von Baer's diagram:

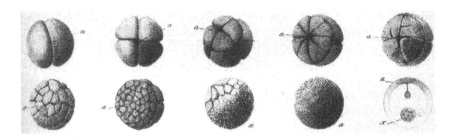

DEMOCRITUS: Let's slow down. It's clear that in this case the ovum is not a completely uniform, symmetrical sphere. Already it has divided into different regions, and the uneven distribution of the yolk determines the location of further cleavage planes and the speed at which the furrows form. Cytoplasm therefore has a decisive role in regulating development, as Morgan asserted.

EXTON: That's true. The uneven distribution of the yolk determines one of the main axes of the embryo. It has an indelible impact on the entire development of the animal.

ALKIMOS: Why? What happens next?

EXTON: The presence of the yolk causes the macromeres to divide much more slowly than the micromeres. The result is a mulberry-shaped embryo with lots of little cells gathered at the animal pole and a few big ones at the vegetal pole. Gradually a fluid-filled cavity, known as a blastocoel, forms in the animal region, and then dramatic changes occur, including one of the most important processes in development: gastrulation. At this point, von Baer's famous germ layers—the ecto-, endo-, and mesoderm—form. During gastrulation a group of cells migrates to the interior of the embryo and the ectoderm cells remain on the surface. The cells migrating inside will form the endoderm and the mesoderm. The invagination of the cells starts at a specific point along the equatorial plane of the blastula to form a slit-like blastopore, the future anus of the animal. The new cavity that forms during gastrulation is called the archenteron. This primitive gut develops into the digestive tract. As the layers fold over one another, the nervous, muscular, and skeletal systems and the other parts of the animal are created. To avoid confusion, I think we can leave out the details.

ALKIMOS: Good. I'm having a hard enough time understanding this all.

EXTON: After dinner, we'll observe the various stages through the microscope. Until then you'll have to be satisfied with von Baer's drawing.

COOK: Gentlemen, shall I serve the soup?

ALKIMOS: Yes, please!

DEMOCRITUS: There's something I still don't understand. What determines where the blastopore forms? The uneven distribution of the yolk affects only one axis of the animal. Although it's clear that invagination begins at just one point, otherwise we'd have embryos with several guts! But at which point?

EXTON: Excellent observation! But I won't reveal the answer yet. First let's have a look at the experiments conducted by Spemann and his student, Hilde Mangold, in the early 1920s. Using amazing surgical techniques, they demonstrated that in the early gastrula, the upper, or dorsal, lip of the blastopore possesses an unusual characteristic. When they surgically removed this section of a newt gastrula, and then transplanted this piece into another gastrula on the opposite side of the blastopore, then this tissue began to work independently. Like at the original blastopore, the embryo at the place of the transplanted piece began to undergo gastrulation. The tissue invaginated and began to form a mesoderm and a nervous system at the vegetal pole, and with such effectiveness that an entire second embryo formed, lying atop the other one. The little double-embryos, lying atop one another, shook their head in indignation at Spemann's clever trick.

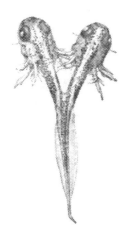

ALKIMOS: They should have been excited about the fame it brought them.

EXTON: Since no other part of the embryo was capable of this degree of reorganisation, Spemann named the tissue around the dorsal lip of the blastopore the *organiser*. This tissue determined the development not only of the back and belly, but the head and tail, too.

DEMOCRITUS: So, one tiny piece of tissue controls development of the two main axes of an embryo.

EXTON: Spemann and Mangold experimented with two newt species, the dark coloured *Triturus taeniatus* and the light-coloured *Triturus cristatus*. The transplant was made between the two species; thus the scientists were able to follow the fate of the transplanted tissue by observing the difference in pigmentation. What they discovered was that the second embryo did not develop exclusively from the transplanted tissue. The host cells also participated considerably in the formation of the animal.

DEMOCRITUS: The vegetal cells of the host embryo became the nervous system!

EXTON: That's what's most extraordinary about the entire experiment. The foreign cells were able to redirect the host cells onto a completely different developmental pathway. This means that different parts of an embryo can affect each other, and modify each other's development. This phenomenon is called *induction*. Transplants are not necessary for parts of the embryos to engage in this kind of dialogue. As the germ layers fold onto each other, there are more and more opportunities for induction.

ALKIMOS: The newly induced tissues may come into contact with others, inducing newer and newer tissue types. This is how embryos differentiate!

DEMOCRITUS: And very likely genes themselves carry on this sort of dialogue. If we mix the chromosomes, as Boveri did, the dialogue becomes confused, and the wrong tissues are induced at the wrong time.

ALKIMOS: In that situation, the embryo differentiates in a way that brings about its own death.

EXTON: This diagram of the newt's secondary organisers shows how induction processes become more and more complicated as the embryo develops. As you can see, when the eye develops, the formation of the lens is also induced.

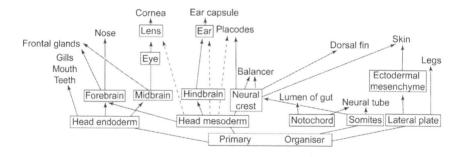

DEMOCRITUS: How interesting. But Exton, there's something you must tell us. Where does it begin—this dialogue that brings about these complicated inductive actions? How does the gastrula know where to fold in?

EXTON: That's one of the most exciting questions; yet the answer is very simple. We've seen that the ovum is not a uniform sphere, with its belly side far richer in yolk. But the distribution of the yolk determines just one axis. If we rotate the ovum along this axis, it will appear the same from every side. But then some small, jumpy thing comes along and breaks the symmetry and the peace.

ALKIMOS: The sperm!

EXTON: That's it! When the sperm penetrates the ovum the cytoplasm completely reorganises. The place opposite the sperm's point of entry begins to take shape and it is here that the blastopore will eventually open. As we've seen, the blastopore on its own is capable of determining the axes of the animal. What plays a key role in this whole process of reorganisation is the sperm's centrosome.

ALKIMOS: What a noble role the sperm has. Not only does it trigger the development of the organism and transport chromosomes and the centrosome, but it determines where the head and tail develop. Nature is so wise.

DEMOCRITUS: Nature is. You, on the other hand, are a fool.

COOK: Did you call me? I'll be right there with the next course, Thracian roast.

EXTON: Thank you.

DEMOCRITUS: So the genes talk with each other. Some speak up and others are silent.

ALKIMOS: And the sperm starts the conversation.

DEMOCRITUS: Yes, but the ovum decides the subject. Remember, the yolk establishes the axis in the early stages of cleavage.

ALKIMOS: But the sperm decides the location of the blastopore.

DEMOCRITUS: And the organiser surrounding the blastopore is capable of a long monologue. But eventually it agitates its neighbouring cells.

ALKIMOS: Induction.

DEMOCRITUS: I suspect an orderly dialogue is very important. If every gene spoke at once, they wouldn't hear each other. There'd be quite a din, like a chorus of frogs.

COOK: Indeed, sir, that's the next course.

DEMOCRITUS: Excuse me?

EXTON: Oh, sautéed frog's legs! Wonderful.

ALKIMOS: They look pretty scrawny to me. Imagine if Spemann had done his experiment on frogs? He could have raised ones with four thick, meaty hind legs instead of two. We'd be enjoying a much heartier meal.

EXTON: Alkimos, wouldn't it be easier to just catch more frogs?

ALKIMOS: But then how would science progress? By the way, Exton, would the experiment work with frogs?

EXTON: It even works with Hydra, as shown by Ethel Browne in 1909, before Spemann described the vertebrate organiser.

COOK: Gentlemen, if you're finished, some ambrosia pudding?

Zeus's Beard

HERMES: I must say, the pudding was delicious.

COOK: I've had some practice.

HERMES: It shows. I'm glad there was some left for me.

COOK: Well, I made it for you.

HERMES: What a relief this day is finally over. I'm ready for bed.

COOK: Hermes?

HERMES: What?

COOK: Did you forget something?

HERMES: No, it's not going to happen.

COOK: You promised.

HERMES: Not today.

COOK: This is the perfect opportunity. Come on!

HERMES: Oh, for the gods' sake ...

COOK: They're asleep. We can sneak over there.

HERMES: If the boss finds out.

COOK: You should have thought about that while you were gulping down the pudding.

HERMES: All right, but quietly.

COOK: Wow, would you look at that.

HERMES: Slide in there.

COOK: On Zeus's beard, that's a lot of buttons!

HERMES: Don't touch that! I said, no!

COOK: Hey, what's happening? Ohhh!!

HERMES: Hooooold onnnn!

COOK: Whoa!

HERMES: What do we do?

COOK: What an odd feeling! What's it doing?

HERMES: Something wild.

COOK: I like the sound of that!

Evolutionary Synthesis

DEMOCRITUS: Alkimos, wake up! It's late.

ALKIMOS: Hmmm…

EXTON: Leave him. He'll get up on his own.

DEMOCRITUS: He's a sluggish one, isn't he?

EXTON: Come to the kitchen. Let's have breakfast.

DEMOCRITUS: All right, I suppose.

EXTON: Last night those two halfwits, Hermes and the cook, went to Tübingen instead of us. They've completely disrupted my plans for today. Now they're fast asleep.

DEMOCRITUS: They used the Deus ex Machina? They must have had quite a fright.

EXTON: Luckily Hermes knows how to operate it.

DEMOCRITUS: Who did they visit?

EXTON: They went to a conference, in 1929.

ALKIMOS: Morning! Ah, how nice to wake up to the smell of toast. But why aren't we being served?

DEMOCRITUS: I think you've become too accustomed to comfort. If we are too well accommodated, our minds grow lax.

EXTON: We're not going to make it to Tübingen today thanks to those two birdbrains, but we can still make good use of our time. Let's examine the special connection between Darwinian evolution and Mendelian genetics in the first decades of the 20th century.

ALKIMOS: Isn't it a little early for that?

EXTON: Not at all. We've reached the 1920s and both of you are now familiar with the fundamental concepts of both theories. It's exactly the right time to join the two together.

ALKIMOS: Do you mind if I take a poetic approach to the question?

DEMOCRITUS: This isn't going to be difficult, Alkimos. We know that Darwin's biggest problem was that he couldn't find a satisfactory explanation for inheritance.

EXTON: Darwin's theory of pangenes was considerably different than the theories of heredity devised by Mendel, Weismann, and Morgan. Darwin's notion that pangenes migrate from every part of the body into the germ cells allowed for the inheritance of acquired traits. In this respect his theory most closely resembled Lamarck's. In fact, many naturalists had made similar assumptions. Their ideas are known as the theories of *soft inheritance*. The foundations of this set of theories were the observation of perfect adaptations in nature and the very slight variations observed within species. As an example, have a look at this drawing depicting slight variations in the abdomen patterns of a wasp species.

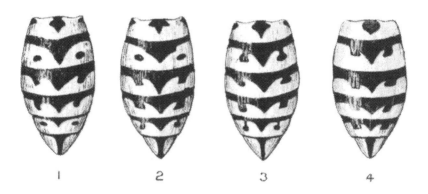

Naturalists like Lamarck imagined that the hereditary material was influenced by small differences in the environment and the usage or nonusage of body parts, with the end result of gradual evolutionary change. Some believed the transformations might be *orthogenetic,* in other words a function of internal efforts.

DEMOCRITUS: Mendel, however, was not interested in minute differences. What he focused on were the obvious, almost certainly inherited, pairs of alternative traits. The same is true of Morgan.

EXTON: That's right. You've just highlighted the fundamental difference in approach between naturalists and experimental biologists, whether they were taxonomists, comparative embryologists, or students of evolutionary theory.

ALKIMOS: I suspect the prominent members of the two camps were good at talking past each other.

EXTON: Indeed they were. But it's not surprising. Their distinct ideas about the nature of inheritance well illustrate the huge chasm between the two camps. As Democritus noted, while the naturalists were focused on subtle changes, the experimental biologists focused on conspicuous ones. Mendel's pea varieties and Morgan's mutants differed considerably from the so-called wild types. Morgan's *white* mutant fruit flies had white eyes, but the wild type had red. Mendel's short-stemmed pea mutants were always short-stemmed in the homozygote form. These genes were passed on in a well-known, predictable ratio and always provided the same categories of phenotypes. Changes in Mendelian factors produce easily distinguishable, discontinuous variations. This is known as *hard inheritance.*

DEMOCRITUS: Then we're in trouble, because the variation in nature that so inspired Darwin is far more nuanced than what we observe in a laboratory. Our laws of hard inheritance, however, are based on changes we've studied in the vegetable patch or a test tube. The variations under these circumstances are not nearly as continuous as the variations in nature.

EXTON: The difference between continuous and discontinuous variation, like the difference between soft and hard inheritance, caused a lot of problems, and in the early decades of the 20th century researchers sought to resolve them not through a thoughtful exchange of ideas but through bitter attacks. Both sides fiercely criticised the other and ignored valuable new information that had accumulated outside their own field.

ALKIMOS:

> *Out in the wild, overgrown forest, as you study soft genes,*
> *The endless variety of life lies exposed before you;*
> *Back in the sombre lab, jamb-packed, brimming with microscopes,*
> *You strive to establish assured, hard-as-rock, laws of science.*

EXTON: Ah, and geneticists indeed attempted to apply the hard-as-rock laws of genetics to evolution. Since the more subtle changes at first could not be observed in the laboratory, scientists relied on the mutations that arose to explain the origin of species. Their thinking was for the most part *essentialist*. They imagined a species to be a group of homozygous wild type individuals—to use the language of geneticists—without any kinds of variations. Among identical individuals occasionally a few mutants will appear. These mutants sometimes differ so greatly from the wild type that they create a new species. In 1906, Hugo de Vries wrote, 'the theory of mutation assumes that new species and varieties are produced from existing forms by sudden leaps.'

ALKIMOS: How simple. One mutation, a new species!

EXTON: Naturalists did little more than smirk at such an approach. How could new species come about simply as a result of a mutation? Yet among geneticists, this kind of evolutionary thought—one that presumes *saltations*, or leaps—had gained considerable currency. Their stalwart belief in the theory of mutations led them to question the governing role of selection in evolution. The most they were willing to concede to natural selection was that it was responsible occasionally for discarding harmful mutations. In their minds evolution was driven by the pressure of mutations. According to Morgan, "natural selection does not play the role of a creative principle in evolution".

ALKIMOS: So all those microscopes made geneticists short-sighted.

EXTON: Yes, that's indeed what happened. John B. S. Haldane, a British population geneticist, stated, 'I believe that the opposition to Darwinism is largely due to a failure to appreciate the extraordinary subtlety of the principle of natural selection.' The truth is those geneticists—in particular Bateson, de Vries, Johannsen and Morgan—who were interested in evolution, unfortunately didn't understand it at all. They had no idea about the nature of biological diversity and natural species. The only way to navigate the essentialist categories they'd created was by leaps. For this reason, they questioned the role of selection and attributed every creative force in evolution to mutations.

DEMOCRITUS: If we ignore the diversity of the natural world, the subtleties of life, then we can easily develop false notions.

EXTON: It's amazing for example, how long geneticists failed to recognise the role of recombination in evolution. Allow me to quote Morgan: 'The kind of variability on which Darwin based his theory of natural selection can no longer be used in support of that theory because ... selection of the differences between individuals, due to the then existing genetic variants, while changing the number of individuals of a given kind, will not introduce anything new. The essential [element] of the evolutionary process is the occurrence of new characteristics.'

DEMOCRITUS: But it was Morgan and his team that discovered recombination! The mixing of various alleles of various genes can result in none other than the appearance of new characteristics.

EXTON: Even without mutations. And the random segregation of chromosome pairs during meiosis can also lead to the formation of variants without the occurrence of mutations.

DEMOCRITUS: But the mutations Morgan observed truly did result in sudden, discontinuous changes. So even if we can mix these various alleles, we still cannot explain the subtle diversity of nature.

EXTON: Many thought soft inheritance was the mechanism to generate subtle variation. It was geneticists who finally managed to answer these contentious questions of soft versus hard inheritance, and continuous versus discontinuous variation. The answer is so simple and elegant; yet for decades it failed to be incorporated into evolutionary theory. Because they were too specialised, geneticists themselves were unable to appreciate the true importance of their new discoveries. Naturalists in turn could not keep pace with the increasing amount of detailed genetic literature and therefore had no idea that many of their quandaries had been resolved by genetics. Bateson's fervent statement of 1922 underscores the friction between the two sides as they refused to listen to one another: 'If we cannot persuade the systematists to come to us, at least we can go to them. They have built up a vast edifice of knowledge which they are willing to share with us, and which we greatly need. They too have never lost that longing for the truth about evolution, which to men of my date is the salt of biology, the impulse which made us biologists.'

ALKIMOS: It's as if I'm listening to my master.

EXTON: The tension between the two sides finally began to abate in the 1930s. In the decades that followed, most major misunderstandings were cleared up and the knowledge of naturalists came to match that of their experimental counterparts. The intellectual achievement of the years 1937–1947 became known as the *evolutionary synthesis*, after Julian Huxley's book of the same name. A major contribution to the shaping of the synthesis was Theodosius Dobzhansky's book of 1937, *Genetics and Origin of Species.*

DEMOCRITUS: What was the essence of the evolutionary synthesis?

EXTON: The first substantial achievement was liberation from essentialism, which was replaced by *population thinking*. For this to happen, the views surrounding soft and hard inheritance needed clarification. Genetic experiments quickly revealed that only one kind of inheritance exists. The discrete differences that some believed were transmitted through soft inheritance were shown to be the results of hard inheritance. What really happens is that minute changes are caused by micromutations, whose effects on phenotype are small, or they are created by the interaction of several genes.

ALKIMOS: Galton's theory of inheritance!

EXTON: Exactly. From 1908 to 1911 Herman Nilsson-Ehle, a Swedish geneticist, demonstrated that the hybridisation of white and red grained wheat yielded nearly continuous phenotypic variation. This peculiar ratio could be interpreted according to Mendelian principles, if we consider that three independently inherited genes are involved. In other words, the laws of Mendelian genetics can explain continuous variation if several genes with different effects work together. This is demonstrated in Ronald Fisher's brilliant theoretical model which he presented in his article of 1918, '*The Correlation between Relatives on the Supposition of Mendelian Inheritance.*' His article was one of the important forerunners of evolutionary synthesis.

ALKIMOS: So soft inheritance can be interpreted according to the principles of hard inheritance?

DEMOCRITUS: Most likely. But a lot of genes are necessary or mutations that have little effect on the phenotype.

EXTON: Morgan and associates were the first to experimentally investigate mutations that caused only small changes in the phenotype. For some genes, for example, they could isolate entire series of alleles. That means that all kinds of mutations could occur in the same gene. Some are fatal, while others have only a moderate or slight impact.

ALKIMOS: And some probably have no effect at all.

EXTON: That's quite possible.

DEMOCRITUS: But those would be difficult to demonstrate genetically.

EXTON: It was also discovered that some mutations, in given circumstances, could even be beneficial.

DEMOCRITUS: So continuous and discontinuous variations arise from the same source. Some mutations have simply a small impact, leading to small gradations in phenotype. Mendel's pea varieties occupied the extremes of the mutation scale.

ALKIMOS: So Mendel could have worked with peas with medium-length stalks?

EXTON: Yes, he could have. But he would have needed to find the appropriate allele. Mendel's decision to use pairs of conspicuously different alternative traits was a conscious one.

DEMOCRITUS: This way he could easily identify the different categories.

EXTON: The concept of a mutation was altered significantly during Morgan and associates' activities. While studying the genetics of *Drosophila pseudoobscura*, Dobzhansky realised that natural populations were full of genetic variations and their frequencies can change rapidly. Therefore evolution does not advance through leaps but through changes in the frequency in minor mutations. According to Dobzhansky' very simple definition, evolution is the 'change in the genetic makeup of a population.'

DEMOCRITUS: If there can be so many kinds of mutations, then it's easy to see how discrete variations can arise in nature.

EXTON: That's right. Hugo de Vries' interpretation was replaced by the notion that every allele is in fact a mutation of the gene. We can only use the term 'wild type' when we're talking about the inbred strains created by geneticists. In nature there's no way to decide what's wild and what's mutant. Nature doesn't work in vials. The main idea of population thinking is that every individual's genotype is unique. If we don't accept this, then the principle of natural selection isn't worth much.

ALKIMOS: We'd only have mutant monsters to choose from. There'd be no adaptation.

EXTON: The correct understanding of population thinking and inheritance also helped to finally lay to rest Lamarckism—the belief that acquired traits can be inherited—and orthogenesis. This is another of the great achievements of the evolutionary synthesis.

DEMOCRITUS: The synthesis then meant mostly unloading erroneous theories: orthogenesis, Lamarckism, essentialism and saltation theory.

EXTON: That's right. In the 1930s, a more refined interpretation of mutations and a better understanding of both variations in natural populations and the differences between individuals gave rise to a new branch of science: population genetics.

DEMOCRITUS: But there's something we have not explained sufficiently. If changes continuously follow one another, and the transformation is gradual, then how do new species arise?

EXTON: The most wonderful accomplishments of the evolutionary synthesis provide the answers to that question, but scientists first needed to clarify the concept of the species, and for that, they needed the help of naturalists. The idea of a biological species replaced the essentialist theory that considered all types fixed and immutable. The greatest evolutionary biologist of the 20th century, Ernst Mayr first elaborated on it in his book *Systematics and the Origin of Species*, published in 1942. According to the biological species concept, a species is a group of actually or potentially interbreeding populations. The crucial determinant is not the totality of its external characteristics, but rather those factors that inhibit reproduction with other species. The essential element in the definition of a species is *reproductive isolation*.

DEMOCRITUS: That means the most important step in the origin of a species is the appearance of limits to reproduction.

EXTON: Of course, geneticists knew this. They searched fervently for those mutations that prevented an offspring from reproducing with the parent generation but allowed them to reproduce with each other. But they never managed to discover such a mutation. It's no surprise. Nature is actually much simpler. Species evolve on the level of populations, not genes. It begins when a part of the population is separated. Once these populations are cut off from one another, tiny genetic changes lead to reproductive isolation. In Ernst Mayr's words: 'A new species develops if a population which has become geographically isolated from its parental species acquires during this period of isolation characters which promote or guarantee reproductive isolation when the external barriers break down.'

DEMOCRITUS: Geographic isolation prevents genes from mixing.

EXTON: Meanwhile, the two populations embark on separate evolutionary paths, and gradually lose the ability to reproduce with each other.

DEMOCRITUS: Later, if they encounter one another, they won't be able to produce offspring.

EXTON: And thus a new species is born.

DEMOCRITUS: Why one and not two?

ALKIMOS:

> *Our two species cannot jumble and mix our precious genes.*
> *But dear, no cause exists for your heart to yield to grief*
> *No cause to spill your tears upon my tender wound,*
> *While your legs are swathed in the sweet pollen of my love, my rose,*
> *With your buzzing desire, little bee, return to my betrothed*
> *Deliver my answer of love upon your wraithlike wings.*

EXTON: I see you've understood everything.

The Casting Moulds of Genes—*Hermann Joseph Muller*

ALKIMOS: Master?

DEMOCRITUS: Yes, Alkimos?

ALKIMOS: What's all this knowledge good for?

DEMOCRITUS: It's better to know things than not to.

ALKIMOS: Really?

DEMOCRITUS: The endless quest to better know the world is Prometheus' gift to us. Knowledge gives us power.

ALKIMOS: Who do we use this great power against?

DEMOCRITUS: Ourselves.

ALKIMOS: There'd be no point in battling the gods anyway …

EXTON: Everything's finally ready. Thanks to the night-time adventure of those scallywags I had to reprogram the Deus ex Machina, but I think it's working now.

ALKIMOS: Haven't we travelled enough?

EXTON: No. We need to unravel the mystery of genetic variation. It's one of the most important aims of genetics and the key to understanding evolution. But if you'd like to stay behind, Alkimos, that's fine with me.

ALKIMOS: No, no. I'm interested, truly I am. It's just the trip there … So where are we going?

EXTON: The University of Texas at Austin in the year 1922. We're going to visit one of Morgan's students, Hermann Joseph Muller.

ALKIMOS: Not fruit flies again!

EXTON: No. Grasshoppers.

ALKIMOS: Well that's better.

EXTON: Hermes, are you ready? We're leaving.

ALKIMOS: Aren't you coming, Hermes?

HERMES: Never again!

EXTON: All aboard!

ALKIMOS: Agghh … here we go!

EXTON: A New Yorker by birth, Hermann Joseph Muller began his studies at Columbia. The writings of Edmund Wilson had already inspired him to pursue biology, and later Wilson's classes at Columbia had a great impact on Muller. As a student he became acquainted with Calvin Bridges and Alfred Sturtevant. In 1910, he became an active participant in Morgan's lab, but he only began experimenting in 1912. His first investigations were into the matters of linkage and crossing over. Later, in Texas, he turned his attention to the causes of mutations. At the end of 1926 he discovered that X-rays could induce them.

ALKIMOS: Didn't you say we're going to 1922?

EXTON: Yes, so please, Alkimos, don't say a word about X-rays to Muller.

ALKIMOS: Why would I do that?

EXTON: We're here!

ALKIMOS: Wow, it's like an oven.

EXTON: That's Texas for you. Hot and dry. Just be happy it's not raining. Now follow me. We're close to the university.

ALKIMOS: I think I'd prefer rain.

EXTON: Maybe Muller was interested in the effect of heat on gene mutation.

ALKIMOS: But it turned out X-rays were better at it.

EXTON: Alkimos, I said not a word about X-rays.

MULLER: Mr. Exton, is that you?

EXTON: Yes. We are so honoured to meet you, Professor Muller.

MULLER: Oh, I'm only an Assistant Professor.

ALKIMOS: Then we're doubly honoured.

MULLER: Excuse me?

EXTON: Don't mind him. Now, I must tell you we are fascinated by your work. The Morgan laboratory has played such a tremendous role in expanding our understanding of genetics; thanks to the Morgan lab's experiments, genetics has become an integral part of the fabric of mainstream experimental science. We think the time has come to ask: what is a gene?

MULLER: I find myself preoccupied by that question too. What's the secret of how genes operate? How do they function in living things? It's undeniable that the basic discoveries of genetics have greatly increased our understanding of the physiology of cells. Yet, physiologists have not managed to integrate these findings into their own science.

ALKIMOS: Ah, like naturalists, who for a long time closed themselves off to genetics.

MULLER: The most important recognition of genetics is that in addition to the material we've known for some time was in cells—fats, sugars and proteins—there are several thousand particles that can't be seen with a microscope: genes. Despite being invisible, genes, however, have an effect on the entire cell, determining the matter, structure and function of the cell, and thus the entire life of the individual.

ALKIMOS: When we look at chromosomes under a microscope, then we should see the genes, too, since they line up on the chromosomes.

MULLER: You're right, except that genes are not large enough to be seen. Just think, one larger chromosome may have several hundred genes, and the length of a chromosome is one-thousandth of a millimetre. Now you see what a tiny particle a gene is.

ALKIMOS: It's incredible that something that small can impact an entire organism.

MULLER: Those pathways by which the genes exert their influence on an individual and manifest themselves in a particular character are the end result of extraordinarily subtle, precisely balanced chemical reactions. The interaction of genes governs these reactions. If the careful balance of the reactions is disrupted by a mutation of one of the genes, then every part of the individual will feel the effects of the change, and can grow, shrink, or disappear entirely.

DEMOCRITUS: So there's no direct relationship between genetic mutations and a particular phenotype. A gene doesn't determine a trait, but is part of reaction, whose end result is a certain characteristic.

EXTON: Now it's easier to understand pleiotropy and polygenic inheritance.

DEMOCRITUS: In the case of pleiotropy, one gene is involved in several reactions, thus a change will affect more than one characteristic. In polygenic inheritance, on the other hand, several genes take part in a reaction, and so a change in any one of them will have a similar outcome.

MULLER: But the most important distinguishing feature of genes is not that they take part in reactions. Other building blocks of life are capable of the same thing. What's special about genes is that they can replicate.

ALKIMOS: Like the individuals themselves.

DEMOCRITUS: We must search for the foundations of life in this process of replication on the atomic level!

MULLER: But what does that mean in terms of genes? Genes must be capable of creating copies of themselves out of the materials present in the complicated environment of the protoplasm. Chemists call this ability *autocatalysis*; physiologists refer to it as growth; and geneticists, assuming it takes place over several generations, consider it to be inheritance itself.

ALKIMOS: And after several thousand generations we have evolution.

MULLER: You'll often hear that evolution rests on two basic pillars: heredity and variation. But this definition is too vague. Inheritance doesn't lead to change, and variation doesn't lead to permanent change. What is most important is that variation is passed on. To put it simply, the necessary conditions for evolution are not inheritance and variation, but rather the inheritance of variation. This is a consequence of the general principles of the construction of genes. The autocatalytic capacity of genes is retained even if the gene occasionally changes.

DEMOCRITUS: What a strange contradiction. How can one characteristic of a gene remain immutable, the ability to reproduce, while another, its power to affect the cells, changes?

MULLER: This is the heart of the problem, and perhaps what can shed some light on it is the molecular structure of genes. Because when a gene undergoes random mutation, its outward catalytic qualities change accordingly, but its autocatalytic qualities are unaffected.

DEMOCRITUS: That's the greatest secret of mutations, of the random variation that arises from a change in individual genes.

MULLER: That's correct. The most fundamental challenge of genetics is to understand the nature of the structure of a mutable gene capable of autocatalysis. It seems that the autocatalytic capability resides within the structure of genes, and the job of the protoplasm that surrounds the genes is merely to supply the building material necessary for the genes to replicate.

DEMOCRITUS: How's that possible?

MULLER: According to the physicist Leonard Troland the growth of crystals can be explained by an attraction that arises between the assembled crystals and those in the solution caused by the identical patterns of their force fields. It's similar to the attraction of two magnets—the north pole exerts a pull on the south, and vice versa. Troland postulated that essentially the same mechanism is at work during the autocatalysis of hereditary particles. If he's right, then different sections of the gene—just as we've seen in crystals—attract similar material from the protoplasm, thus creating a second gene made up of particles in the same arrangement as the original gene next to it. The particles then bond together and the copy is completed.

DEMOCRITUS: Incredible! So genes use themselves as a mould! That's how they can make a perfect replica of themselves.

ALKIMOS: Master, I've been to several bronze casting workshops in Abdera, but the pattern on the mould is never the same as the pattern on the bronze statue. They're like mirror images. They fit together, but they're not identical.

DEMOCRITUS: Alkimos, how right you are! I didn't even think of that. It's similar to the way the north pole of the magnet attracts the south. The original gene creates a complementary image of itself.

ALKIMOS: Then the complementary gene creates a copy identical to the original, which then creates another complementary image and so on. This process can be repeated over and over again.

DEMOCRITUS: This is what lies behind the inheritance of genes. Now all we need is to find these moulds, which are surely constructed of atoms. The mould metaphor is helpful in understanding how genes that have undergone mutation can still replicate. If a mould is scratched, this scratch will appear on the statue. The changes are inherited but the principle and technique of bronze casting is unchanged!

MULLER: Gentlemen, your reasoning is brilliant. The model of complementary forms does indeed explain the many aspects of a gene's functioning. But sadly we have no proof that this is how things work. I think we need to discuss another phenomenon that will bring us closer to the solution. It's known as the d'Hérelle phenomenon.

EXTON: Allow me, Professor, to explain the d'Hérelle phenomenon to my students.

MULLER: Of course.

EXTON: Félix d'Herelle was a Canadian microbiologist. Ever since Van Leeuwenhoek, microbiologists have dealt mainly with microscopic forms, in particular bacteria. These creatures have no nuclei, and are known as prokaryotes. Organisms that have nuclei—plants, animals, fungi, and protists—are called eukaryotes. Ferdinand Cohn and Louis Pasteur laid the groundwork for bacteriology in the second half of the 19ᵗʰ century. Cohn developed a system of categorizing bacteria according to their morphology, while Pasteur discovered their role in illnesses.

ALKIMOS: I remember Pasteur. He disproved spontaneous generation using his swan-necked flasks.

EXTON: In Félix d'Hérelle's time bacterial cultivation was a well-established procedure. They could now grow bacteria not only in liquid cultures like Spallanzani's broth, but in petri dishes containing agar, made form sea algae, or gelatine. The bacteria grow into a confluent lawn across the substrate surface. While travelling in Mexico in 1910, d'Hérelle discovered a hitherto unknown epidemic among grasshoppers. He was able to isolate large quantities of the bacteria responsible and then used it to infect healthy grasshoppers. As he was growing the bacteria in a petri dish, he made an interesting observation. Every so often strange patches appeared in the bacterial lawn. He later wrote, 'The anomaly consisted of clear spots, quite circular, two or three millimetres in diameter, speckling the cultures grown on agar.' In 1915, while searching for the cause of an outbreak of dysentery among French soldiers, again he discovered it was a bacterial infection, but obviously a different type than what caused the illness among grasshoppers. Yet, the same strange circular patches appeared in the cultures of the *Shigella* bacteria responsible for dysentery. Whatever caused the patches was much smaller than the bacteria, since it easily passed through the porcelain filter that trapped the bacteria.

ALKIMOS: But why were the spots clear?

EXTON: Because something killed the bacteria in those places. When d'Hérelle mixed the material produced in the clear spots with a thick bacterial culture, he experienced something incredible. This is how he described his discovery: 'The next morning, on opening the incubator, I experienced one of those moments of intense emotion which reward the research worker for all his pains: at first glance I saw that the broth culture which the night before had been very turbid, was perfectly clear: all the bacteria had vanished; they had dissolved away like sugar in water. As for the agar spread, it was devoid of all growth and what caused my emotion was that in a flash I had understood: what caused my clear spots was in fact an invisible microbe, a filterable virus, but a virus which is parasitic on bacteria.' D'Hérelle named this virus *bacteriophage*, or bacteria eater.

MULLER: Mr. Exton, you are well prepared for our discussion. Your knowledge is impressive.

EXTON: Thank you. I've always liked bacteriology.

MULLER: These observations of d'Hérelle's were all quite interesting, but the most important was yet to come. D'Hérelle demonstrated that a drop of the material found in the clear spot could be transferred to a pristine bacterial colony and this second colony would be killed. A drop from this spot could be used to kill another colony and so on.

ALKIMOS: So the bacteriophage must be able to reproduce.

MULLER: That's right. This strange material multiplied on the bacteria—it self-replicated. This means it meets the criteria of being an autocatalytic material. Even if its makeup and the mechanisms by which it works are totally different than what we find in genes, it still corresponds in every respect to the definition of a gene. In fact the two have even more similarities. André Gratia discovered that a bacteriophage will mutate if handled in a certain way; yet, like genes, it retains its ability to replicate.

DEMOCRITUS: That means two different materials, d'Hérelle's bacteria-eating parasite and genes, both possess the extraordinary characteristic of inherited variation.

MULLER: If we take into consideration the complexity of the protoplasm, however, genes and bacteriophages must work in different ways. The possibility arises that two fundamentally different kinds of life exist on Earth. Nevertheless, the similarity between chromosomes and these invisible d'Hérelle particles, the bacteriophages, is startling and peculiar. If these phages really are a sort of gene, fundamentally like the genes on chromosomes, then this would open a whole new set of possibilities in genetic research. The phages can be filtered and even placed in test tubes. Although it's too soon to call them genes, our present knowledge suggests, as I've already mentioned, that they correspond in every way to genes. We cannot exclude the possibility that one day we'll be able to grind them up and boil them in a beaker.

DEMOCRITUS: What a heretical thought! That would mean the ultimate triumph of genetics. If genes worked in beakers the same way that they do in the special environment of the protoplasm, then the notion of *vis vitalis* would be laid to rest.

MULLER: But at present, genetics can only progress through investigation of changes in the individual genes; for example, the investigation of crossing over revealed to us the linear arrangement of genes on the chromosome. Therefore to understand the true nature of genes, we need to know a lot more about their variations. But the situation is the same in other branches of science. Chemists, for example, were only able to discover the makeup of molecules based on how they changed during chemical reactions. Similarly, if we want to uncover the structure of genes, then we need to observe as carefully as we can the changes caused by genetic mutation.

DEMOCRITUS: Mutations are therefore our greatest aid in unravelling the secret of genes.

MULLER: Without variation there would be no study of genetics, and without mutations there is no variation.

ALKIMOS: Unless it takes place through crossing over or sexual reproduction.

MULLER: That's true, but those processes can only mix those genetic differences that have already been produced by mutations. For example, during crossing over new allele combinations are created, many have never existed before, but the alleles themselves are a result of mutations. The crossing over of completely identical chromosomes is in vain, since no new variations will ever be produced.

ALKIMOS: So geneticists need as many mutations as possible. Well, that's easy!

MULLER: Unfortunately, the frequency of mutations in laboratory strains is very small. Do you know of a good method of causing mutations?

ALKIMOS: Me?

EXTON: Alkimos!

MULLER: You said it was easy. Yet, I've experienced exactly the opposite in my lab.

ALKIMOS: Oh, well I was just ….

EXTON: Professor, please excuse us but we're in a hurry.

MULLER: But our discussion is just getting interesting. Tell me your thoughts on mutations.

EXTON: I'm afraid we can't. Our data is awaiting publication. Let's go!

DEMOCRITUS: The mutation of the casting mould is the key to variation.

ALKIMOS: Let's stay, Exton!

EXTON: Impossible!

MULLER: Gentlemen, we can revolutionise genetics. Let's not delay!

DEMOCRITUS: The true revolution will be if we can uncover the details of attraction and repulsion that allow a gene to create a complementary replica ...

MULLER: Wait!

EXTON: Everyone aboard!

ALKIMOS: But Exton, I wasn't going to mention X—

EXTON: Alkimos!

ALKIMOS: Aaagh … here we go!

Fronts on the Wings of a Moth—*Alfred Kühn*

ALKIMOS: Master? We're here.

DEMOCRITUS: I see that.

ALKIMOS: It was an interesting trip.

DEMOCRITUS: Indeed it was.

ALKIMOS: Master …? Do you really think that someday we'll be able to grind up genes and put them in a beaker?

DEMOCRITUS: I have no reason to think not.

ALKIMOS: And do you think the genes will behave the same way they do in a living organism?

DEMOCRITUS: Maybe not exactly the same way, but I'm sure we'll be able to uncover something about their characteristics in a beaker. The d'Hérelle phenomenon shows us it's possible.

ALKIMOS: Master, I don't understand. Will the fruit fly's *white* gene be white in a test tube?

DEMOCRITUS: I doubt it.

ALKIMOS: And the gene for short-stemmed peas will be short? The shortness of the gene won't mean anything if it has no stem to act on.

DEMOCRITUS: No, it won't.

ALKIMOS: Doesn't this all mean that genes only have meaning when they're among other genes inside the living thing? Without the organ whose trait the gene determines, the gene will have nothing to do.

DEMOCRITUS: You're right, Alkimos.

ALKIMOS: Maybe it doesn't even exist. But then what would we find in the beaker?

DEMOCRITUS: Son, although your reasoning is sound, I must call your attention to a rather important consideration. The physical existence of genes is certain. The chromosomes, which carry genes, have an undeniable materiality. My theory of atoms leads to the same conclusion. My entire view of the world is based on atoms and the forces that hold them together and set them in motion. No observations, no arguments, have as yet challenged this.

EXTON: Democritus is a materialist to the core. That's what I like about him.

ALKIMOS: What do you mean by materialist?

EXTON: Someone who dares to believe.

EXTON: Believe what?

EXTON: In the infinite capabilities of matter.

ALKIMOS: I don't understand.

EXTON: The world is extraordinarily complicated. That's why we take such delight in it. Creatures creep and crawl, do battle and devour each other, go into heat, and produce offspring similar to themselves. Human beings have desires, fall in love, have doubts and hopes, engage in trade and food production, and sometimes boil things in test tubes. And sometimes they contemplate the world. They ask, 'How is it possible that so many things take place around us, and that they occur in exactly that way?' Usually no one can answer that question, but it doesn't stop people from trying. Some assert, 'There can be only one solution. Somebody is directing this company of actors. I know this because I can see everything teeming and swarming, rutting and copulating.' To help them understand they call upon *vis essentialis* and the gods of Olympus. But another group says, 'Why would I think all of this is being controlled by some outside force, when I don't even understand what all this is? There's no reason for me to think this couldn't happen on its own. All my life I've been a hard worker, why would I now choose the lazy thinker's solution?'

ALKIMOS: Master, then you're a materialist?

DEMOCRITUS: No. I'm a sophist.

ALKIMOS: My Zeus, and I'm a hedonist!

EXTON: I see you've understood the general idea.

HERMES: Gentlemen, may I …?

EXTON: Not now, Hermes. So, we've been thinking about the role of genes in living things. I think it's time to investigate the question in further detail. In the 1930s, nobody thought as clearly about the working of hereditary material or its role in development than the German geneticists. In America, Morgan and associates had created an entirely new school, and the problems of development and evolution slipped to the periphery of their enquiries. Genetics was essentially divorced from embryology, the field that had given birth to it. But the situation in Germany was different. Because the German university system was far more conservative (and still is), numerous older professors continued to teach in the language of 19th-century comparative embryology. This atmosphere had an impact on young researchers, spurring some of them to strike a brilliant balance between classical knowledge and the most advanced thoughts on genetics at the time. Not surprisingly, the field of developmental genetics, a product of the marriage of Roux's *Entwicklungsmechanik* to genetics, was born in Germany. We can find the forerunner of this in the work of Boveri, but the real master of this new field were Karl Henke and Alfred Kühn, who both worked first in Göttingen and later in Berlin.

ALKIMOS: So we're about to take a trip, aren't we?

EXTON: Yes, direction Berlin, 1937! Hermes, prepare the Deus ex Machina!

ALKIMOS: But we just got back.

HERMES: And dinner is ready. I've been trying to tell you this.

EXTON: Don't worry, it's just a quick trip. We'll be back soon enough. All aboard!

ALKIMOS: But I'm hungry!

EXTON: Alkimos?

ALKIMOS: What?

EXTON: Don't yell!

ALKIMOS: Bbbbut ... Agggghh ... hhhere wweeee gggooooo!

EXTON: Kühn conducted groundbreaking experiments in developmental genetics on the moth *Ephestia kühniella*. He was not satisfied with collecting and mapping mutants. What excited him was the role of genes in development. He wanted to know what genes do, in what chemical reactions they participate, and when they are active during development.

ALKIMOS: But he didn't want to know what they are?

EXTON: That's a question we'll leave to Democritus. We've arrived!

ALKIMOS: Exton, can we have a look around the city?

EXTON: If we have time. But for now, come with me. The Institut für Biologie der Kaiser-Wilhelm-Gesellschaft zur Förderung der Wissenschaften. It shouldn't be far.

ALKIMOS: This maddening fog!

EXTON: Here we are!

KÜHN: Von Eschtand!

EXTON: Alfred!

KÜHN: I'm so happy to see you again, dear friend.

EXTON: I am too!

KÜHN : I haven't seen you since our paths crossed in Lisbon.

EXTON: Ah, but what unforgettable days we spent together.

KÜHN : Indeed. The zoologists didn't understand my lectures, so we had to find other ways to spend the time.

EXTON: Yes, such fine wines.

KÜHN : And fine food! And, yes…

EXTON: Well… What memories! Oh, but I'm afraid we haven't got time for reminiscences.

KÜHN: Well then, my friend, what brings you here?

EXTON: I'd like to finish our discussion, which was interrupted unfortunately. I've brought two colleagues with me, Mr. Alchumbach and Mr. Demiescher.

KÜHN: I'm glad to meet you.

EXTON: Let's go over the main points of your lecture in Lisbon.

KÜHN: My pleasure, but will it interest your colleagues?

ALKIMOS: Oh, we live for science.

KÜHN: I'm happy to hear that.

EXTON: Let's get started.

KÜHN: All right. In the first thirty years of research on heredity, we managed to elucidate the laws of genetic transmission. Nowadays, research centres increasingly on the genetic-developmental question of how the hereditary material works.

EXTON: Your research does, but not everyone's.

DEMOCRITUS: We've already spoken a lot about Mendel and Morgan. We're also fairly well versed in embryology. We'd love to combine the two — genetics and development.

KÜHN: A gene, as a unit that can mutate and be passed on, can be considered an entirely independent agent. However, if we view the gene from the perspective of developmental biology, then this definition doesn't work. After all, during development, no independent, simultaneously occurring reaction chains exist that connect certain genes to certain characteristics. The assortment of genes hidden within the nuclei of germ cells in no way corresponds to the assortment of characteristics seen in a fully developed individual.

ALKIMOS: Too bad because the explanation would be so simple!

EXTON: That was the beauty of the theory of preformation. It provided an explanation the human mind could easily grasp.

KÜHN: The explanations for life's phenomena, however, are never simple. The completed individual is the outcome of a very complicated set of processes. In these processes, various genes play only partial roles in specific reactions. The question is what role genes play in these reactions and what happens to the reactions when an individual gene carries a mutation. Richard Goldschmidt thought the effect of mutations was to change the speed of the reactions. To better understand this, we need to examine individual examples—it's the only way to delve deeper into the question of genetic interaction. My studies of *Ephestia kühniella* can help.

ALKIMOS: Another strange creature.

EXTON: No, not that strange. *Ephestia* is just the latin name for the meal moth. Here is a picture from Richard Hertwig's book.

KÜHN: The development of patterns on moth wings gives us some insight into the question of morphogenesis and the determination of various cell types. So, what are we talking about? The first two wings of a fully developed animal have a distinctive pattern. We find two white transverse bands, one closer to the base of the wing, the other toward the tip. Each white band is bordered on both sides by black stripes that enclose a middle field. This pattern can vary considerably in different strains of meal moth. For example, in the black-winged line the entire wing is black except for the two white bands. By the way, you don't see these bands in the line that Hertwig illustrated, so have a look at my drawings:

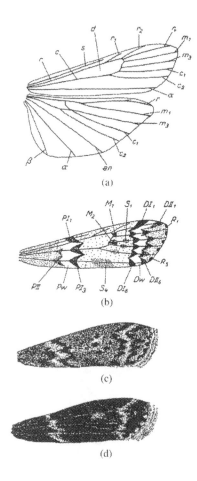

(a)

(b)

(c)

(d)

ALKIMOS: The wings really are striped.

DEMOCRITUS: If we're interested in morphology, then we need to ask why they're striped.

KÜHN: To be more precise we need to ask how and when the stripes arise. We might at first think that the stripes just appear where they are. But that's not at all that happens. The truth is breathtaking. The entire process takes place during the pupal stage, between 36 and 72 hours after spinning the cocoon. Initially the stripes don't appear as stripes. They're the result of two *determination fronts* that run the length of the wing's surface.

ALKIMOS: A determination front? What's that?

KÜHN: On the second day of the pupal stage, two fronts begin, one along the anterior and the other along the posterior edge of the wing. They spread and spread until they merge in the centre area of the wing surface. The melded fronts result in two transverse stripes on the wing. After, they slowly begin to distance themselves from one another perpendicular to the original direction of progress, creating an increasingly wide middle section.

DEMOCRITUS: Amazing. If D'Arcy Thompson could see this! It's none other than mathematics on the wing of a moth.

ALKIMOS: Professor Kühn, there's something I don't understand. How can you follow those fronts?

KÜHN: I was expecting that question. It's not a simple matter. If we open the cocoons at different stages of development, we won't see anything. The stripes become visible only much later. That's why we say a determination front spreads across the wing and not a stripe. Where the front finally arrives is where the stripes form. The cells in the area grow scales of different colours and shapes in response to signals given by the front.

EXTON: So these cells will be different than the others of the wing. It's called a determination front because it directs a group of cells to do something different.

ALKIMOS: But how do we know where the front has spread?

KÜHN: We did experiments that involved burning the cocoons.

ALKIMOS: Cocoon burning?

KÜHN: If we use a hot needle to burn the developing wings in different places and at different stages we can stop the spread of the fronts. The front can't pass through cells killed by the heat. Then we just had to wait for the stripes to appear where the front was stopped or redirected. With this method we could determine the origin of the front, in what direction it spread, and how fast. For example, if we injured the anterior edge of the wing, the front can't begin its spread. The result is a stripe that begins on the opposite side and halts in the middle. These drawings should help. The small dots mark the spot where the wing was burned.

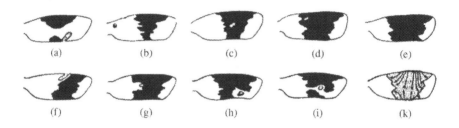

DEMOCRITUS: Fantastic! So the patterns we so often see in nature—the patches on a butterfly's wing or the stripes and dots on a flower petal—are not some pre-planned, divine, or genetic blueprint, but the interplay of fronts spreading here or there?

ALKIMOS: In other words there are no stripe genes, only front genes?

EXTON: Yes, we'll have a better understanding if we think in those terms.

DEMOCRITUS: The triumph of epigenesis! If everything is created rather than growing or developing from something smaller, then the concept of a preformed *animalculum seminalia* or a miniature design loses its meaning. Genes do not determine a characteristic, but rather a developmental process: the spread of a front, the speed of a reaction, the creation of a pigment.

EXTON: That's right. There are no designs for living things, only recipes. A recipe sounds like this: take a fertilised ovum with an intact genome, leave it at the right temperature, with the necessary nutrients, and wait. Nine days, nine weeks, or maybe nine months. During this time the zygote begins to cook itself. The cytoplasm activates the genome, the genome changes the cytoplasm, and from this point on the embryo reads the recipe. If necessary it activates new genes, creates new fronts, and differentiates cells. If a gene is mutant, then the recipe changes a little. If the deficit is serious then a leg will be missing, or the eyes will be white, or a front won't spread as it does in individuals of the wild type. The most important, most variable, and most susceptible to injury is the genome, but without the environment of the cytoplasm and metabolic networks, the zygote would have no chance. But let's return to the wings of the moth. Alfred, tell us, what does this process teach us about genes?

KÜHN: Well, as we know, the role of genes can be studied by examining their mutations. Fortunately we've found several genes that affect the pattern and colour of moth wings. The *Sy* mutation affects the distance between the stripes. If we have an *SySy* homozygote, then the individual dies, but we can study the mutation in *Sysy* heterozygotes. Have a look at this drawing.

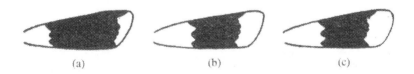

(a)   (b)   (c)

In panel (a) we see the *sysy* "wild type." Panel (b) shows the heterozygote *Sysy*, which has a much narrower centre section. The difference is the speed at which the determination front spreads. If we apply heat at the same point in time, the expansion in the *Sysy* individuals is always interrupted at an earlier stage than in the wild types (c). This means the front spreads more slowly in the mutants.

DEMOCRITUS: This corresponds to Goldschmidt's idea, which we discussed earlier. One gene influences the speed of a certain developmental process. The *Sy* gene is responsible for ensuring the appropriate speed of expansion of the front across the wing.

ALKIMOS: I still don't understand what a front is. Is it a gene?

KÜHN: Unfortunately, we still can't say what the front is composed of. It may be physical matter or an impulse. We also can't say what the *Sy* mutation influences: the characteristics of the front, or the characteristics of the cells which the front passes through. But we do know where and when the gene operates.

ALKIMOS: In the wings of the pupa between the hours 36 and 72.

EXTON: And we can also be sure that the *Sy* gene has other tasks besides this.

KÜHN: This is true. We mustn't forget that the *SySy* homozygotes can't survive, so the gene clearly participates in a much more essential, vital reaction.

ALKIMOS: And we can't examine that because the experimental subjects all die. Too bad.

DEMOCRITUS: So there are fronts upon which the individual's life depends.

KÜHN: Yes, those fronts that take part in the basic chemical reactions in the cell.

EXTON: Metabolism, cell division, maintenance of the cell membrane, chromosome splitting.

ALKIMOS: There must be a lot of mutations that are lethal!

EXTON: There certainly are. At this moment, Muller is using X-rays to investigate this.

DEMOCRITUS: The X-rays damage the genes, which damage the fronts, which damage the organs, which kill the individual.

EXTON: Exactly. If we think about it, we can see what a long road lies between the genotype and the phenotype.

KÜHN: And a very circuitous one.

ALKIMOS:

> *You dreamed of life, but were conceived in bitter, unrelenting cold,*
> *You'll take no part in existence, because your gene, from which*
> *Your heart's front expands, filling your cavernous chest*

Limps along, mangled, defeated, bearing the X-rays badge
Neither heavenly nor earthly word echoes from within.
You're not to this world born, nor will you age and whither,
On a dusty road, your remains shall never scatter.

DEMOCRITUS: Genes can destroy us, but they can never rule us.

ALKIMOS: You think so?

DEMOCRITUS: You said it yourself, my friend.

EXTON: Alfred, we must be going.

KÜHN: Already?

EXTON: I'm afraid so.

KÜHN: Well, I'm honoured by your visit.

EXTON: Take care!

KÜHN: You too, especially in that old clunker.

EXTON: Democritus, Alkimos, come on! Here we go!

DEMOCRITUS: Thank you, Alfred!

ALKIMOS: Aggghhh … here we go!

The Birth of Patterns

ALKIMOS: We're finally back!

EXTON: You didn't like the trip?

ALKIMOS: Oh, I did, I just missed my good old friend.

HERMES: Exton, sir, may I serve dinner?

EXTON: Yes, or course.

DEMOCRITUS: There's something that confuses me.

EXTON: What's that?

DEMOCRITUS: How do you know Alfred?

EXTON: I met him when I was planning this trip.

DEMOCRITUS: In Lisbon?

EXTON: Yes.

DEMOCRITUS: And where's that?

EXTON: In Portugal.

ALKIMOS: So why didn't we go there?

EXTON: I didn't want to go back there, so I reprogrammed the journey.

ALKIMOS: Aha. Was a woman involved?

EXTON: Of course not! Now let's return to our discussion of morphogenesis.

ALKIMOS: I'm more interested in the other topic.

DEMOCRITUS: Alkimos, please apply the same enthusiasm to the question of how living things develop shape.

ALKIMOS: I'll try, master.

EXTON: Our discussion with Alfred showed how important the spreading of fronts in various directions is during development. That's why we need to understand the concept of *diffusion*.

HERMES: Gentlemen, flamingo soup.

ALKIMOS: That sounds delicious!

EXTON: Diffusion is the movement of materials down a concentration gradient. If we salt the soup, the salt will eventually distribute evenly throughout the liquid even if we don't stir it.

ALKIMOS: I hope you don't mind if I start eating.

EXTON: If you put drops of ink in milk, they'll also spread evenly throughout it. Liquid molecules are in constant motion.

DEMOCRITUS: Dear Exton, please tell us what molecules are. D'Arcy Thompson used the word and so did Muller, but we still don't understand what they are.

EXTON: Molecules are created by the joining of atoms.

DEMOCRITUS: Joining? Why that's in my theory of atoms!

EXTON: Chemists so far have identified more than a hundred different atoms, and millions of molecules that arise from various atomic reactions.

ALKIMOS: There are that many kinds of atoms? And they can all join to each other?

EXTON: Yes, but they can't join at random. There are specific laws which determine which ones can join to which. The number of atoms in a molecule can range from two or three to a hundred, and in some types of molecules, the number of atoms that can join is nearly endless.

ALKIMOS: Those must be huge molecules.

EXTON: They are indeed. We call these kinds of molecules *polymers*.

ALKIMOS: *Poly* means "many" and *mer* "parts."

DEMOCRITUS: So polymers are enormous molecules composed of many units. I didn't realise my theory allowed for so many possibilities.

EXTON: Now let's get back to *diffusion*. Let me explain why it happens.

ALKIMOS: Mmm, I've never tasted anything as good as this soup!

EXTON: As I said, the molecules in a fluid are in constant motion. They float here and there, knock into each other and spin. If a group of foreign molecules is added to the mix, they're immediately bombarded. The pushing and shoving goes on until they are fully mixed into the liquid.

ALKIMOS: And then they stop?

EXTON: No, the pushing continues, but by that time the two materials are so well combined that the concentration is uniform. The motion caused by the flurry of activity is known as Brownian motion after the man who first discovered it, Robert Brown.

ALKIMOS: Is he the same man who discovered the cell nucleus?

EXTON: Yes. Naturally diffusion takes place in a similar manner in organisms composed mostly of liquid.

DEMOCRITUS: I'm not sure what you're trying to say, Exton. We were discussing morphogenesis, the process by which an organism assumes its shape. Now you've brought up diffusion. If I understand correctly, the process doesn't create shapes. In fact, it completely breaks up existing forms.

EXTON: Yes, I know it seems that way. But under certain conditions, the opposite occurs. Alan Turing proposed a *reaction-diffusion* model that is able to do exactly that.

ALKIMOS: Does that mean that while the molecules diffuse, they react with each other and join together?

EXTON: Not together. Apart. They disconnect. To be more precise they help the other molecules to disintegrate.

DEMOCRITUS: How does this lead to the development of form?

EXTON: As strange as it sounds, it's actually possible, and Turing's model shows this. Although his model is purely mathematical, we can understand the essence of it without going into the details of his calculations.

HERMES: Gentlemen, some chanterelles with cream?

ALKIMOS: Of course. Could we have some wine too?

EXTON: In his article "The Chemical Basis of Morphogenesis," published in 1952, Turing said this about his model: 'It is suggested that a system of chemical substances, called morphogens, reacting together and diffusing through a tissue, is adequate to account for the main phenomena of morphogenesis. Such a system, although it may originally be quite homogeneous, may later develop a pattern or structure due to an instability of the homogeneous equilibrium, which is triggered off by random disturbances.'

ALKIMOS: So the genes that create form are morphogens themselves?

EXTON: That's exactly what Turing wanted to suggest with that expression. Morphogens, for example, can be materials that diffuse in a tissue and can persuade the cells they come in contact with to develop differently than they would when not in the presence of the morphogen.

ALKIMOS: We saw this with the moth wings. The spreading front determined the location of the stripes.

EXTON: Where there was no front, there was no stripe. We can think about morphogens in this way: they affect tissue differentiation.

DEMOCRITUS: Something's not clear to me. Fronts and morphogens diffuse in the tissue, but we know that genes are attached to chromosomes, and chromosomes are in the nucleus of the cell. I would think that genes are therefore not capable of diffusion.

EXTON: Weismann had similar thoughts. Because he didn't believe in the permanence of chromosomes, he postulated that the hereditary material in the nucleus disintegrates when the cell differentiates, and certain parts seep into the cytoplasm through pores in the nucleus and then begin to operate.

ALKIMOS: If the chromosomes are permanent forms, then they can't disintegrate, and the genes can't flow out through the pores.

DEMOCRITUS: There's only one solution. Transmitter molecules must exist. They're the messengers of genes. They can escape from the nucleus and act as intermediaries between the genes and the cytoplasm. They move about freely in the cytoplasm, creating fronts, and taking part in reactions.

EXTON: Yes, something like that. But I'd like to talk more about Turing's model. You see, it's the triumph of epigenesis, the perfect mathematic formulation of Caspar Friedrich Wolff's *vis essentialis*, the first working model of the creation of shape from nothing. Turing showed that spatial inhomogeneity, or inequality, can arise in an evenly distributed system.

ALKIMOS: So something can arise from nothing?

EXTON: Maybe not from nothing, but from a perfectly uniform system. The extreme theory of epigenesis needs to explain exactly this. Imagine a spherical ovum, uniform in every direction, filled with uniform cytoplasm. What could break this perfect symmetry? How will the embryo develop a head and tail, a back and belly, a right and left half?

ALKIMOS: We already saw with the frog ovum that the sperm and the uneven distribution of the yolk are responsible.

EXTON: Well, that may be true with the frog sperm, but think back to Hans Dreisch's experiments with sea urchins.

ALKIMOS: I'll try …

EXTON: Boveri mentioned Driesch's experiments; in fact, his experiments with quadripolar eggs are based on Driesch's work.

DEMOCRITUS: I remember, dear Exton. Driesch discovered that when the blastomeres of a four-celled sea urchin embryo were separated, each of the four cells developed into an intact larva. This means that in the four-cell stage of the embryo, nothing has been decided yet. The asymmetry of an embryo can arise even without penetration of the sperm, as it did in the four separated cells.

EXTON: Driesch conducted these experiments in 1894. He concluded that development cannot be explained by the laws of physics. According to Driesch, embryos must possess some kind of wisdom or spirit, and were capable of development despite the interference of scientists. Eventually he abandoned his experimental work and became a Professor of Philosophy. Until his death in 1941, he taught and defended his theory of a life force. He was the last genuine vitalist.

DEMOCRITUS: The sea urchin egg and similar malleable, regulative ova require an epigenetic explanation free of any kind of initial inhomogeneity.

EXTON: Turing supplied this. His simplest model relied on two morphogens. He posited that their production, disintegration and diffusion occur at different speeds. Moreover, each morphogen affects the production of the other. At first the two morphogens distribute uniformly, their production and disintegration are balanced, and the concentration of neither changes. But, at the atomic and molecular level, balance doesn't exist, nor does quiet. Those terms, like the concept of temperature, can only be used for systems comprised of a large numbers of units.

ALKIMOS: So individual morphogens dash here and there, spin and coil?

EXTON: That's right. At the molecular level, thickening and thinning can take place. This is not apparent at the macroscopic level if no reactions take place. The ink and milk mixture always remains a dark colour. Morphogens, however, are continually being produced, breaking down, and entering into reactions. Because of this, at the atomic level, the instability of the equilibrium can increase. If in one place a few more of one kind of morphogen accumulates, more of the others will break down. What arises is a patch in which one morphogen completely crowds out the other. However, this morphogen is not able to flood the entire system, because its production is dependent on the other morphogen. The patches caused by the interaction of morphogens can have an effect not only at the atomic level but at the cellular level, too.

DEMOCRITUS: This is how order can arise from chaos. By chance.

ALKIMOS: If D'Arcy Thompson could hear this!

EXTON: Turing probably relied a lot on D'Arcy Thompson's work. His article had six references and one of them was D'Arcy Thompson's *On Growth and Form*.

DEMOCRITUS: Turing's model is epigenesis in the language of numbers.

EXTON: That's right. This kind of explanation for patches, stripes, and patterns is the most abstract approach to morphogenesis. Pure mathematics. Turing was able to create all kinds of patterns by changing the parameters to describe the characteristics of morphogens. He carried out the difficult calculations on the world's first computer at the University of Manchester.

ALKIMOS: Exton, do you mean that Turing's morphogens trigger development in the ovum of a sea urchin?

EXTON: No, we can't say that. What's important about Turing's model is not that it explains a specific case, but that he discovered a principle that no one before him had ever dreamed of. He didn't show that patterns and forms were made in this way, but rather that they could be. He established the theoretical foundations of epigenesis. Perhaps only he could have persuaded Driesch to abandon vitalism.

One problem with Turing's model was that it did not create stable patterns. In the 1970s, the German theoretical biologist Hans Meinhardt, together with Alfred Gierer, developed a new set of biologically inspired reaction-diffusion models. The Meinhardt-Gierer model was able to produce stable patterns that also scaled with tissue size. Their model included a short-range activator with self-enhancing capabilities, and a long-range inhibitor that suppressed the activator. However, like Turing, Meinhardt and Gierer also had no idea of the molecular nature of morphogenes.

DEMOCRITUS: To truly understand development, we need to find these morphogens and the genes that set them on their path.

ALKIMOS: But master, Kühn already found the morphogens of the moth's wings. After all, determination fronts, which spread like a wave and give orders to cells, are surely none other than morphogens.

EXTON: Moreover, Kühn found mutations which affect the spread of morphogens.

ALKIMOS: Master, can we now say we know something about development?

DEMOCRITUS: I'm afraid we can't. We still have no idea what kind of molecules or atoms make up morphogens or their genes. I'm afraid our journey continues until we discover this.

EXTON: Democritus, do you realise that you've just formulated the goals of *molecular biology*? I agree we shouldn't delay. It's time for us to descend to the atomic level. As we do, the world of frog eggs, embryos, sea urchins, and cypress stands will fade in the distance, but we'll eventually come back to them.

HERMES: Gentlemen, if I may, some honey wine …

Part Five

The Atoms of Life

Hormones in Larva Blood

ALKIMOS: We could have slept a little more …

DEMOCRITUS: I had a strange dream. A fruit fly sat on my nose and started to laugh at me. But it wasn't the Drosophila we know so well, but a completely new species. It held its wings in a steeper fashion, its body was narrower, and its legs longer. It just laughed, scornfully, and said I was a despicable scientist, while it was life itself. But I didn't back down. Do you know what I said? 'My little friend, you're seriously mistaken. I'm a philosopher!' I could see it was surprised; it had never heard this word. The fly just looked at me with its large, compound eyes. 'Hey little fly, do you understand?' I asked. It just stared. I must have swatted myself, because I woke up with a throbbing nose.

ALKIMOS: I hope you got the fly.

DEMOCRITUS: Hmm, let's see … ah yes, I did. Here's the casualty.

ALKIMOS: Aha.

EXTON: You're awake! I've been up for hours. You shouldn't have drunk so …

ALKIMOS: Yes, Exton, but that wasn't the problem. My master was plagued by nightmares, and his incessant tossing and turning kept me awake.

DEMOCRITUS: Yes, it's true, Exton.

EXTON: Well, I hope breakfast will invigorate you.

ALKIMOS: Oh, it will. Come on, I'm starved.

EXTON: Over breakfast we can discuss *gynanders*.

ALKIMOS: Gynanders? What a bizarre word. I know that *gyno* means female and *andro* male, so a gynander is both male and female?

EXTON: Something like that.

DEMOCRITUS: I've heard of that kind of thing. Hermaphroditos, the child of Hermes and Aphrodite, was both male and female.

ALKIMOS: Did you dream that too, master?

DEMOCRITUS: Tiresias, the old prophet, was once transformed into a woman for seven years.

ALKIMOS: But he wasn't a man and a woman at the same time.

DEMOCRITUS: No, he was a man and woman in succession. He had the opportunity to experience love from both sides.

ALKIMOS: As an old person?

DEMOCRITUS: He wasn't old at the time. Later, when he had to arbitrate the debate between Zeus and Hera, he proclaimed that women enjoy far greater sexual pleasure than men ...

ALKIMOS: And he knew from experience.

DEMOCRITUS: ... Hera in her anger struck him blind. Zeus compensated Tiresias with long life and the gift of foresight.

EXTON: *Hermaphroditism*, the state of being both male and female, is common in the living world. Gynandromorphism, however, is different. For example a hermaphrodite snail is male and female throughout its entire body. It can produce eggs and sperm. In a gynandromorph only some parts of the body are male and others are female.

Gynandromorphs occur at random among species that don't normally produce hermaphrodites. This drawing shows a gynandromorph moth.

The left side is female and the right is male. In nature gynandromorphs are very rare, but scientists have been able to make thorough studies of them in fruit flies.

ALKIMOS: So what happens is that a male and a female merge?

EXTON: No. From a genetic point of view, it's far more exciting than that. Remember, male fruit flies have an X and a Y chromosome. The females have two X chromosomes. Keep this in mind as I explain: if an animal has only one X chromosome, it will be male.

ALKIMOS: Even if there's no Y?

EXTON: Yes. Gynadromorph fruit flies occur when a female zygote begins developing and during the earliest stages, it somehow loses one of its X chromosomes. The cells that contain just one X chromosome will produce daughter cells that are male. Those that have both X chromosomes will produce daughter cells that are female. Depending on how early, and which cells suffered a loss of the X chromosome, body parts of various size and location will change from female to male. The animal is what we call a genetic mosaic.

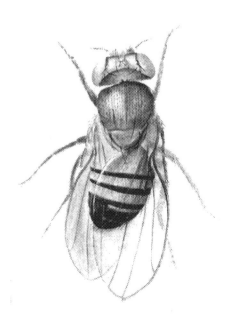

ALKIMOS: Incredible.

EXTON: Now think about what happens if the embryo is a heterozygote for a mutation on the X chromosome, and the wild type allele was located on the lost chromosome.

DEMOCRITUS: If the normal allele is lost from half of the animal's cells, then these cells will all be mutant. In a genetic sense, the male part is mutant.

EXTON: And the female part continues to be heterozygous. A recessive mutation will only have its mutant effect on the male part.

DEMOCRITUS: I see.

ALKIMOS: Exton, what are you trying to say?

EXTON: Just wait a moment. It's extremely important that you understand this puzzle, if you want to understand the rest.

ALKIMOS: Okay …

EXTON: The boundary between the male and female parts of a gynandromorph is well defined by those recessive, X-linked mutations that clearly affect the phenotype. There are a number of these. For example, *yellow*, which causes a yellow body colour, and *forked*, which causes bent and shortened bristles. These particular mutations reveal exactly which parts of the animal are male, because they're what we call *cell autonomous*. Let me put it another way: in a genetic mosaic, cells that are mutant for these traits will show this in their phenotypes and cells that are not mutant will be unchanged.

ALKIMOS: But I thought that's how it worked. Only a cell that has the mutation will have the corresponding phenotype.

EXTON: It's reasonable to think that. But strangely enough there are genes that are not at all cell-autonomous. These genes affect not only those cells that contain them, but also other cells around them.

DEMOCRITUS: Exton, please explain!

EXTON: In 1920, Sturtevant discovered that the *vermilion* (*v*) mutation, which resulted in bright red eyes, worked rather strangely in gynandromorphs. If the animal's head was entirely male and contained the *vermilion* mutation, the eyes were frequently dark red, as is typical in wild types, and not bright red, as we'd expect based on its genetic makeup.

DEMOCRITUS: If its eyes were wild type, how did Sturtevant know the head was male?

EXTON: Because other recessive mutations were present and clearly showed the head to be male. Only the *vermilion* mutation acted so strangely. The eyes were definitely mutants; yet their colour was the wild-type red. It was as if something diffused from the female region to the male, and coloured the eyes of the mutant.

ALKIMOS: The female's intact *vermilion* gene.

DEMOCRITUS: Yes, but we've seen that genes can't diffuse. But the intact *vermilion* gene may set off a front. Then the molecules produced by the genes and necessary for a certain eye colour may diffuse, even if the genes can't.

EXTON: At the end of the 1930s, our friend Alfred Kühn and two other scientists, the Russian-born Boris Ephrussi and the American George Beadle, all decided simultaneously to try and isolate this diffusing molecule.

ALKIMOS: Isolate a molecule?

EXTON: It sounds incredible, but it wasn't an extravagant idea. By that time, chemistry was already very advanced. Thousands of the carbon-based, or organic, molecules necessary for life had been discovered; moreover scientists clearly understood how they reacted with each other. A lot of organic compounds had been stored in laboratory vials in their completely pure, crystal forms. Reliable methods of identifying new materials had also been devised. In theory, all scientists needed to do was remove the material that affected eye colour from the animal and hand it over to an organic chemist.

ALKIMOS: Master, we're getting close to understanding the atoms of life!

DEMOCRITUS: I can hardly wait!

EXTON: Let's see what Kühn and the others based their experiments on. As you recall, Kühn worked with the *Ephestia* moth, whose stock contained mutations that affected body colour. One gene was labelled *A* by Kühn.

Its mutant *a* allele resulted in red eye colour instead of black in the homozygote form. The dark colour of the epidermis and testes of the larva were also made lighter by the mutation.

ALKIMOS: Why did Kühn call the gene *A*? Why not *red* or *testes-lightener*? Where was his imagination? The fruit fly geneticists were much more creative.

EXTON: Yes, Alkimos, the fruit flies' genes all have colourful names. *Decapentaplegic, zerknüllt, escargot, cactus.* The fruit fly geneticists enjoyed naming them. But let's continue with our discussion of the meal moth's *A* gene. It may have a dull name, but the experiments were exciting. Ernst Caspari worked in Kühn's lab and started experimenting with transplantation in the early 1930s.

DEMOCRITUS: If I understand correctly, his experiment was similar to Hans Spemann's with the frog embryo.

EXTON: That's right. Tissue transplantation. Caspari said, 'It has not been tried yet to approach the problem of the mechanism of gene action by the method that is very frequently used in developmental physiology, operative experiments.' Caspari used the *a* mutation in his experiments. He transplanted the testes of moth larvae homozygous for the red eye mutation (*aa*) into wild type (*AA*) larvae and vice versa.

ALKIMOS: Moth-testicle transplantation. Oh, what will scientists come up with next?

EXTON: Caspari was far more enthusiastic than you, Alkimos. The majority of animals he operated on died, but those that survived exhibited some very interesting characteristics. Here's how he described his observations: 'When *aa* testicles are transplanted into *AA* animals, one observes … an influence of the host on the implanted organ … When *AA* testicles are implanted in *aa* animals, the opposite effect obtains, that is the implanted organ has an influence on the host.'

DEMOCRITUS: In every case, the wild type tissue will influence the mutant tissue.

ALKIMOS: What kind of influence?

DEMOCRITUS: Probably colouration. Exactly as we saw with Sturtevant's gynandromorphous fruit flies.

EXTON: Yes. The mutant moth's eyes became black not red when it was implanted with the wild-type testes. In the opposite case, mutant testes implanted in the wild-type animals turned dark, a colour typical of the wild type.

DEMOCRITUS: So something can travel from the wild-type tissue to the mutant tissue and can induce pigment production.

EXTON: In Caspari's own words: 'The *A* testicle secretes a substance into the blood that has the capacity to induce intensive formation of pigment in testicles or eyes in whose cells gene *A* is lacking.' There had to be a substance capable of diffusion between the gene and the pigment, like this:

'A' Gene → 'A' Substance → Pigment

Kühn and associates called the 'A' substance a *hormone*, because only small amounts of it were needed to produce an effect, as was the case with the other hormones that were well known at the time among physiologists and biochemists.

ALKIMOS: So the 'A' gene sends the 'A' hormone to colour the eyes.

EXTON: Now, do you see what it means if a gene is not cell-autonomous? The genes of the wild-type cells can affect those of the mutant cells. Later, scientists were able to dissect this chain into smaller parts. During the transplantation experiments, they'd discovered that the quantity of transplanted material influenced the strength of the pigmentation. If they transplanted more, the colour was darker. Therefore, the more hormone that reaches the mutant cells, the more pigment they can produce.

DEMOCRITUS: That makes sense.

EXTON: But the degree of pigmentation didn't change if the genotype of the implanted tissue was heterozygous *Aa* instead of *AA*. Thus, the *A* gene does not directly determine the quantity of hormone, because that would mean the heterozygote would have only half as much hormone. But since it has the same amount, Kühn and associates concluded that the gene's immediate impact is on the *primary reaction* that produces the hormone instead of the hormone itself.

DEMOCRITUS: So the *A* allele either triggers a reaction in the cell or it doesn't. The cell that contains this allele will then produce the appropriate quantity of hormone, that can diffuse.

EXTON: That's right. This reaction works the same way whether there's one *A* or two *A* alleles in the cell.

DEMOCRITUS: That's why the scientists could only introduce more of the hormone into the larva if they implanted more of the tissue.

EXTON: Right. Whether the tissue was heterozygous (*Aa*) or homozygous (*AA*) didn't matter. This is how Kühn and associates explained the operation of the *A* gene:

'A' gene → primary reaction → 'A' hormone → pigment

The genes are in the cell nucleus and the primary reactions take place in the cytoplasm; the hormones circulate in the blood, and colour the organs.

DEMOCRITUS: If we could somehow trap this mysterious hormone and give it to a chemist, then we'd know a lot about these primary reactions governed by genes.

EXTON: This seemed increasingly possible. Scientists had managed to induce pigmentation in mutants using extracts from the larva instead of transplants.

DEMOCRITUS: So the hormone was present in the extract, and this affected pigment production.

EXTON: Meanwhile, another startling discovery was made. The '*A*' hormone in the extract made from *Ephestia* larvae also had an effect on the *vermilion* mutation in fruit flies and vice versa. Later, it was shown that the '*A*' hormone and the *vermilion* material in fruit flies were one and the same.

DEMOCRITUS: Two animals exhibit kinship on the atomic level too!

ALKIMOS: If von Baer could hear this! Not only do embryos, cells and chromosomes resemble each other, but so do the atoms of life!

EXTON: Now you understand why the transplantation experiments were so important and why Kühn's lab was in tight competition with Beadle and Ephrussi to isolate the hormone.

DEMOCRITUS: They all wanted to know something about the chemistry of genes. If they could shed light on this, we'd know something important about fruit flies and meal moths. But if that something is true for both, if the same hormone circulates in the blood of both species, then this must be true of countless other species.

ALKIMOS: It must be true of all of them. But then, wouldn't all living creatures have red eyes?

EXTON: Ephrussi and associates tried to use Caspari's transplantation techniques on Drosophila larvae to further study the genes that influence the animal's eye colour. It was a difficult task, since the fruit fly larva is much smaller than the meal moth's. But in the end they too were successful. Instead of the testes, they transplanted the eye rudiments, called imaginal discs, and wound up with three-eyed fruit flies. After they developed the technique, they tried transplanting imaginal discs with the *vermilion* mutation. The transplantation of this mutant *vermilion* eye tissue into wild type larvae resulted in wild-type eyes, as was expected.

DEMOCRITUS: We can interpret this observation as we did Sturtevant's and Caspari's. Something diffuses from the wild-type tissue into the mutant tissue, and participates in the production of normal, wild-type eye colour.

EXTON: Ephrussi and Beadle examined all twenty-six fruit fly mutations that affect eye colour. Only the *cinnabar* (*cn*) mutation behaved the same way as the *vermilion* mutation.

ALKIMOS: Then an intact *cinnabar* gene in the wild-type tissue also synthesizes a material that can diffuse into the *cinnabar* mutant eye tissue and produced the wild-type colour?

EXTON: That's one possibility. We could also describe it as: *vermilion* $\rightarrow v^+$ *material* and *cinnabar* $\rightarrow cn^+$ *material*.

DEMOCRITUS: So each gene produces a material that's necessary for normal eye colour, and if either material is missing the eyes will be a bright cinnabar red. But what's the other possibility?

EXTON: The other is that both genes participate in the production of the same material. After all, the phenotypes of the *cinnabar* and *vermilion* mutations are almost identical. In both cases, the eye lacks the brown pigment, and only the bright red remains; thus the eyes are bright cinnabar red. We can show it like this:

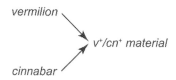

DEMOCRITUS: We can easily decide between the two possibilities. We just have to think of the appropriate experiment. Hmm ... if I'm thinking about this correctly, then the mutant *cinnabar* eye tissue needs to be transplanted into the larva with the *vermilion* mutation and vice versa. If both genes are needed to produce the material that makes the brown pigment, then neither animal will have the wild-type eye colour, because the material can't be made in the absence of one or the other of the genes.

ALKIMOS: And what happens if the genes produce two different materials?

DEMOCRITUS: Then in both cases we'll end up with wild-type eye colours, because the material missing from the *vermilion* mutant can be produced by the *cinnabar* mutant, since it's not a *vermilion* mutant. Meanwhile the material lacking from the *cinnabar* mutant can be produced by the *vermilion*. Each mutant compensates for what the other lacks, and the situation is the same regardless of which mutant is the host and which the donor.

EXTON: This is what Beadle and Ephrussi reasoned too.

ALKIMOS: See master, your mind is as sharp as a razor.

EXTON: They performed the necessary experiments, and much to their surprise, their results matched neither of their hypotheses.

ALKIMOS: How's that possible?

EXTON: When mutant *cinnabar* imaginal discs were transplanted into a mutant *vermilion* larva, the eye colour was cinnabar. But in the opposite case, when mutant *vermilion* imaginal discs were transplanted into a mutant *cinnabar* larva, then the animal had wild-type eyes.

ALKIMOS: Oh forget it—I can't follow that.

DEMOCRITUS: I'm afraid I can think of no solution either.

EXTON: Because it's not a simple one, but it's exciting. Ephrussi and his team were able to explain this puzzling result by positing the existence of two materials capable of diffusion, and that one material is produced from the other, that is, the first material is the precursor of the second.

ALKIMOS: So the material produced by the *vermilion* gene is the precursor of the material produced by the *cinnabar* gene?

EXTON: You've got it right, Alkimos. Both materials represent a stage in the reaction chain that leads to production of the eye's brown pigment.

$$precursor \rightarrow v^+ \ material \rightarrow cn^+ \ material \rightarrow pigment$$

DEMOCRITUS: That means both the v^+ and the c^+ materials are diffusible and can therefore work in a non-cell-autonomous manner to rescue a mutant eye from its mutant state.

EXTON: At least on the phenotypic level. The rescued eyes may appear as wild types, but they are in fact mutants.

DEMOCRITUS: Let's continue our thinking on this question. In mutants, the reaction chain is broken, because the gene responsible for a certain material doesn't work.

EXTON: That's correct. Certain reaction chains are interrupted in mutants, and the materials that should be produced as the reaction moves down the chain aren't. If one material is not produced then the next one isn't either, because one material is the precursor of another.

ALKIMOS: And in the end no pigment can be produced.

EXTON: In the chain that leads to pigment production, the *vermilion* gene is responsible for the first reaction and the *cinnabar* gene for the second reaction. Mutant *vermilion* eyes in a mutant *cinnabar* host are capable of producing the pigment, because the v^+ produced in the mutant *cinnabar* host diffuses into the transplanted mutant *vermilion* eyes, where it can transform into c^+ and thus into the pigment, since the mutant *vermilion* eye tissue contains the wild-type *cinnabar* gene needed for the second reaction. In the reverse, however, the mutant *cinnabar* eyes cannot complete the reaction, because the mutant *vermilion* host can't produce the v^+ material needed at the start of the reaction chain. The mutant *cinnabar* eyes also cannot use their own v^+, since they lack the *cinnabar* gene that transforms the v^+ into c^+.

ALKIMOS: One moment, I need to think this through slowly...

DEMOCRITUS: So genes' role in creating the material needed for pigmentation is indirect.

EXTON: That's right. During a series of chemical reactions, materials can transform into others, and genes are necessary for these transformations. If a gene is missing the reaction chain breaks and the pigment isn't produced. If a chain is very long, then many genes will be responsible for its operation.

DEMOCRITUS: So a chain can break because of any number of mutations that may occur at different places along it.

EXTON: But the end result is always the same. The production of the final material never happens.

DEMOCRITUS: For example the production of a pigment.

EXTON: And so the phenotype of the various mutations that could break the chain can be the same.

ALKIMOS: So the *vermilion* and *cinnabar* mutations both result in a bright cinnabar red. I got this. But if the phenotype of the *vermilion* and *cinnabar* mutations is the same, how did Beadle and Ephrussi know that the red-eyed flies were caused by mutations in two different genes?

EXTON: I am sure you can figure that out, Alkimos.

ALKIMOS: Maybe... Well, I guess the two genes mapped to different regions on the chromosomes.

EXTON: Indeed. But the simplest way to tell if two mutations with the same phenotype occur in the same or two different genes, even without genetic mapping, is to do a *complementation test*. If homozygous *vermilion* and *cinnabar* mutants are crossed to each other, the resulting progeny is wild type. The two mutations complement each other therefore they occur in two different genes. As an exercise, you can draw the crossing scheme.

ALKIMOS: Let me see...

EXTON: Anyway, in 1940, chemists finally found and defined the 'A' hormone, the same molecule what we've been calling the $v+$ material. Kühn had the help of Adolf Butenandt and Beadle the help of Edward Tatum. The first stage in the reaction path leading to the brown pigment was *kynurenine*. Its precursor is an amino acid called *tryptophan*. By the way, the race to identify the material was won by Kühn and associates by almost a year. The material was well known to chemists, who already knew exactly what atoms it was composed of. Democritus, I think you'll be interested in this *structural formula* I've drawn. It's a model of the relationship of the atoms within the molecules.

Soon enough the scientists discovered other elements in the reaction chain.

DEMOCRITUS: Exton, I think we need some time to digest this information.

ALKIMOS: Did you say digestion?

EXTON: I wouldn't mind a break. We've immersed ourselves in the topic and completely forgotten about …

ALKIMOS: Breakfast!

DEMOCRITUS: First, finish your crossing scheme, son.

One Gene, One Enzyme—*George Wells Beadle*

ALKIMOS: What were you humming?

HERMES: Just a little song. You don't know it?

ALKIMOS: No, I've never heard it.

HERMES: Where are you from?

ALKIMOS: Abdera. And you?

HERMES: Arcadia.

ALKIMOS: A lot of shepherds around there?

HERMES: That's right.

ALKIMOS: So what's the song about? Sing it!

HERMES: About a clever child.

ALKIMOS: And?

HERMES: He was so clever he stole magnificent cattle from Apollo's herd.

ALKIMOS: Impressive. How'd he do it?

HERMES: At night he crafted sandals out of the bark of a fallen oak tree and attached them to the feet of the cattle. In the dark he drove the cattle away. Because of the sandals they left no prints.

ALKIMOS: Apollo must have been furious.

HERMES: Oh, he was.

ALKIMOS: Let me hear the song.

HERMES: In the evening. I think you need to go.

EXTON: Alkimos! Come on! Let's continue our conversation.

ALKIMOS: I'm coming! Hermes, my friend, let's meet tonight.

HERMES: As you wish.

DEMOCRITUS: Son, I'm shaking with excitement. Exton and I have been discussing the secrets of atoms and what we've learned so far—how we've discovered the molecules that cause one of life's chemical reactions, the creation of eye colour; in fact, we've discovered the entire chain!

ALKIMOS: Yes, I know, master.

DEMOCRITUS: We can be certain that genes control individual steps in the process, because if a gene is ruined by a mutation, the reaction comes to a halt. So you see, on one side we have the reactions and on the other the genes.

ALKIMOS: Yes, I see.

EXTON: This is exactly the scheme proposed by Kühn. On one side we have genes, *Ephestia's A* and Drosophila's *vermilion*, *cinnabar*, and several other genes. On the other side are the stages of the reaction that lead to pigment production: tryptophan, kynurenine, and finally the pigment. The primary reactions that take place in the cytoplasm can be inserted into the scheme. For some time, scientists suspected that these primary reactions were directed by *ferments*, or *enzymes* that are determined by genes.

ALKIMOS: What are enzymes?

EXTON: Huge molecules made up of several hundred atoms, and they act as biological *catalysts*—this is a material that speeds up chemical reactions by several orders of magnitude. Studies of them began in the late 19th century, mainly by microbiologists. A certain enzyme is the catalyst of a well-defined reaction. This means that it comes into contact with the given chemical and accelerates a step in the reaction. Even though it is a participant in the reaction, the enzyme remains unchanged. So it can play its role over and over again.

ALKIMOS: So eye colour results from the influence of enzymes?

DEMOCRITUS: Most likely. And enzymes are under the influence of genes.

EXTON: Following Alfred Kühn's example, we can draw a more elaborate diagram of the process *gene → primary reaction → 'A' material → pigment*. Since we now know about many more genes and reactions, we can reshape the chain into a two-dimensional network. On the left side we have genes, in the middle enzymes, and on the right the elements in the reaction chain.

DEMOCRITUS: One gene determines one enzyme, and one enzyme one reaction.

EXTON: Kühn's scheme already contained the famous 'one gene, one enzyme' concept.

ALKIMOS: Famous? I've never heard of it.

EXTON: Kühn never had an opportunity to test his hypothesis. It was the 1940s, and his lab was barely able to function after many of his younger colleagues were conscripted and perished in the war. For Kühn this was the second time his work was affected by external realities. He'd already experienced the dismissal of his colleague Caspari years earlier in Gottingen by the powers that controlled Germany at the time. They had little respect for science, and none for humanity. Kühn paid dearly, although others paid even more.

ALKIMOS: What kind of powers? What kind of war? Not those Persians again?

EXTON: Pardon?

ALKIMOS: I knew it.

DEMOCRITUS: The Persians are excellent spear throwers. Not as adept with swords, however. I do hope Alfred had an opportunity to avenge his colleagues.

EXTON: My good friend, I suppose he did what he could. After the war he immediately contacted his colleagues who'd been forced to leave Germany and resumed collaborative work.

ALKIMOS: My Zeus, I was sure Alfred wouldn't give up!

EXTON: I suggest we now have a closer look at the one gene, one enzyme concept.

ALKIMOS: We're ready, Exton.

EXTON: Then let's visit George Wells Beadle. Hermes, the Deus ex Machina! Direction Stanford University, 1946!

DEMOCRITUS: Right away?

EXTON: Yes. All aboard!

ALKIMOS: Me too?

EXTON: Come on!

ALKIMOS: All right, all right. Aaggh ... here we go!

EXTON: We've heard a lot about Beadle and the eye pigments of fruit flies. At first he dealt with plant genetics, and then for a while he worked in Morgan's lab on the problem of recombination. In 1933, he made the acquaintance of Boris Ephrussi. Beadle approached the subject of genetics by conducting numerous transplantation experiments, which also established his reputation. He believed that questions of genetics should be examined in terms of chemistry, which is why he began working with a chemist, Edward Tatum. But let's hear the rest from Beadle. I think we've arrived.

ALKIMOS: It's about time.

EXTON: Come on. We're in the middle of the university campus. I think we need to go that way.

ALKIMOS: What a quiet, relaxing place. Beautiful trees.

EXTON: There it is! Let's go in.

ALKIMOS: Exton, wait!

DEMOCRITUS: Come on, son.

EXTON: This is Beadle's lab.

ALKIMOS: Ugh, look at all the mould!

EXTON: Dr. Beadle, allow me to introduce myself. I'm Professor Exton, from the University of Sussex.

BEADLE: Dr. Exton, welcome!

EXTON: These are my students, Mr. Alsmith and Mr. Dimatthews.

BEADLE: Nice to meet you. Please come in.

EXTON: Thank you.

DEMOCRITUS: Dr. Beadle, your research on the eye pigment of fruit flies is inspiring. We know that you faced many challenges, but I trust you have since managed to overcome them?

EXTON: In fact, that's why we're here. We'd like to hear about your latest achievements.

BEADLE: Of course, I'd be happy to discuss them with you. It's true that we had many challenges. Our work to isolate the eye pigments of fruit flies was slow and laborious. Tatum and I realised that, in most cases, we would have the same problems investigating other mutations and the chemical background of their effects.

ALKIMOS: So you knew you had a daunting task ahead of you.

DEMOCRITUS: But a promising one. After all, if you found the reactions that lead to pigment production, you must have been confident of finding the developmental reactions that lead to growth and form!

BEADLE: Gentlemen, I hate to disappoint you, but with pigments, we just got lucky. We had no reason to believe we'd have the same luck uncovering the chemistry of other reactions.

EXTON: You mean your research on eye pigment was successful because you made some lucky choices? Like Archibald Garrod, who was able to discover the cause of *alkaptonuria* only because chemists already had a thorough understanding of the chemical reactions behind it?

BEADLE: Yes. Homogentisic acid was identified years before Garrod's work, and so were the chemical pathways that lead to it.

ALKIMOS: What is homogentisic acid?

BEADLE: It's an intermediate product of the degradation of the amino acid tyrosine. It appears in large quantities in the urine of people with alkaptonuria. When their urine is exposed to air it turns black.

ALKIMOS: Like the bile of the Lernaean Hydra!

EXTON: In 1902, Sir Archibald Garrod, a British doctor and the founder of biochemical genetics, noted in his article on alkaptonuria that inheritance patterns of the disease correspond to Mendel's laws, assuming the mutation is recessive. According to Garrod, patients with alkaptonuria lack one of the enzymes that helps to break down tyrosine. As a result, homogentisic acid, one of the stages in the degradation of the amino acid is excreted in the urine.

ALKIMOS: Again we have a gene that affects a chemical reaction, so in the mutant this reaction doesn't work.

BEADLE: That's right, Mr. Alsmith.

EXTON: Garrod believed the diseases albinism and cystinuria were also the result of a recessive mutation that causes an enzyme deficiency.

BEADLE: These cases have helped us gain a basic understanding of how genes operate. My problem, however, is that these traditional approaches involve gruelling, painstaking efforts, before any results are achieved.

ALKIMOS: Is there another way?

BEADLE: Yes. You choose the reverse approach. Instead of taking well-known mutations and searching for the chemical reactions they affect, you start with well-known reactions and search for the mutations.

ALKIMOS: It sounds simple ...

BEADLE: All Tatum and I had to do was find the appropriate experimental system.

ALKIMOS: Fruit fly, moth, frog ovum, or intestinal worm?

BEADLE: None of them. We sought a living thing that did not have great nutritional demands; in other words, that was capable of carrying out reactions necessary to produce essential vitamins and amino acids. If we used mutagenic agents to induce mutations, we'd produce mutants unable to execute one or more of these basic, well-known reactions.

DEMOCRITUS: So what critter did you choose?

BEADLE: A sac fungus known as *Neurospora*.

ALKIMOS: Excuse me, but making mutants isn't a problem. Just ask Muller!

BEADLE: Indeed, you're right. We took advantage of the mutagenic effect of X-rays, which Muller discovered.

EXTON: Allow me to explain to my students about the life cycle of *Neurospora*.

BEADLE: If you think it's necessary ...

EXTON: I do. Gentlemen, please listen: neurospora is a bread mould. This fungus emerges from spores located in a sac known as *ascus*. If the spores land in an appropriate place they form a colony of threadlike structures. The fungus spends the majority of its life in the haploid state, which means it contains just one copy of every chromosome. From the spores, two types of mating colonies can develop, and if they are in close proximity they can fertilise one another. Both types of colonies produce male and female reproductive cells, so the fertilisation is mutual. After fertilisation, the nuclei fuse, and diploid cells are created. Then through meiosis and further mitosis, eight haploid spores are created in every ascus.

ALKIMOS: And then the process begins all over again.

EXTON: Yes. The huge advantage of *Neurospora* is that the products of a meiotic division are present in one ascus. Because it's easy to retrieve the spores, the separation of the alleles can be directly observed.

DEMOCRITUS: This means that we don't need to think about probability as we did in Mendel's and Morgan's experiments. We can see the alleles for ourselves.

ALKIMOS: Sounds promising.

EXTON: If for example we have a mutant mould, and we cross it with a wild type, then exactly four of the spores in the ascus will be mutant and four will be wild type. Moreover, the mutants and the wild types will line up next to each other on either side of the ascus.

BEADLE: Yes, this was a great help to us when we examined our mutants.

EXTON: And there's yet another big advantage. The spores and the *mycelia* that develop from them are haploid, so the mutants can be studied directly.

DEMOCRITUS: So, a recessive mutation exhibits its effect in a haploid mycelium too, since the chromosome carrying the mutant gene has no wild type companion.

BEADLE: That's why it was so easy to find mutations. We exposed the mould to X-rays or ultraviolet rays, and then allowed them to produce haploid spores. If a mutation occurred, we could identify it immediately in the first generation.

ALKIMOS: And what kind of mutations occurred?

BEADLE: We wanted to investigate the genetic background of metabolic reactions. After irradiating the mould, we waited for the spores to arise through meiosis, and then we placed them on substrate amply supplied with vitamins and amino acids. The spores happily grew. Then the offspring of individual—possibly mutant—spores, were transferred to a nutrient-poor medium.

EXTON: If I'm not mistaken this is known as a *minimal medium*. It contains only the sugars, salts, and vitamins that are absolutely essential for the fungus to grown, but nothing more.

BEADLE: Exactly. If a culture can't grow on a minimal medium, this means that it's not capable of producing one or more of the materials present in the enriched medium.

ALKIMOS: The reaction can't take place because the enzyme needed to speed it up is lacking. The mutation damaged the gene responsible for this enzyme.

BEADLE: After that, it was easy to determine which mutants had lost the ability to produce which material. We just had to grow the cultures in a variety of substrates, each containing an additional supplement of one nutrient.

DEMOCRITUS: If the supplemented nutrient restored the colony's ability to grow, then we knew the mutation damaged the gene responsible for its production.

ALKIMOS: So it's similar to the experiment with eye pigmentation.

EXTON: The principle is the same, but mould is much easier to use and leads to quicker results.

BEADLE: That's right. In less than five years we discovered one hundred mutations. Each one influenced the biosynthesis of just one vitamin, amino acid, or nucleic acid.

ALKIMOS: So it's not just one gene, one enzyme; it's one hundred genes, one hundred enzymes.

DEMOCRITUS: What a momentous discovery! There can no longer be any doubt that genes govern chemical reactions.

BEADLE: We were able to verify fairly easily that the mutations occurred in the genes. We paired the strains with wild-type colonies of the opposite mating type and examined the ascospores that arose.

DEMOCRITUS: If one gene is responsible for death by starvation, then according to Mendel's laws, four of the eight spores must be mutant, and four wild type.

ALKIMOS: And they neatly line up: four on one side, four on the other.

EXTON: If no recombining occurred.

BEADLE: Gentlemen, you are masters of genetics. I don't know if there's anything new I can teach you. Here, have a look at our ingenious culture tubes.

EXTON: Excellent!

DEMOCRITUS: Maybe we now understand the metabolic steps, but we still have no idea about the structure of the genes that regulate them. If we could clear up this mystery, we'd discover the true essence of life.

ALKIMOS: Dr. Beadle. I have one more question.

BEADLE: Of course Mr. Alsmith. I'm listening.

ALKIMOS: How did it feel when Alfred won the race?

BEADLE: Excuse me?

EXTON: Alkimos!

ALKIMOS: He found the eye pigments, and he was the first to recognise the one gene, one enzyme relationship.

BEADLE: Why, I don't know what you're talking about.

EXTON: I'm sorry Dr. Beadle. We must be going.

BEADLE: Already?

ALKIMOS: But he never had an opportunity to work with mould … you know, because of the Persians.

BEADLE: Why, I've no idea …

DEMOCRITUS: … you see, if we knew the structure, then we might understand how mutations damage the atoms of life …

EXTON: Alkimos, Democritus! Let's go!

BEADLE: Gentlemen, I'm still confused …

DEMOCRITUS: We've come this far. All we need to do is assemble a living thing. From atoms, we can make embryos; from embryos, we can make …

ALKIMOS: Wait! Dr. Beadle, don't you know about the Persians …

EXTON: Alkimos, now!!!

ALKIMOS: Aagghh, here we go!

The Protein–Genes

ALKIMOS: Master, my head hurts.

DEMOCRITUS: Rest a little.

ALKIMOS: Master? We've been gone for more than a week. They must be wondering where we are back home.

DEMOCRITUS: I'm afraid it's been more than a week.

ALKIMOS: And when are we going home?

DEMOCRITUS: I have no idea. Why don't you ask Exton?

ALKIMOS: Do we really have that much more to learn? I feel extraordinarily knowledgeable already.

DEMOCRITUS: I'm afraid now that we've started, we can't stop.

EXTON: Gentlemen, allow me to quote Erwin Chargaff: 'In science there is always one more Gordian knot than there are Alexanders.'

ALKIMOS: Then let's decide on a set goal, and it we achieve it, we give ourselves a longer rest.

DEMOCRITUS: Good idea. Our goal should be to uncover the secret of the atoms of life.

EXTON: In the early 1940s, Erwin Schrödinger, one of the greatest physicists of the 20th century, set the very same goal for himself.

ALKIMOS: A physicist?

EXTON: There were many great physicists among the researchers of the molecules of life: Schrödinger, Delbrück, Crick ...

DEMOCRITUS: As we get closer to the atoms, I suspect we'll have an increasingly greater need for physicists.

EXTON: In his book *What is Life?* published in 1944, Schrödinger formulated the question in this way: 'How can the events *in space and time* which take place within the spatial boundary of a living organism be accounted for by physics and chemistry?' To put it another way: what's the secret of the order and stability of living things?

ALKIMOS: You mean: why doesn't a frog dissolve in water like a sugar cube?

EXTON: The laws of statistical physics apply to large numbers of atoms or molecules. The movement, or kinetic energy of certain atoms can vary greatly, and their collective properties are described by thermodynamics. Therefore, the behaviour of individual molecules would be very difficult to describe precisely (in fact according to quantum mechanics it's not even possible). But a lot of atoms together behave according to an expected pattern.

DEMOCRITUS: And so the law of populations is born.

EXTON: The laws of diffusion are a good example. If we drip ink into milk, the individual ink particles will move randomly throughout the milk, going wherever they're pushed. But if we conduct a statistical study of the movement of all these particles together, we'll discover that they flow from areas of high concentration to areas of low concentration with such precision that it can be quantified.

DEMOCRITUS: What about genes? We know that they remain unchanged as they're passed on from generation to generation, even though in each individual cell only two copies of the gene are present. Why don't genes disintegrate in response to random thermodynamic forces? What holds genes together?

EXTON: Schrödinger stated that over time only crystals or atoms that form Heitler–London type chemical bonds can resist thermodynamics. The Heitler–London theory describes how atoms can form molecules by covalent bonds that are powerful enough to hold them together despite the destructive forces of thermodynamics. The molecules are thus stable like crystals.

ALKIMOS: The chromosomes are the stable crystals of genes?

EXTON: Yes, but not according to the general definition of crystals. The crystals that were known at that time all had generally regular, repeating structures; that is to say, specific atoms were located at the various points on the grid. The crystals of the inorganic world are periodic.

However, according to Schrödinger, the molecules of genetic material are *aperiodic crystals.*

DEMOCRITUS: Chromosomes are aperiodic crystals?

EXTON: In Schrödinger's words: "A small molecule might be called 'the germ of a solid.' Starting from such a small solid germ, there seem to be two different ways of building up larger and larger associations. One is the comparatively dull way of repeating the same structure in three directions again and again. That is the way followed in a growing crystal. Once the periodicity is established, there is no definite limit to the size of the aggregate. The other way is that of building up a more and more extended aggregate without the dull device of repetition. That is the case of the more and more complicated organic molecule in which every atom, and every group of atoms, plays an individual role, not entirely equivalent to that of many others (as is the case in a periodic structure). We might quite properly call that an aperiodic crystal or solid and express our hypothesis by saying: We believe a gene—or perhaps the whole chromosome fibre—to be an aperiodic solid."

DEMOCRITUS: The aperiodic structure contains the possibility of endless variation. So aperiodicity ensures the variety of the hereditary material, while the crystal ensures the stability of inheritance. Astounding!

EXTON: Indeed. If every section of the crystal is unique, then we can construct a never-ending genetic text.

DEMOCRITUS: Language works in a similar way. The arrangement of letters and words also offer an almost endless number of possibilities.

EXTON: Moreover, the genetic text in a crystal can occasionally change, thus creating a mutation.

ALKIMOS: Ah, it makes the genetic text mumble–jumble.

DEMOCRITUS: Or just alters its meaning.

EXTON: The idea that a chromosome is an aperiodic crystal containing a code is fascinating. Schrödinger's book, *What is Life*, is among the most important scientific works of all times. It inspired numerous scientists to attempt to uncover the molecular secrets of inheritance. It's time for us, too, to examine the chemistry of genes, or to be more precise, the chromosomes that carry them, to determine the molecular structure of genes.

DEMOCRITUS: Is it even possible? We saw what a great problem it was for Kühn and associates to isolate just one pigment molecule. What chance do we have of discovering the crystal structure of aperiodic chromosomes?

EXTON: Indeed, only meticulous work carried out over generations can help us accomplish the task.

DEMOCRITUS: But I'm afraid we're still a long way from that.

EXTON: Not at all! The work has already been done! Let's have a look at the history of the chemistry of inheritance.

ALKIMOS: Master, your atoms of life! That's what you've been waiting for!

DEMOCRITUS: Have we arrived at last?

EXTON: Obviously, the nucleus and its chromosomes had to be extracted from the cells and subjected to chemical analysis. The Swiss chemist Friedrich Miescher experimented with this in 1869. He sought a tissue that contained lots of nuclei and little cytoplasm. As you'll see, his choice was clearly based on scientific and not aesthetic criteria. He began to collect and study the pus that accumulated on surgical bandages. He extracted the nuclei from the white blood cells in the bandages by bathing them in alcohol and hydrochloric acid. Then he cleaned away the proteins in the cytoplasm with an extract from the stomach of a pig.

ALKIMOS: That sounds like Paracelsus' recipe!

EXTON: Miescher's procedure was based on far more sound scientific principles. The extract from the pig's stomach, for example, contained digestive enzymes that broke down the proteins.

ALKIMOS: What are proteins?

EXTON: We'll discuss that in a minute. Meischer dissolved the nuclei in a dilute alkaline solution. At the end of the procedure he had sediment consisting of fine flakes. He called this material *nuclein*, derived from the word nucleus. The slightly acidic material contained not only carbon, hydrogen, and oxygen atoms, but also nitrogen (14%), phosphorous (6%), and sulphur (2%). Miescher believed he'd found a new material never before known to science. He was right.

ALKIMOS: Nuclein is the material of genes?

EXTON: Oskar Hertwig certainly thought so. Here's what he said: 'I believe that I have at least made it highly probable that nuclein is the substance that is responsible not only for fertilisation but also for the transmission of hereditary characteristics.'

DEMOCRITUS: Aren't we jumping to conclusions? After all, the pig stomach extract may have washed away all the other material from the isolated nuclei leaving only the nuclein. How do we know the genes survived the rather harsh treatment they were subjected to?

ALKIMOS: We can't know …

EXTON: You're right. The best we can do is further examine the materials in the nucleus. In 1889, Richard Altmann showed that the nuclein contained a small amount of protein in addition to the acidic material which he called nucleic acid.

DEMOCRITUS: Miescher's extract was not entirely pure?

EXTON: It's no surprise given the difficulty of the circumstances and the rudimentary nature of his methods. Meischer's greatest achievement was laying the groundwork for the chemical analysis of nuclei.

ALKIMOS: At least he had a method that could be perfected.

EXTON: Yes, and this process of perfecting it ultimately revealed that the main components of the nuclei are protein and nucleic acid.

DEMOCRITUS: And what is the nature of these mysterious substances?

EXTON: They can be found in all living things. By the middle of the 18th century, scientists had begun to study proteins. The first extracts were made from flour, eggs, blood, and other accessible sources. By the 19th century, scientists knew that proteins were enormous molecules containing nitrogen.

ALKIMOS: That means proteins contain lots of atoms bound together.

EXTON: Thousands of them. In 1885, Oscar Zinoffsky determined the atomic makeup of one of the main proteins in blood, *hemoglobin.* According to his measurements, this molecule was composed of 712 carbon atoms (C), 1130 hydrogen atoms (H), 214 nitrogen atoms (N), 245 oxygen atoms (O) and one iron atom (Fe). Its molecular formula is $C_{712}H_{1130}O_{245}N_{214}S_2Fe$. Chemists slowly realised that proteins are made up of smaller building blocks called *amino acids*. It took more than one hundred years to discover all twenty of them. Nevertheless, it was immediately apparent that proteins from difference sources had different amino acid components. When certain protein extracts were broken down, the resulting samples revealed a variety of amino acids in varying ratios. Heinrich Ritthausen posited that these kinds of differences could not be attributed to a random event. Clearly such results were obtained because the proteins used at the start were different from one another.

DEMOCRITUS: Proteins are large molecules composed of a variety of amino acids.

EXTON: The amino acids all have similar structures. NH_2–CHR–COOH. The letter 'R' indicates the acid's side chain. This is how the various types differ from each other.

ALKIMOS: Theme and variation.

EXTON: According to Franz Hofmeister, a protein scientist born in Prague, chains are constructed of amino acids with specific chemical bonds. In the building of these chains, always the same atoms of the amino acids take part. The NH2 (amino) group always links to the COOH (carboxyl) group, creating a peptide bond with the structure –CO–NH. A piece of this chain looks like this:

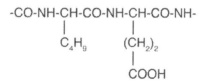

DEMOCRITUS: If I understand correctly, the chain can continue infinitely.

EXTON: At least for very long. That's how giant molecules are made.

ALKIMOS: My Zeus, that's incredible.

DEMOCRITUS: If the atoms join to form chains, then they can resist the forces of thermodynamics. The varying amino acid makeup of proteins suggests that the chains are also varied. Or aperiodic, as Schrödinger would say.

EXTON: The importance of proteins becomes even more obvious if we consider that enzymes, the biological catalysts of chemical reactions, are also proteins.

ALKIMOS: So the gastric juices of a pig contain digestive proteins?

EXTON: Yes. Protein digesting proteins, in fact. Digestion, the breaking down of food, is also a catalytic process.

DEMOCRITUS: So the digestive enzyme breaks down proteins, and it is a protein itself. So indeed, it's as you say: proteins that digests themselves.

EXTON: Yes, that can happen.

ALKIMOS: An enzyme digests an enzyme.

DEMOCRITUS: The snake bites its own tail.

ALKIMOS: Exton, what about the other component of the nucleus, the nucleic acid?

EXTON: The study of nucleic acids progressed more slowly. Their structure at first appeared rather simple and 'monotonous', which is how Phoebus Levene, a Russian chemist working in New York, described them after conducting many experiments that gradually shed light on their internal order. In the early 20th century, it became clear that nucleic acids were made up of four different types of components: a sugar, purine bases, pyrimidine bases, and phosphate groups. In nucleic acids, the phosphate groups, the sugar and one base link together in a specific order. Levene called the phosphate–sugar–base trio a *mononucleotide*. When nucleic acids were carefully broken down, these mononucleotides were obtained. This suggested that nucleic acids were constructed of these units to create *polynucleotides*.

ALKIMOS: New molecular chains!

EXTON: The backbone of the chain is made of alternating phosphate and sugar molecules. The purine and pyrimidine bases attach to the sugar. There are two kinds of purine bases: *adenine* and *guanine*. And two kinds of pyrimidine bases: *cytosine* and *thymine*. Levene tried to discover the proportions of these materials in the nucleic acid. Based on some early investigations, it seemed that the molecules contained even amounts of the four kinds of bases. Levene discovered that the sugar was none other than *deoxyribose*. Levene published a model that corresponded perfectly to reality at the level of the monomers and named the molecule *deoxyribonucleic acid* (DNA). In his *tetranucleotide model* the four kinds of bases formed a tetramer. These identical tetramers then were repeated over and over again. In addition to DNA, another nucleic acid form with a similar structure also appeared in great quantities. This molecule, however, had *uracil* in place of thymine and the sugar was *ribose*. Levene called this form *ribonucleic acid* (RNA). I've drawn the structure of the five nucleotide monomers for you.

Adenine	Thymine	Guanine	Cytosine	Uracil

DEMOCRITUS: DNA and RNA are chains consisting of four different kinds of nucleotides in even amounts. These four nucleotides repeat in a regular pattern over the entire length of the chain. So these molecules aren't aperiodic. This means we'll have to search for our genes elsewhere, maybe in the area of self-producing molecules capable of autocatalysis.

ALKIMOS: Self-producing or self-digesting molecules. Enzymes!

EXTON: When discussing crystals, Leonard Troland—he was mentioned by Muller—argued that enzymes work as autocatalytic molecules. In his article, 'The chemical origin and regulation of life,' published in 1914, he wrote the following: 'We have said that enzymes and catalysts in general have the power to assist in the production of specific chemical substances. Now there is no reason why the same enzyme should not aid in the formation of more than one and also why one of these substances should not be identical with the enzyme itself. A process of the last mentioned variety, in which the presence of a catalyst in a chemical mixture favours the production of the catalyst itself is known as autocatalysis.'

ALKIMOS: The snake doesn't just eat itself, but also creates itself.

DEMOCRITUS: Like genes! Muller showed us that because genes are capable of replicating, they must certainly possess the power of autocatalysis.

ALKIMOS: The autocatalysis of genes means that a gene, using itself as the mould, can create more of itself?

DEMOCRITUS: This is what all our evidence suggests.

ALKIMOS: Master, so a gene must be an autocatalyst.

DEMOCRITUS: Yes, so it seems. We've seen that enzymes work as catalysts.

EXTON: And viruses, too. At least that's how it appeared initially. In 1935, Wendell Stanley managed to crystallise the tobacco mosaic virus. The crystal material was capable of infecting even when diluted a hundred million times, and it seemed to be made up entirely of proteins. According to Stanley, '[the] tobacco mosaic virus is regarded as an auto-catalytic self-reproducing protein which, for the present, may be assumed to require the presence of living cells for multiplication.'

DEMOCRITUS: An autocatalytic protein capable of infecting and repro-ducing. The d'Hérelle particle, the gene itself! Moreover in a pure and crystalline form!

ALKIMOS: Incredible! Life is engraved in crystals!

EXTON: Later the virus crystals were found to contain some ribonucleic acid …

ALKIMOS: Just some contamination.

DEMOCRITUS: Yes, most likely. It's beyond a doubt that genes themselves are proteins. The complex makeup of proteins, the infinite variety of their structure, their ability to catalyse, and the protein composition of viruses all clearly point to that. We must consider them the true bearers of life. Within each one, *vis vitalis* fuses with infinite variety, inheritance with the chemical reactions of development, replication on the molecu-lar level with catalysis. This fact is the greatest discovery of genetics!

ALKIMOS:

> *How all our proteins seethe, pulling the yoke of life,*
> *Cast from its own mould, the enzyme copies itself.*

DEMOCRITUS: Autocatalysis is the essence of life. This is why like things produce like things. This is how life creates itself.

EXTON: Troland asserted the same thing. According to his argument, the origin of life can be traced back to the appearance of the first autocatalytic molecule. If such a molecule ever comes into being, it can produce a tremendous number of like copies. These molecules can reproduce, have inheritance and can also change because of mutations that occur during replication.

DEMOCRITUS: An autocatalytic molecule can correspond to the three basic requirements of evolution: reproduction, inheritance, and variation.

ALKIMOS: So life is an enzyme? Or an enzyme is life? Interesting …

EXTON: Interesting indeed. Later I'll have a few words about that. You see, it seems we've forgotten about the nucleic acids.

DEMOCRITUS: Is that so surprising? You said it yourself—those molecules are extraordinarily monotonous. They have only four components and all in the same proportions. No, their structure offers little reason to think they have any significance. Yes, they make up a portion of the nucleus, but all our evidence suggests these molecules are nothing but the framework of the chromosomes.

ALKIMOS: Yes, they're as simple as a Thracian peasant.

DEMOCRITUS: Indeed. Strong and resistant.

ALKIMOS: Master, do you suppose that the nucleic acids hold the chromosomes together in one thread?

DEMOCRITUS: Yes, that's exactly what I think. They form the thread upon which life fastens its pearls.

ALKIMOS: Protein, genes!

DEMOCRITUS: This framework keeps the genes together, and the copies that are created through autocatalysis break from the chain and make their way into the cytoplasm, where they direct chemical reactions. And during cell division, too, the chromosomal proteins replicate through autocatalysis.

EXTON: You have no idea how close you are to the truth.

ALKIMOS: Exton, are you complimenting our reasoning skills?

EXTON: Indeed I am.

The Principle of Transformation—*Oswald Theodore Avery*

DEMOCRITUS: Isn't science wonderful? We've made so much progress in understanding the chemistry of life. Just a few generations earlier scientists couldn't have dreamed it was possible. But now we've uncovered the atomic structure of genes. The incredible discoveries of chemistry have revealed that proteins contain the variation required by life— Schrödinger's aperiodicity. This resolves the question of gene replication, because enzyme-genes are capable of producing copies of themselves and thus perpetuating the germplasm.

EXTON: In the 1940s, the general view was similar in scholarly circles. It seemed obvious that proteins and genes were identical. The greatest question had been answered.

ALKIMOS: Finally! Now we can rest.

DEMOCRITUS: Let's apply it to what we know about living things—let's devise a theory that explains inheritance on the atomic level.

EXTON: That's fine with me. But first we need to address a few questions.

DEMOCRITUS: Such as?

EXTON: Although enzymes seem to resolve the question of how genes can be autocatalytic, another possibility exists in principle.

ALKIMOS: Why should we bother with it?

EXTON: You'll see. It's conceivable, after all, that genes don't actively copy themselves, but serve as models, and the copying is performed by other enzymes.

DEMOCRITUS: But there's no need for that. The principle of Occam's razor is that the simplest solution is always the best.

EXTON: Let's rehash what needs to happen when a gene replicates.

ALKIMOS: Oh, I suppose, if you insist ...

EXTON: Trust me, Alkimos, it'll be worth it. One thing is clear: a gene is an enormous molecule made up of a chain of subunits. In his article of 1936, Hermann Muller refined the views he shared with us on how genes replicate. Like Schrödinger later, he considered the gene to be a macromolecule built from various subunits. According to Muller, when the chain replicates, every single building block draws to itself an identical substance present in the cytoplasm. These elements that are attracted to it then form links, creating a new chain, a copy of the gene. J. B. S. Haldane, in his article of 1938, slightly modified this explanation. He believed that the building blocks of genes did not attract exact copies to themselves, but rather their negatives. The copy of the gene was therefore an anti-gene.

ALKIMOS: Like a bronze statue and its mould.

EXTON: Something like that. One of the greatest chemists of the 20th century, Linus Pauling, argued the same thing. He reached this conclusion after pondering the nature of the interactions that weakly hold molecules together. He introduced the expression *complementarity* to describe the relationship between molecules. The copy of a gene is in fact its complement.

DEMOCRITUS: So genes produce complementary genes, and the complementary genes then duplicate to form a complement of themselves …

EXTON: Which are identical to the original gene.

DEMOCRITUS: So if genes are proteins, than both the genes and their complements must work as enzymes? Interesting.

EXTON: Quite so. But how could the mould and the statue both be capable of catalysis?

ALKIMOS: That's a problem.

DEMOCRITUS: Maybe we shouldn't expect so much from atoms …

ALKIMOS: Master, does that mean Occam's razor doesn't work?

EXTON: We might do well to reject the theory of the autocatalytic nature of genes, in other words the notion that genes and enzymes are one and the same. We need a new theory. Since it's highly unlikely that genes and their complements work as enzymes, it would be simpler if we considered genes to be passively replicating aperiodic chains.

DEMOCRITUS: The copiers, however, are certainly enzymes. They assemble and transport the building blocks, and they link them together into chains.

EXTON: Since genes are just passive participants in replication, the process is not autocatalytic.

ALKIMOS: Exton, you're confusing us! Then what was the point of the whole autocatalytic theory?

EXTON: What's the point of our entire journey through the history of genetics? My task is not to give you knowledge set in stone, but to show you the sweat and toil that led us to the knowledge. If you follow the path to its end, you will far better understand everything and develop a great admiration for the methods of science. That's my only goal.

DEMOCRITUS: It's true we've learned a lot from our mistakes, just as all respectable scientists have. What tangles and thorns we've struggled with, but you're right; our appreciation of the discovery of the one and only path out of the morass is all the greater. After all, we've done it! There it is! The aperiodic chain is the guiding principle of life! It determines what an organism becomes. An exact copy of the chain's aperiodicity is made when the chromosomes replicate thanks to the attraction between the chain's subunits and their complements. The new chain contains the negative of the original chain's regulatory content, but another copy is produced, a negative of the negative, which is thus identical to the original. Enzymes are needed for the construction of the chain and the generation of subunits. These set the regulatory chain in motion.

EXTON: The chain's regulatory content is encoded in its aperiodicity.

ALKIMOS: This aperiodicity is what is needed to create the different types of organisms.

DEMOCRITUS: Aperiodicity is the secret to an organism's form. We've finally found the key! All our evidence tells us that genes are giant molecules composed of chains of proteins.

EXTON: Indeed all our evidence points to this. The time has come for us to visit Oswald Theodore Avery. He was the first who managed to isolate and identify the chemical bearer of this hereditary information.

ALKIMOS: I can hardly wait!

EXTON: Hermes, prepare the Deus ex Machina! Direction New York, 1944!

ALKIMOS: Exton, we were just in the year 1946.

EXTON: Chronology doesn't matter. Rest assured, we're progressing in a logical order.

ALKIMOS: Whose logical order?

EXTON: Why yours of course. Come on, squeeze in.

ALKIMOS: Aagghh, here we go!

EXTON: On the way, let's discuss the forerunners of Avery's experiments. We're going to talk about *bacterial transformation*.

ALKIMOS: Good. Keep talking! It distracts me from all the shaking.

EXTON: Avery worked with the bacteria *Diplococcus pneumoniae*, which causes pneumonia. At the time, tens of thousands of people were dying of it every year in America, so he was trying to develop a vaccine against it. But in the mid-1930s, some other scientists developed sulfonamide antibiotics which were very effective against the disease; his vaccine research was considered unnecessary. Avery, however, refused to abandon his studies of pneumococcus, but he did gradually take his investigations in another direction.

ALKIMOS: Genes?

EXTON: The *transforming principle*.

ALKIMOS: Excellent!

EXTON: His studies were based on the research of Frederick Griffith, a London doctor, in the 1920s. Griffith began to examine two already known strains of the pneumococcus bacteria. One strain, known as the *R* (rough) type, forms rough-surfaced colonies on the growth plates, while the other, the *S* (smooth) type, grows into much larger, smoother colonies. The reason for the difference in the surface texture is that the *S* type forms a polysaccharide capsule which ensheaths the cell. What was particularly interesting was that only the *S* strain was capable of infecting the host. It appeared that the polysaccharide capsule helped the bacteria adhere to the lung tissue.

ALKIMOS: Then the *R*-type must be a mutation. Let's call it *Sugar-free*!

EXTON: As you wish. Griffith wasn't interested in the name. He began to study how each strain could transform into the other in mice. He mixed live bacteria with bacteria killed by heat and then injected the mixture into the animals. A couple of weeks later he examined which bacteria had proliferated. Much to his surprise, when the mixture he injected into the animal contained dead *S* bacteria and live *R* bacteria—remember, the *R* is incapable of causing infection—the *S* bacteria proliferated.

ALKIMOS: The *S* bacteria were resurrected?

EXTON: Well, that's one explanation. Perhaps the living *R*-type bacteria were able to revive the dead *S* bacteria in the mouse. The other possibility is that the presence of the dead *S* bacteria made the *R* bacteria capable of producing the polysaccharide necessary for it to spread in the animal. He soon realised this second possibility was correct. The *R* bacteria transformed under the influence of some material contained in the dead *S*.

DEMOCRITUS: Ah, so that's the transforming principle!

ALKIMOS: But how is it possible?

EXTON: That's exactly what Avery wanted to know. He resolved to identify the transforming material contained by the dead bacteria. Oh, we're here!

ALKIMOS: Finally!

DEMOCRITUS: This is New York?

EXTON: Yes, but we're already inside the Rockefeller Institute. There was too much traffic outside.

ALKIMOS: But where exactly are we?

EXTON: We're in the chemical storage room of Avery's lab. The exit's over there.

ALKIMOS: I've never seen so much equipment.

EXTON: No surprise. Avery's lab was among the best equipped of its time. Don't touch anything!

ALKIMOS: We wouldn't think of it.

EXTON: Professor Avery, greetings.

AVERY: Welcome, gentlemen. Please come into my office. Excuse me, it's a little cramped in here …

EXTON: Oh, that's all right.

AVERY: I suspect you're here to discuss my experiments with the transforming material.

EXTON: That's right, as I wrote in my letter.

AVERY: Well then, let's start at the beginning. I'm sure you're already familiar with the basic experiments on the transformation of pneumococcus.

DEMOCRITUS: We were just talking about Griffith's enormously exciting experiments.

ALKIMOS: How the *S* bacteria died when exposed to heat, but the property that makes them *S* didn't!

AVERY: Yes, that's at the heart of transformation. In the transformed *R* cells, synthesis of the characteristic polysaccharide capsule takes place under the influence of the transforming material from the dead *S* cells.

EXTON: We know that you and your colleagues, Colin MacLeod and Maclyn McCarty, managed in the end to identify the material necessary for transformation.

AVERY: That's right. Our research focused on producing a pure extract of the transforming material. After that, we were able to induce transformation *in vitro*. This made our job much easier.

ALKIMOS: So you didn't need to inject mice?

AVERY: No. We were able to do everything in an appropriate bacterial culture.

EXTON: I imagine you were able to work much more quickly.

AVERY: Yes, we were.

EXTON: Professor Avery, please tell us how were you able to produce and identify the material?

AVERY: Making the pure transforming agent from the bacterial culture was a long and challenging process. We had to subject the material to heat and treatments with alcohol, enzymes, and chloroform. The end result was a colourless and viscous solution.

DEMOCRITUS: If I understand correctly, you managed to extract an uncontaminated material from the *S* bacteria, and this all by itself was able to transform the *R*-type?

AVERY: To be precise, it turned the *R* back into *S*. The *S* bacteria that we obtained from transforming *R* corresponded in almost every respect to the *S*-type from which the transforming agent was retrieved.

ALKIMOS: What must happen is the *S* bacteria's capsules move to the *R* bacteria, and these turn into *S* bacteria.

DEMOCRITUS: Yes, that's the most obvious answer.

AVERY: Except that it's not supported by the evidence. When we cleansed the transforming material of impurities we used an enzyme that broke down the polysaccharide capsule completely. When we subjected the extract to serological tests, we found no traces whatsoever of the capsule.

ALKIMOS: Serological study?

EXTON: Serology is the study of antibodies in the blood plasma. Antibodies are the proteins or groups of proteins that are produced by the *immune system*, which protects the body against disease. These molecules recognise the chemical properties of certain molecules and bind to them very specifically. You could say, antibodies are fastidious, each one recognising only one kind of molecule or molecular unit and binding only to that.

ALKIMOS: Blood plasma is full of those?

EXTON: Yes.

ALKIMOS: But what do they bind to exactly?

EXTON: Foreign molecules. Proteins, polysaccharides, whatever. This is how the immune system protects the body from invaders: from bacteria, fungus, viruses, and of course serologists, since they have a particular fondness for injecting foreign molecules into bodies. Usually they inject a substance to which they'd like to find an antidote. The immune system can be depended on to fiercely resist its attackers by generating copious amounts of antibodies.

ALKIMOS: Let me guess, as soon as the animal starts to defend itself, the serologist comes along and chops its head off.

EXTON: Or to be more precise, draws some blood. What the serologist obtains is a valuable serum containing large quantities of antibodies specifically produced by the animal to bind the injected protein or polysaccharide.

AVERY: That's right, Exton. This is exactly how we acquired the serum with the specific antibodies to recognise the capsule's polysaccharides. The antibodies worked so well, that even when we diluted the serum 6 million times, they were still effective.

DEMOCRITUS: Therefore we can safely conclude the extract of transforming material was truly free of polysaccharides. This means the capsules couldn't be responsible for the transformation of *R*.

AVERY: It also suggests that the *S* bacteria responsible for the transformation of the *R* bacteria can pass on their characteristics.

DEMOCRITUS: Aha, so the *S* bacteria don't pass on their capsule, only their ability to make the capsule.

AVERY: That's what we think.

ALKIMOS: Okay, so if the capsule is not passed on, then what is? An enzyme that makes the capsule?

DEMOCRITUS: That seems the most likely explanation. We know that enzymes act as catalysts; that is, they speed up chemical processes, but they are never changed by the reactions. Therefore, very little of them is needed, since they can be used over and over again in the reactions.

EXTON: Griffith argued something similar. He thought a protein from the dead bacteria migrated to the live one and began producing the capsule there.

AVERY: We were inclined to believe that too, but using the Biuret test, we determined that the pure preparation of the transforming material contained no proteins.

ALKIMOS: Master, you said only a small quantity of enzymes is needed for the chemical reactions.

DEMOCRITUS: That's true. Perhaps the Biuret test was not sensitive enough to show the presence of a miniscule amount. This means we still can't rule out the possibility of a protein causing the capsule formation.

AVERY: Of course we weren't satisfied with our results either. We decided to treat the extracts with two enzymes, trypsin and chymotrypsin, which digest proteins, but this had no effect on the transforming material's ability to induce the creation of capsules.

DEMOCRITUS: How? The results should have been the opposite. If the transforming principle is a protein, the two enzymes should have destroyed its transforming capabilities.

AVERY: But that's not what happened.

ALKIMOS: But couldn't the transforming principle be a protein that is resistant to those two enzymes? After all, if it's clever enough to transform, why wouldn't it be clever enough to outsmart some enzymes?

AVERY: Okay, but how would you explain the following observation: when we treated the extract with an enzyme that we'd proven digests deoxyribonucleic acid, the transforming effect vanished.

DEMOCRITUS: Deoxyribonucleic acid?

EXTON: Yes. And the enzyme is *deoxyribonucleic acid depolymerase*, or DNase.

AVERY: This means the transforming agent is none other than deoxyribonucleic acid. We were able to verify this using several chemical methods. The atomic makeup of the extract, its molecular density, and migration in an electric field all point to this. Our tests done with enzymes allow no other interpretation.

DEMOCRITUS: So, deoxyribonucleic acid is the transforming principle? It's what makes the cells of *R* capable of producing a polysaccharide capsule? How?

AVERY: I'm afraid, I don't know either. But the conclusion we've drawn based on the experiments is that the deoxyribonucleic acid extract from the *S*-type pneumococcus bacteria can transform bacteria without capsules into *S*-type polysaccharide encapsulated bacteria.

ALKIMOS: What a simple formulation.

DEMOCRITUS: DNA, however, no longer appears to be as simple as we thought.

AVERY: Oh, it's not at all. We were astounded to discover that nucleic acids possess a variety of qualities. The very specific effect of the transforming principle, of deoxyribonucleic acid, in the pneumococcus bacteria proves this.

DEMOCRITUS: Specificity and the ability to produce a capsule, in other words this regulatory force, demonstrate that the transforming principle is the gene itself.

AVERY: In his recent book, Dobzhansky wrote something along the same lines. He also compared the transforming principle to a gene, and the polysaccharides of the capsule to the product of a gene.

ALKIMOS: I don't understand, master. Didn't we swear not too long ago that genes were enzymes?

DEMOCRITUS: We'll have to reinterpret nucleic acids and the role of proteins, now that we know that nucleic acids ...

AVERY: At least the deoxy form of them ...

DEMOCRITUS: ... have different specificities. Phoebus Levene's model must therefore be erroneous. We need to find the key to the information bearing abilities of the seemingly simple nucleic acid.

EXTON: So you see, gentlemen, today was worth the effort. We've discovered the bearer of biological information.

ALKIMOS: DNA.

DEMOCRITUS: The principle that regulates proteins, and, in so doing, life.

ALKIMOS: I think we really have earned a rest.

EXTON: We can rest on the way home.

DEMOCRITUS: Are we leaving already?

EXTON: Professor Avery, thank you for your time.

AVERY: You're very welcome.

EXTON: Now, let's go.

DEMOCRITUS: Could we take a jar of the atoms of life? Or DNA. I'd like to finally see them with my weary eyes.

ALKIMOS: Professor, could we have just one test tube of ...

AVERY: Hey, what are you doing? We're using that to develop a vaccine. It contains a virulent strain!

EXTON: All aboard!

ALKIMOS: Aagghh! Here we go ...

The Triple Helix—*Linus Pauling*

ALKIMOS: Master?

DEMOCRITUS: Yes, son?

ALKIMOS: It's been a long day.

DEMOCRITUS: Yes, it has.

ALKIMOS: An aperiodic day.

DEMOCRITUS: Somewhat, yes.

ALKIMOS: It's left me confused.

DEMOCRITUS: How so?

ALKIMOS: We've learned a lot about DNA, but I can't make sense of it. We were convinced genes were proteins, and now we see that …

EXTON: Avery's contemporaries weren't sure either what to make of his results. Some simply ignored them, while others sought reasons to reject them, and they had a fairly easy time of it. After all, no one knew whether his findings were generally applicable. Maybe DNA was the stuff of genes only in the pneumococcus, but in other organisms proteins were still the masters. It's also possible that the tiny amount of protein contamination in the extracts was responsible for transformation.

ALKIMOS: But Avery and his team performed every possible experiment to verify their results.

EXTON: Yes, their work was flawless. That wasn't the reason scientists objected to Avery's conclusions. The problem was the concept of naked, protein-free DNA didn't fit into the scientific world view of the time. Nevertheless, more and more scientists began to grapple with the puzzle of DNA.

ALKIMOS: Like my master and me.

EXTON: Other observations also helped to excite interest. Max Delbrück's 'Phage Group' which examined bacteriophages, had a great impact. A theoretical physicist by training, Delbrück turned his attention to biology in hopes of discovering the nature of life through the investigation of very simple, replicating particles.

ALKIMOS: Oh yes, Muller also thought the study of gene-like, bacteria-eating particles had a lot of promise. Delbrück made a good choice.

EXTON: It seemed that way at first. But he wasn't the one to unravel the mystery of genes. He quickly discovered, however, that bacteriophages weren't nearly as simple as he thought. Under the electron microscope, which has powers of magnification far greater than the light microscope, bacteriophages looked like complex viruses with a head, tail, and legs.

ALKIMOS: So genes have a head, tail, and legs?

EXTON: No, genes don't. But phages do. In other words phages are not merely naked genes. They're coated in proteins, and their DNA is located within this coating.

DEMOCRITUS: If phages are not as simple as we thought, then we aren't going to find an answer quickly to the questions of gene structure and functioning.

EXTON: At least not by studying phages. Still, an interesting study of bacteriophages finally managed to direct the attention of scientists to DNA. During infection, the phage, which resembles a syringe, adheres to the bacteria. It then injects the bacteria with some sort of substance, but the protein coat remains outside the cell.

ALKIMOS: It injects the DNA encased in the protein coat!

EXTON: That's right. Alfred Hershey and Martha Chase showed this experimentally in 1952. The protein-coated bacteriophage injected its DNA into the cell. The DNA then acted alone in directing the cell to create copies of the bacteriophage. This new understanding of DNA finally persuaded the scientific community that DNA was a hereditary material.

ALKIMOS:

> *DNA-shooting phage able to generate its lordly self,*
> *And shamefully defile the bacteria's cold gutted body.*

EXTON: That's a slight exaggeration. In any case, the hearty bacteriophage needs only half an hour for its offspring to devour a bacterium. But enough about phages. Their time hasn't come yet to unlock the secrets of genes.

ALKIMOS: Too bad. They're a great subject for a poet. Proliferation, murder, and DNA injections. It couldn't be better!

EXTON: Yes, well, DNA itself is a great subject—for research anyway. The Austrian-born, American biochemist Erwin Chargaff realised this after learning about the nature of the transforming principle. In admiration of Avery's work, he declared, 'Avery gave us the first text of a new language, or rather, he showed us where to look for it. I resolved to search for this text.'

ALKIMOS: The text of aperiodic DNA!

EXTON: Chargaff realised Avery's discovery was incompatible with the tetranucleotide model proposed by Phoebus Levene in 1933. If DNA was the bearer of genetic information, then its structure could not be as monotonous as Levene asserted: 'If different DNA species exhibited different biological activities, there should also exist chemically demonstrable differences between deoxyribonucleic acids.' Chargaff repeated the measurements upon which Levene's model was built. He also calculated the exact quantity of the four nucleotides in several samples of DNA from various species. His results contrasted sharply with Levene's: he discovered their quantities varied significantly from species to species.

DEMOCRITUS: So the tetranucleotide model can't be universally applied, because that would mean the nucleotides always occurred in equal amounts.

EXTON: That's right. At the same time, Chargaff noticed a surprising regularity. No matter what tissue he examined, the quantity of adenine (A) matched that of thymine (T), while the quantity of guanine (G) matched that of cytosine (C).

DEMOCRITUS: So DNA always has the same amount of adenine and thymine, and also the same amount of guanine and cytosine, but the ratio of one pair to the other can change. Therefore, we need to modify Levene's tetranucleotide model. We can assume that adenine and thymine are always in pairs, and so are guanine and cytosine. So a DNA molecule could look like this: AT–GC–GC–GC–AT–AT–GC–AT–AT.

ALKIMOS: Master, why AT and GC? Why can't it be TA and CG?

DEMOCRITUS: You're quite right—why not? The pairing of bases, whether we call them AT or TA, or GC or CG, are responsible for the regularity observed by Chargaff. The ratio between the two pairs, however, is arbitrary, which is why it differs in the various samples.

ALKIMOS: We have a new DNA structure. We can call it Democritus' nucleotide-pair model!

DEMOCRITUS: Naturally, the chain can be aperiodic. This is why it contains information that determines capsules made of sugar and other characteristics.

ALKIMOS: Amazing!

EXTON: It truly is. Except, unfortunately, it's not entirely accurate.

ALKIMOS: Even my master can make mistakes, I suppose …

EXTON: Ah, but he pointed out something extremely important. Democritus, you're right about DNA being aperiodic. Chargaff proved this in a rather ingenious experiment. He subjected DNA to slow, partial digestion. He discovered that the freeing of the individual nucleotides during digestion was followed no regular pattern.

DEMOCRITUS: In an aperiodic chain, the sequence of nucleotides is irregular.

EXTON: That was Chargaff's greatest discovery. He showed that the DNA chain is not repetitive and monotonous, but infinitely varied, and thus possibly extraordinarily specific. This discovery should inspire us in our search for the detailed molecular structure of DNA, of this mysterious substance capable of both replicating and bearing information.

ALKIMOS: Exton, you aren't trying to mislead us again, are you?

EXTON: Of course not! But we're going to encounter a few flawed models of DNA.

ALKIMOS: I thought so.

EXTON: But you'll see. You will appreciate the truth even more!

DEMOCRITUS: Then off we go! Another adventure!

EXTON: Hermes, the Deus ex Machina!

ALKIMOS: But Exton, I thought we were done for the day.

EXTON: Relax, Alkimos, we'll be quick. Just a short chat with Linus Pauling.

ALKIMOS: But we've already talked about the theory of complementarity.

EXTON: Yes, we have. Now we're going to discuss macromolecules. Come on! Direction Pasadena, the California Institute of Technology, 1953!

ALKIMOS: Aaggghh … here we go!

DEMOCRITUS: Alkimos, enough screaming! You should be used to it by now!

ALKIMOS: I know, master, but it shshshsakes ssssso mmmmuch!

EXTON: We've stabilised. Now settle down, both of you, and listen here. Linus Pauling was the greatest chemist of the 20th century. In the 1920s and 1930s, he elucidated the nature of the chemical bonds between atoms, for which he will be forever remembered. His book, *The Nature of Chemical Bonds*, is one of the cornerstones of chemistry. He also did pioneering work in uncovering the structure of complex macromolecules. In 1951, he described one of the basic elements in the complicated spatial structure of proteins: the alpha-helix. Here's the drawing Pauling published in his article on the alpha-helix:

You can see the actual model he constructed in his office.

ALKIMOS: I can hardly wait! But Exton, you said we were visiting him to talk about DNA.

EXTON: Well, among other things. Pauling's research involved primarily proteins. For a long time, he wasn't even interested in DNA.

ALKIMOS: I guess we'll have to fill him in on Avery's studies.

EXTON: That won't be necessary. By early 1952, he began to develop a minor interest in discovering the structure of DNA. But two young scientists were far more determined.

DEMOCRITUS: I think we've arrived.

EXTON: Okay, everyone out.

ALKIMOS: Finally some sunshine!

EXTON: Let's find Pauling!

DEMOCRITUS: My friends, we're finally going to learn how the atoms of life are capable of organizing themselves and replicating!

ALKIMOS: Come on, master! Let's find out how Pauling solved the mystery of the structure of genes!

EXTON: Don't assume he did! He certainly was the most likely one to do it; yet surprisingly, he missed the mark. Many think his progress was impeded because his request for a passport was denied in 1952. As a result he wasn't able to take part in the conference on protein structure organised by the Royal Society. Thus, he was unable to meet with some English colleagues who had already done considerable work on determining the structure of DNA.

ALKIMOS: Poor Pauling.

EXTON: But I don't think this was the reason.

ALKIMOS: Then what was?

EXTON: Pauling didn't take the problem seriously.

ALKIMOS: Master, did you hear that?

DEMOCRITUS: Yes, I did. But tell us, dear Exton, why couldn't Pauling travel to London?

EXTON: The authorities suspected he was a communist.

ALKIMOS: A communist? What's that?

EXTON: An extremist group of Persians.

ALKIMOS: Ah, now I understand.

EXTON: Pauling wasn't a communist, of course. Nevertheless, Mrs. Ruth B. Shipley, head of the passport office of Washington State, declared that Pauling's trip would not be 'in the best interest of the United States.'

ALKIMOS: Hmm, maybe all protein researchers are dangerous Persian criminals in disguise!

EXTON: I don't think we'll ever understand the motivations of Mrs. Shipley. And we don't have time to think about it, anyway. We're here. This is Pauling's office.

ALKIMOS: Oh, look at all the colourful balls!

EXTON: Alkimos, don't touch anything!

PAULING: How can I help you, gentlemen?

ALKIMOS: The atoms of life!

PAULING: Excuse me?

EXTON: Professor, please excuse my students ...

PAULING: Oh, it's all right.

EXTON: I'm Exton, from the University of Sussex, and these gentlemen are ...

PAULING: Of course, Exton! Welcome. Have a seat.

EXTON: Thank you.

ALKIMOS: Is that what a protein looks like?

PAULING: Some proteins, yes. Others only have parts that look like this. It's called an alpha-helix.

ALKIMOS: Look how the atoms are twisting!

DEMOCRITUS: How do you build a structure like this?

ALKIMOS: Master, it's a single, coiled chain!

DEMOCRITUS: Yes, son, we've learned that proteins are made up of long polypeptide chains. But why do they twist like that and fold onto each other?

EXTON: Professor Pauling, could you explain?

PAULING: Of course. My colleague Alfred Mirsky and I developed a representation of native protein molecules—that means ones that exhibit typical characteristics. As you said, every protein is an unbroken chain of polypeptides. But that's not all. This chain spirals into a unique, fixed, spatial configuration.

ALKIMOS: As we see with this alpha-helix.

PAULING: Exactly.

DEMOCRITUS: But what holds the chain in this spiral form?

PAULING: Weak hydrogen bonds.

ALKIMOS: What are those?

PAULING: The electrostatic attraction between oxygen and nitrogen atoms and some hydrogen atoms.

EXTON: These forces are much weaker than the covalent bonds that hold the chain together. Imagine that the chain is full of negatively or positively charged atoms, and they attract or repel each other. The closer two oppositely charged atoms come to one another, the stronger their attraction to each other. If they move away from each other the bond is broken.

PAULING: According to Mirksy—and I agree with him—the weak bonds are broken when we denature a protein.

ALKIMOS: '*De*' means to remove something, so denaturation can't be good for proteins.

DEMOCRITUS: Since when do you know Latin, Alkimos?

PAULING: We can denature proteins by heating them or treating them with certain chemicals. The denatured proteins lose their uniquely defined spatial configuration.

ALKIMOS: If we boil them in salt water they'll lose their nice spirally, tangled shape?

PAULING: Yes, and at the same time their biological activity.

ALKIMOS: So they stop being enzymes.

DEMOCRITUS: The key to their biological activity lies in their twisty spatial structure. A revolutionary idea!

EXTON: It certainly is. This picture of the complicated spatial structures held together by weak interactions allows us to understand the astounding phenomenon of biological specificity. Professor, could you explain to us your view on the relationship between biological specificity and the spatial configuration of molecules, so my students can better understand what I'm talking about.

PAULING: As you know, the most surprising and characteristic property of these biological materials is their specificity. Because of this high degree of specificity, proteins bind to and influence the behaviour of certain materials, but leave other materials completely alone.

ALKIMOS: For example, Alfred Kühn's gene—the one that darkened the testes of moth larvae—affected only one pigment, and not, for example, its polysaccharide capsule, assuming moth larvae even have one …

EXTON: To be precise, the enzyme produced by the gene impacted the pigment molecule, but it was very selective about which one. That is the essence of specificity.

PAULING: Such unfailing precision is rare in physical or chemical systems.

EXTON: But this quality is pervasive in biological systems.

PAULING: A good example of biological specificity is gene replication.

ALKIMOS: Right. A gene can generate only a copy of itself but not of any other gene.

PAULING: You've also mentioned the great specificity of enzymes. These molecules are able to select the one material upon which they apply their catalytic action from a mix of several thousand. Hormones, medicines and in particular antibodies also exhibit a similarly great degree of specificity.

ALKIMOS: Ah, it's amazing what serologists can draw from experimental mice!

PAULING: What's so interesting about antibodies is their extraordinary ability to recognise invaders. I believe that if we come to understand the physicochemical basis of biological specificity, we will be much closer to a solution to the most basic problem of biology: what is the nature of life?

DEMOCRITUS: And what are your thoughts on the physicochemical basis of biological specificity?

PAULING: The crucial aspects are as follows: the molecules' astounding structural variety, weak chemical bonds and complementarity.

ALKIMOS: Bronze statues and moulds …

DEMOCRITUS: Held together by weak electrostatic bonds.

EXTON: In an endless variety.

PAULING: Statues and moulds. What a wonderful comparison! It's exactly what I think. A stable, spatial molecular configuration has a completely unique surface to which its complement binds. This is the case with antibodies and antigens, also enzymes and substrates. Two complementary molecules, because their surfaces fit together tightly, come in close contact over a much larger area, and as a result, the many weak interactions can work together to create a strong bond.

DEMOCRITUS: The molecule may try to pull other material toward it, but it's all in vain because their surface won't match that of the molecule.

ALKIMOS: An enzyme may spit out a mouthful without chewing it.

DEMOCRITUS: That's the secret of specificity. Precise molecular recognition. But how does it work with genes? What kind of moulds and statues do nucleic acids create?

PAULING: Yes, nucleic acids. Interesting subject.

ALKIMOS: We've known about them ever since Avery identified them!

PAULING: I'm familiar with the debate about Avery's work and the idea that nucleic acids are hereditary material. Until recently I considered it impossible that genes were composed of nucleic acids. Proteins impressed me so much that I was sure only they could comprise hereditary material. But at least year's conference in Royaumont, Hershey shared some very interesting new results.

ALKIMOS: About DNA shooting bacteriophage? We heard about that too!

EXTON: Yes, we've read Hershey and Chase's most recent article in the *Journal of General Physiology* on the role of nucleic acids in the proliferation of phages.

PAULING: The final blow was delivered by my colleague Robley Williams, who produced images of nucleic acids using an electron microscope.

ALKIMOS: We can see genes with a microscope?

PAULING: Yes, using the right techniques. The images were astonishing, even to me. Threads—long, uniform, cylindrical strands. It was clear to me that what I was observing were helices of macromolecules twisting around each other.

DEMOCRITUS: So nucleic acids form long, helical threads, just like certain parts of proteins?

PAULING: That's the most likely explanation. But the threads are not built of one chain, like in protein. Using William's results I calculated that the structures consisted of three polynucleotide chains coiled around each other.

ALKIMOS: Three chains twisted together!

DEMOCRITUS: Held together by hydrogen bonds?

PAULING: Probably. Have a look at this diagram:

DEMOCRITUS: I don't understand. If nucleic acid strands really are genes, then I reckon there can only be two chains. The laws of duality, the symmetry of life, the splitting of chromosomes, demand this.

PAULING: The result surprised me, too. I also expected two chains, but as our understanding of biology grows, perhaps we'll be able to make sense of this configuration.

ALKIMOS: Or if our understanding of chemistry grows, we'll discover the correct one.

EXTON: Alkimos!

PAULING: Oh, I know—it's quite possible I made a mistake. But even if I did, it's not a huge problem. I've never worried about making mistakes, and nor should you. Thousands of scientists without anything better to do are always in the ready to point out your mistakes.

ALKIMOS: What's that chiming sound?

PAULING: The telephone. Please, excuse me.

ALKIMOS: What's a telephone?

PAULING: Hello Max, how are things? ... Really? A letter from Jim Watson? What did he say? ... He doesn't want you to tell me? So why on earth did you call? ... Oh, I see. So you'll tell me anyway. ... Indeed? A new model of DNA ... together with Francis Crick? ... Bring the letter over ... Of course you're not disturbing me. I have some guests, but it's no problem.

EXTON: We're just about ready to leave.

PAULING: What's the rush?

EXTON: We don't want to keep you.

PAULING: But you're not.

EXTON: Forgive me, but we have a lot to do.

PAULING: Well, if you say so.

MRS. SHIPLEY: All right, no one move!

PAULING: Excuse me, ma'am, but where did you come from? What's the meaning of this?

MRS. SHIPLEY: Oh, you'll find out when we take you into custody.

PAULING: That's quite unnecessary.

MRS. SHIPLEY: They're here!

ALKIMOS: Exton, who is this fury?

MRS. SHIPLEY: Fury? Is that what you called me? Why, I'll make you eat your words!

PAULING: Ma'am, please. What is going on?

MRS. SHIPLEY: Lousy Reds!

EXTON: We should really get out of here!

MRS. SHIPLEY: Officer, over here!

DEMOCRITUS: But the structure of DNA …

EXTON: Please excuse us, Professor. We hate to leave you in the midst of such unpleasantness, but you see …

OFFICER: You've got them Mrs. Shipley? Who's the whiskered one? I don't think we have him on file.

DEMOCRITUS: We're here to talk about the atoms of life.

MRS. SHIPLEY: Trotsky! It's Trotsky! Arrest him!

PAULING: Ma'am, you're making a terrible mistake.

EXTON: On board, quick!

DEMOCRITUS: Trotsky?

ALKIMOS: Aaaggh … here we go!

DNA with Ambrosia

HERMES: The ambrosia pudding on Mt. Olympus had a completely different taste.

COOK: Really?

HERMES: Yes, I remember well.

COOK: Hmmm …

HERMES: Are you doing something different?

COOK: Of course not!

HERMES: Are you sure?

COOK: All right, just don't say a word to them. But ambrosia is only for the gods.

HERMES: Well they're starting to think of themselves as gods.

COOK: If I served them real ambrosia, that's what they'd become. Or demigods at least.

HERMES: But then, maybe they'd understand that it's impossible to know everything.

COOK: Oh, that old guy wouldn't admit it in two hundred years!

HERMES: Yes, but your gluttonous friend might.

COOK: Hah, except I'm not willing to cook for them for another two hundred years!

HERMES: Why not? In that time, he might actually become a good poet.

EXTON: Here you are! It's late. We'd like some dinner.

COOK: Humph. I guess we won't be getting to bed anytime soon.

EXTON: Come on!

ALKIMOS: Exton, I'm starved. I could eat three flying horses!

EXTON: Relax, Alkimos. They'll have dinner ready soon. It has been a long day, hasn't it?

DEMOCRITUS: Dear Exton, could you tell us what was in that letter Professor Pauling received?

EXTON: In a minute. But first I need to acquaint you with *X-ray crystallography*.

ALKIMOS: Okay, but just a quick overview, right?

EXTON: Excuse me?

ALKIMOS: Well, it sounds sort of scary …

EXTON: Democritus, you'll be very interested. With this tool we can determine the exact position of the atoms in a crystal.

DEMOCRITUS: How is that possible?

EXTON: A German physicist, Max von Laue, and two English physicists, William Henry Bragg and his son William Lawrence Bragg, developed this ingenious method. In 1912, Laue began to work with X-rays. These mysterious electromagnetic rays permeate everything except metals. In the human body they are blocked only by teeth and bones.

COOK: Gentlemen, soup á la Pegasus wing.

EXTON: The physicists couldn't decide if X-rays were made up of particles or whether they were waves. Laue came up with a brilliant experiment: if we pass an X-ray through a crystal, and the rays are wavelike, then they will be diffracted like light waves when they go through narrow slits. When Laue's colleagues performed the experiment, they discovered that the X-ray beam was indeed diffracted when it passed through the crystal and left characteristic spots on a photographic plate placed behind the crystal.

ALKIMOS: What's a photographic plate?

EXTON: A plate that's coated with a substance sensitive to light or X-rays.

DEMOCRITUS: How does a straight beam cause these spots?

EXTON: The waves are diffracted by every atom in the crystal, and they either strengthen or extinguish each other, and because the crystal has a regular structure, the spots will also create a regular pattern. Laue thus proved that X-rays are waves and that crystals have an organised three-dimensional atomic structure.

DEMOCRITUS: But how were they able to reach that conclusion?

EXTON: Because of the arrangement of the spots. After reading Laue's article, the younger Bragg, Lawrence, realised how one could determine the exact arrangement of the atoms in the crystal based on the spot pattern.

DEMOCRITUS: Amazing!

ALKIMOS: Why didn't we think of that?

EXTON: The method was indeed extraordinary. Father and son very quickly established the atomic structure of numerous crystals including diamond.

DEMOCRITUS: For us, the atoms of life are far more valuable than any gemstone, even a diamond.

COOK: Gentlemen, roasted partridge with tamarisk?

ALKIMOS: Splendid!

EXTON: The Braggs' work eventually led to an investigation of the structure of the building blocks of living cells. The first studies were of cellulose, which consists of a long chain of sugar molecules, and then simple proteins, such as keratin in wool. It turned out that some proteins—including enzymes—are capable of forming crystals.

ALKIMOS: Like the tobacco-mosaic virus!

EXTON: In 1926, James Sumner crystallised the enzyme *urease*, which catalyses the transformation of urea into ammonia and carbon dioxide. He made an extract of wild bean which showed very strong urease activity. After one day, crystals formed in the extract, crystals of pure urease protein. This was the first time a completely pure form of an enzyme had been obtained. Like the crystallisation of the tobacco mosaic virus by Wendell Stanley in 1935, this discovery astounded biologists. If a biologically active agent can be crystallised then it must have a regular structure.

DEMOCRITUS: Proteins therefore must truly have a determined atomic structure as Pauling showed us, and to such a degree that it can form a crystal.

EXTON: The discovery of the enormous molecular mass of proteins and the crystallization of enzymes overturned the colloid theory, which had been generally accepted in the 1920s. Gradually a study of the structure of proteins and finally nucleic acids got underway.

ALKIMOS: So we simply need to deduced the position of the atoms from the spot pattern?

EXTON: No, it was anything but simple. Unlike simple mineral crystal, the crystals of macromolecules leave exceptionally elaborate spot patterns in the X-ray diffraction machine. For a long time it wasn't possible to establish the exact structure of such large molecules. But the molecules got increasingly larger. In 1945, the British Dorothy Hodgkin, a pioneer of the crystallographic study of biomolecules, solved the structure of the antibiotic penicillin (27 atoms). A few years later she solved the structure of vitamin B12 (181 atoms), an achievement as significant "as breaking the sound barrier," according to

Lawrence Bragg. The diffraction patterns of even larger molecules also provided a lot of information.

DEMOCRITUS: Did Pauling's determination of the structure of the alpha-helix also rely on earlier observations of X-ray diffraction?

EXTON: Yes, and his brilliant powers of intuition eventually led him to the correct model.

ALKIMOS: His brilliance is also what helped him develop the three-chain model of DNA.

EXTON: That's right. But it seems his brilliance fell a little short, because the three-chain model was completely wrong.

ALKIMOS: Exton, you led us down the wrong path again!

DEMOCRITUS: Remember, I suspected it wasn't right. If the duplicating gene works according to the laws of complementarity, then it can't be constructed of three chains.

ALKIMOS: Well then, we just remove one chain from the model, right?

EXTON: That would take us closer to the correct solution, but it's not as easy as that. The two strands have to be turned inside out. Now it's time to read the letter James Watson, a researcher in Cambridge's Cavendish lab, wrote to Max Delbrück: 'Our model (a joint project of Francis Crick and myself) bears no relationship to either the original or to the revised Pauling–Corey–Schomaker models. It is a strange model and embodies several unusual features. However, since DNA is an unusual substance, we are not hesitant in being bold.'

ALKIMOS: Great scientists are always bold!

DEMOCRITUS: And much boldness is needed to create a model which encompasses the autocatalytic capability of genes and their complementariness.

ALKIMOS: And Pauling didn't manage to do it.

EXTON: At first, Watson and Crick didn't either. A year before Pauling, they also built an erroneous model of DNA with three strands. In their enthusiasm they showed the model to two colleagues, Maurice

Wilkins and Rosalind Franklin. Wilkins and Franklin both worked at London's King's College and were also trying to decipher the structure of DNA. It was clear to them that the triple helix model was wrong based on their own research in chemistry. In the face of their strident criticism, Watson and Crick did not publish their model of a triple helical DNA.

ALKIMOS: But if they'd published it, Pauling wouldn't have wasted his time constructing his own model.

EXTON: Perhaps he would have been the one to point out the problems with the model. Let me read you more of Watson's letter. He says, 'The main features of the model are: (1) The basic structure is helical—it consist of two intertwining helices.'

DEMOCRITUS: Watson and Crick simply removed a strand—just as I suggested!

ALKIMOS: Master, excuse me, but, I believe I came up with that idea ...

DEMOCRITUS: You?

EXTON: 'The core of the helix is occupied by the purine and pyrimidine bases; the phosphate groups are on the outside.' We can illustrate it like this:

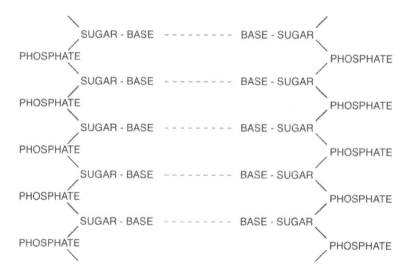

DEMOCRITUS: In Pauling's model, the phosphates were on the inside and the bases on the outside. So they've reversed the chains!

COOK: Gentleman, some honey cake from Akragas?

ALKIMOS: Oh, I'd love some!

EXTON: 'The chains are not identical, but complementary so that if one helix contains a purine base the other helix contains a pyrimidine.' In an earlier model, Crick and Watson had created two identical chains, with purine matched to purine and pyrimidine matched to pyrimidine. But because purine is a larger molecule than pyrimidine, the girth of the double chain varied. Where there were purine pairs the strand was thicker, and where there were pyrimidine pairs, it was thinner.

DEMOCRITUS: But that can't be, because we've seen the DNA with an electron microscope and the strands are of uniform thickness.

EXTON: But if we pair purine with pyrimidine, then this problem disappears. Every purine–pyrimidine pair is of equal thickness, so the phosphate–sugar backbone is smooth. What's more, the base-pairs can appear in any sequence.

DEMOCRITUS: If the bases are in the interior of the spiral, then they certainly must be what keep the two chains together.

ALKIMOS: So they bind to each other. But how?

DEMOCRITUS: I believe by weak hydrogen bonds.

ALKIMOS: I thought so …

DEMOCRITUS: We just need to find which atoms bind together with a hydrogen bond.

EXTON: Watson found it: 'The pairing of the purines with the pyrimidines is very exact and dictated by their desire to form hydrogen bonds— adenine will pair with thymine while guanine will always pair with cytosine.' This diagram illustrates how the base pairs form.

ADENINE   THYMINE

GUANINE   CYTOSINE

0   5Å

ALKIMOS: My Zeus! How'd they figure that out?

EXTON: Watson cut a model of the four bases out of cardboard, and turned them this way and that until he found the pairs. Only A–T and G–C were capable of forming double hydrogen bonds. Later, it was discovered that G–C forms not a double, but a triple hydrogen bond. Watson didn't know that, but he still managed to find the pairs. To match purine to pyrimidine, he had to flip the purine molecule 180 degrees, so that it was upside down.

ALKIMOS: So one chain was much more twisted up than the other.

EXTON: He inverted it—exactly 180 degrees. As one chain runs upward, the other runs downward. The two strands of the double helix are *antiparallel,* as you can see in this diagram after Watson and Crick.

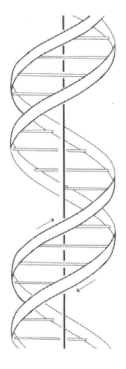

Crick realised this when he saw Rosalind Franklin's unpublished results on X-ray diffraction. Franklin wanted to decipher the structure of DNA using exclusively X-ray diffraction studies and not by building models. Using exact data, it's possible to establish the most essential features of the DNA structure. Rosalind Franklin was very close to discovering the true structure, but not close enough.

DEMOCRITUS: Perhaps she lacked the boldness.

ALKIMOS: Maybe if she'd joined forces with Pauling ...

EXTON: It was too late. In March 1953, Crick and Watson solved the problem. In April 1953, before Watson and Crick's article was published, Pauling visited the Cavendish lab at Cambridge and had a look at the double helix model. It nearly reached the ceiling.

ALKIMOS: So he was issued a passport after all?

EXTON: Several days later, after Lawrence Bragg had shown him the double helix model, Pauling said, 'Although it is only two months since Professor

Corey and I published our proposed structure for nucleic acid, I think we must admit that it is probably wrong ... Although some refinement might still be made, I feel that it is very likely that the Watson–Crick structure is essentially correct.' Pauling, great man that he is, acknowledged his mistake.

ALKIMOS: How can he be so great if he was wrong?

DEMOCRITUS: Only the gods are infallible, son.

ALKIMOS: And maybe they're not either ...

EXTON: The double helix structure of DNA is a perfect example of Pauling's complementariness. It consists of two strands that run in opposite directions and pair with each other; one strand is the other's complement. The order of the bases of one determines the order of bases of the other. If one has A, the other has T; if one has G, the other has C, and of course, the other way around.

ALKIMOS: Bronze statues and moulds!

DEMOCRITUS: A with T and G with C. The spiralling double helix model explains Chargaff's ratios!

EXTON: That's right. The amount of A will always match the amount of T, and the same is true for C and G. But the model also offers an explanation of autocatalysis, the ability of the double helix to replicate. At the end of their article, Watson and Crick stated, 'It has not escaped our notice that the specific pairing we have postulated immediately suggests a possible copying mechanism for the genetic material.'

DEMOCRITUS: Autocatalysis and complementariness. The deepest secrets of life.

ALKIMOS: Master, we've solved the mystery?

DEMOCRITUS: I believe we have.

ALKIMOS: By Hercules! Master! The atoms of life! So can we go home now?

DEMOCRITUS: Oh no, not yet. We have travelled to the innermost recess of living things. We've exposed them, down to their very atoms.

ALKIMOS: Yes, master ...

DEMOCRITUS: We now know the structure of the atoms of life in such great detail that the mystery of autocatalysis is now revealed to us.

ALKIMOS: Complementary strands of DNA ...

DEMOCRITUS: However, I don't think Miescher, Avery or Crick would call DNA life; nor would Pauling call his alpha-helix life.

ALKIMOS: Then what is life, master? Exton?

EXTON: In the words of Chargaff: "Life is what's lost in the test tube."

ALKIMOS: So why did we put it in a test tube? Why'd we bother with Spallanzani, Boveri, Pauling, and all the others? Why didn't we just stay home? We could be fishing in the Abdera Bay.

DEMOCRITUS: Now that we've taken it apart, we need to put it back together again. We're staying until life is reborn in the test tube.

EXTON: You might not have time for that ...

DEMOCRITUS: But we can try!

ALKIMOS: Master, I guess I trust you.

DEMOCRITUS: But first we should have a rest.

ALKIMOS: Oh, what a novel suggestion.

EXTON: Yes, it's late. Time to turn in.

COOK: Can we finally clean up here?

EXTON: Yes, of course. We couldn't eat another bite.

ALKIMOS: Well actually, I could ...

COOK: Some wine?

ALKIMOS: Do you have anymore of last night's ambrosia pudding ...?

Part Six

Codes and Links

The Central Dogma—*Francis Crick*

ALKIMOS: Master, are you awake?

DEMOCRITUS: Hmmm ...

ALKIMOS: I didn't sleep well. It was too much.

DEMOCRITUS: Yes, it's a lot of knowledge to absorb.

ALKIMOS: No, I meant the ambrosia pudding. I feel like I swallowed a rock.

DEMOCRITUS: Oh, Alkimos, I'd rather not hear about it.

ALKIMOS: Sorry, master. In that case, maybe you can tell me what comes next.

DEMOCRITUS: I think what's next is perhaps the most difficult to ...

ALKIMOS: Then let's ask Exton for help. Exton!

EXTON: Yes? Ah, you're up already? Then come on.

ALKIMOS: I have this heavy feeling ...

EXTON: All that science. I know, Alkimos. But here, have a little tamarisk jam, and let's think about what the discovery of DNA's structure means for the study of biology.

ALKIMOS: So early in the morning?

EXTON: Why not? Now, so that you better understand the complementariness of DNA strands as it relates to gene replication, let me read to you from Watson and Crick's second article on DNA. Here they discuss the structure as it pertains to genetics.

DEMOCRITUS: I have a feeling I know what they have to say.

EXTON: The article appeared in May 1953, just one month after their first article on the structure of DNA. It begins with a discussion of the importance of DNA, and then continues with this: 'Until now, however, no evidence has been presented to show how it might carry out the essential operation required of a genetic material, that of self-duplication.'

DEMOCRITUS: Because we didn't know the structure of the atoms of life. But the double helix model makes it absolutely clear how duplication is possible.

ALKIMOS: How, master?

DEMOCRITUS: Exton has already explained it. One strand is the complement of the other. If the sequence of bases is provided by one strand, then you can immediately know the sequence of the other strand because of the specific pairing.

ALKIMOS: A cast bronze statue and its mould.

DEMOCRITUS: Right, son. This tells us how genes self-duplicate, and this time on the atomic level.

EXTON: That is exactly what Watson and Crick wrote: "Previous discussion of self-duplication have usually involved the concept of a template or mould. Either the template was supposed to copy itself directly, or it was to produce a "negative," which in turn was to act as a template and produce the original "positive" once again. In no case has it been explained in detail how it would do this in terms of atoms and molecules."

ALKIMOS: Master, we tried.

DEMOCRITUS: Let some others have some glory too, Alkimos.

EXTON: Let's take for example this sequence: GCATGAAACTAGG. Alkimos, what's the sequence of the complementary strand?

ALKIMOS: A with T, and G with C, then … one moment … it would be CGTACTTTGATCC.

EXTON: Exactly. The upward and downward running strands of the double helix will match up like this:

G	C	A	T	G	A	A	A	C	T	A	G	G
C	G	T	A	C	T	T	T	G	A	T	C	C

DEMOCRITUS: Between each G–C pair are three hydrogen bonds, and between each A–T pair are two.

EXTON: And all in a spiral.

ALKIMOS: I don't understand how the helix can duplicate.

EXTON: This is what Watson and Crick had to say about DNA duplication: 'Our model for deoxyribonucleic acid is, in effect, a pair of templates, each of which is complementary to the other. We imagine that prior to duplication the hydrogen bonds are broken, and the two chains unwind and separate. Each chain then acts as a template for the formation on to itself of a new companion chain, so that eventually we shall have *two* pairs of chains, where we only had one before. Moreover the sequence of the pairs of bases will have been duplicated exactly.'

ALKIMOS: In other words …

EXTON: The two strands unwind like this:

G	C	A	T	G	A	A	A	C	T	A	G	G

C	G	T	A	C	T	T	T	G	A	T	C	C

Then each strand produces its complement from the single nucleotides, or *monomers*, roaming freely in the cell. The end result is:

G	C	A	T	G	A	A	A	C	T	A	G	G
C	G	T	A	C	T	T	T	G	A	T	C	C

G	C	A	T	G	A	A	A	C	T	A	G	G
C	G	T	A	C	T	T	T	G	A	T	C	C

ALKIMOS: So we'll have two genes instead of one!

EXTON: But we have to assume that the free nucleotides are in large quantities in the cell, and that after the chains unwind, they line up in the right places and attach using hydrogen bonds. Then the individual nucleotides also need to bind to each other.

ALKIMOS: How?

DEMOCRITUS: That part's simple! They may link simply because they're so close together, or an enzyme might help them. Those are just the details.

ALKIMOS: Details?

EXTON: Crick also loved to say they'd work out the details later.

DEMOCRITUS: First, they needed to establish the big picture.

ALKIMOS: And the biochemists would fill in the gaps.

EXTON: More or less. There were a number of unanswered questions, but Watson and Crick weren't too concerned.

ALKIMOS: What sort of questions?

EXTON: For example, how could DNA strands of several thousand bases unwind without the whole thing turning into a tangled mess?

ALKIMOS: Using special chain-unwinding enzymes!

DEMOCRITUS: Excellent idea!

EXTON: They said, 'let the details take care of themselves,' and in the end they did.

ALKIMOS: Thanks to biochemists!

EXTON: Crick was what we'd call an intuitive genius, and this is true of very few scientists. Those minds that can see beyond the details, that can dream up new systems, that can simplify and summarise, are the ones that really take science forward.

ALKIMOS: Yes, like us. I mean, we're trying ...

EXTON: Crick is perhaps the most striking representative of a revolutionary new approach. In 1947, he described his area of enquiry like this: 'The particular field which excites my interest is the division between the living and the non-living, as typified by, say, proteins, viruses, bacteria and the structure of chromosomes. The eventual goal, which is somewhat remote, is the description of these activities in terms of their structure, i.e. the spatial distribution of their constituent atoms, in so far as this may prove possible. This might be called the chemical physics of biology.'

ALKIMOS: But that's my master's approach too! He's interested in the biology of the atoms of life!

EXTON: Otherwise known as *molecular biology*.

DEMOCRITUS: Molecular biology. Yes, that sounds much better.

EXTON: Jacques Monod once said, 'No one man discovered or created molecular biology. But one man dominates intellectually the whole field, because he knows the most and understands the most ...'

ALKIMOS: Do you hear that, master?

EXTON: ... Francis Crick.

ALKIMOS: Francis Crick, master.

EXTON: It's time for us to hear him in person.

ALKIMOS: Oh, not again.

EXTON: Oh yes, Alkimos!

ALKIMOS: But ...

EXTON: All aboard! Direction London, 1957!

ALKIMOS: Master, no ...

DEMOCRITUS: Stop fidgeting, Alkimos.

ALKIMOS: Aagghh ... here we go!

DEMOCRITUS: Settle down!

EXTON: Gentlemen, listen here. Crick was an unremitting sceptic. His greatest desire was to shed light into the last gloomy corners in which vitalism still managed to take refuge.

DEMOCRITUS: Kronides! That's exactly what we're trying to do—I couldn't have said it better myself!

EXTON: Once the structure of DNA was established, Crick began to investigate the relationship between genes and proteins.

DEMOCRITUS: But we know that. Kühn, Ephrussi, and Beadle all demonstrated that genes are responsible for the production of proteins.

EXTON: Yes, but how? What's the link between the structure of DNA and the structure of proteins? How does it all work? How does a cell use genes?

ALKIMOS: We should ask Crick.

DEMOCRITUS: That's why we're going to Cambridge.

ALKIMOS: You said 'London.'

EXTON: We're here! We can enter the building through the back door. They won't notice us.

ALKIMOS: Why do we have to sneak in?

EXTON: So we don't cause trouble again.

ALKIMOS: What trouble?

EXTON: It's already started. Be quiet.

DEMOCRITUS: Are we late?

EXTON: Sh. We can slip in now.

CRICK: 'In the protein molecule, Nature has devised a unique instrument in which an underlying simplicity is used to express great subtlety and versatility; it is impossible to see molecular biology in proper perspective until this peculiar combination of virtues has been clearly grasped.'

ALKIMOS: He should ask Pauling!

DEMOCRITUS: If you recall, the structure he built of sticks and balls was not exactly versatile.

CRICK: 'It is at first sight paradoxical that it is probably easier for an organism to produce a new protein than to produce a new small molecule, since to produce a new small molecule one or more new proteins will be required in any case to catalyse the reactions.'

ALKIMOS: The hormones of the moth larva required a couple of proteins. Or genes.

CRICK: '...the main function of the genetic material is to control (not necessarily directly) the synthesis of proteins. There is a little direct evidence to support this, but to my mind the psychological drive behind this hypothesis is at the moment independent of such evidence.'

DEMOCRITUS: The genetic material can recombine too.

ALKIMOS: And replicate.

EXTON: Just listen to him!

CRICK: '...There are a few cases where a Mendelian gene has been shown unambiguously to alter a protein, the most famous being that of human sickle cell anaemia haemoglobin .... Until recently it could have been argued that this was perhaps not due to a change in amino acid sequence, but only to a change in the folding. That the gene does in fact alter the amino acid sequence has now been conclusively shown by my colleague, Dr Vernon Ingram ... It may surprise you that the alteration of one amino acid out of a total of about 300 can produce a molecule which (when homozygous) is usually lethal before adult life but, for my part, Ingram's result is just what I expected.'

ALKIMOS: Who's Ingram?

EXTON: Ingram also worked at Cambridge, in the lab of Frederick Sanger. In 1957, he showed that the haemoglobin protein of people with sickle cell anaemia differed from normal, or 'wild-type,' haemoglobin in just one amino acid.

ALKIMOS: What's sickle cell anaemia?

EXTON: Haemoglobin is a protein in red blood cells that transports oxygen to the cells of the body. Sickle cell anaemia is an inherited disease. The patient's haemoglobin forms long strands that distort red blood cells giving them a crescent shape, like a sickle. As a result red blood cells are unable to operate efficiently. This can lead to trouble breathing and ultimately death. Pauling hypothesised that the distortion of the red blood cells was the result of a change in the structure of haemoglobin in those afflicted with the disorder. In 1949, one of his students, Harvey Itano managed to verify this. He found that the aberrant haemoglobin differed electrophoretically from normal haemoglobin. This means that it moves differently in a gel when an electric field is applied. The disease—the alteration of the haemoglobin—is the result of a genetic mutation.

DEMOCRITUS: So a mutant gene causes haemoglobin to be misshapen? Genes determine the structure of a protein? Is that what's behind the one gene, one enzyme relation?

EXTON: Yes, exactly. Ingram demonstrated that the normal and the mutant protein differed from each other in just one amino acid.

DEMOCRITUS: Genes determine the structure of the protein down to the very last amino acid. What mind-boggling precision!

ALKIMOS: Quiet, master!

EXTON: Yes, we'd better pay attention.

CRICK: 'As I was saying, it is instructive to compare your own haemoglobin with that of a horse ...'

ALKIMOS: Why a horse?

CRICK: '...Both molecules are indistinguishable in size. Both have similar but not identical amino acid compositions. They differ a little electrophoretically, form different crystals, and have slightly different ends to their polypeptide chains. All these facts are compatible with their polypeptide chains having similar amino acid sequences, but with just a few changes here and there.'

ALKIMOS: So I'm a mutant horse?

DEMOCRITUS: I'd say an ass not a horse.

EXTON: Sh!

CRICK: "...This 'family likeness' between the 'same' protein molecules from different species is the rule rather than the exception. It has been found in almost every case in which it has been looked for. One of the best-studied examples is that of insulin, by Sanger and his co-workers, who have worked out the complete amino acid sequence for five different species, only two of which (pig and whale) are the same."

ALKIMOS: They're both fat!

DEMOCRITUS: What's insulin? And how could it be the entire sequence? Why, it's not possible!

EXTON: I'll explain it later. Now let's listen to Crick. He's talking about the relationship between genes and proteins.

CRICK: '…My own thinking is based on two general principles, which I shall call the Sequence Hypothesis and the Central Dogma. The direct evidence for both of them is negligible, but I have found them to be of great help in getting to grips with these very complex problems.'

ALKIMOS: See, master, Crick is continually formulating new theories!

CRICK: 'The Sequence Hypothesis has already been referred to a number of times. In its simplest form it assumes that the specificity of a piece of nucleic acid is expressed solely by the sequence of its bases …'

DEMOCRITUS: Schrödinger's aperiodicity.

CRICK: '…this sequence is a (simple) code for the amino acid sequence of a particular protein.'

DEMOCRITUS: A simple code for a protein's amino acid sequence. Fantastic!

CRICK: "… my second principle is the Central Dogma. This states that once 'information' has passed into protein it cannot get out again. In more detail, the transfer of information from nucleic acid to nucleic acid, or from nucleic acid to protein may be possible, but transfer from protein to protein, or from protein to nucleic acid is impossible."

ALKIMOS: How's that?

CRICK: '…Information means here the precise determination of sequence, either of bases in the nucleic acid or of amino acid residues in the protein.'

ALKIMOS: What do you say to that, master?

DEMOCRITUS: It's quite a theory.

ALKIMOS: Exton, is it true?

EXTON: I'm afraid we need to delve in a bit deeper to answer that question. But we need to go!

ALKIMOS: We're not going to talk to Crick?

EXTON: This isn't the time.

ALKIMOS: But why not? We have so many questions. Dr. Crick, Dr. Crick!

CRICK: If you have a question, please ask.

ALKIMOS: Master, go ahead!

DEMOCRITUS: Me? Um, well, um. So you see, if I understand correctly, according to the Sequence Hypothesis, the nucleotide sequence of genes is in some way responsible for the amino acid sequence in proteins.

CRICK: Yes, that's what I think.

DEMOCRITUS: So let's imagine that the sickle cell mutation occurs in the middle of the gene responsible for haemoglobin. In this case, will the altered amino acid be in the middle of haemoglobin? What's your opinion on that?

CRICK: Yes, I think so. Of course to prove the Sequence Hypothesis, we need to show that the DNA chain and the protein chain are *colinear*. To do this, of course, we need to …

ALKIMOS: Leave the details to the biochemists!

CRICK: Excuse me?

EXTON: Alkimos, let Dr. Crick finish his thought!

ALKIMOS: But Exton, I have something really important to ask. Dr. Crick, who discovered DNA? You or Watson?

CRICK: To my knowledge, it was Friederich Miescher.

ALKIMOS: No, I mean the double helix!

CRICK: I'm afraid …

EXTON: Alkimos, please come here.

DEMOCRITUS: Son, instead ask him what his thoughts are on the inheritance of acquired characteristics.

CRICK: Watson or me?

EXTON: What we mean is …

ALKIMOS: Who's smarter? You or him?

EXTON: Alkimos!

CRICK: Perhaps you should ask Watson.

EXTON: Okay, we really need to go.

DEMOCRITUS: You see the central dogma places the concept of acquired characteristics in a new light. Proteins can change, but their changes can't be written into the DNA. Therefore …

ALKIMOS: Dr. Crick, I really loved it that you ran rings around that snooty Pauling.

CRICK: Pauling? What are you talking about?

EXTON: Come on already!

DEMOCRITUS: … only the DNA can be passed on to the next generation, but not the proteins. Unless the ovum …

EXTON: Alkimos, Democritus! Now!

CRICK: What did I run?

EXTON: Alkimos!

ALKIMOS: Exton, let go of me!

EXTON: We're boarding.

DEMOCRITUS: But I still have questions about the Sequence Hypothesis …

ALKIMOS: Aagghhh … here we go!

The Diamond Code of Proteins

ALKIMOS: Master, what did you want to ask Crick?

DEMOCRITUS: Oh, it doesn't matter. Your question was much more important. Besides, I think I know the answer to mine.

ALKIMOS: Then tell me! What was your question? What's the answer?

DEMOCRITUS: The central dogma sheds light on the question of the inheritance of acquired traits. If it's true that information from the proteins cannot be transferred back into the genes, then changes in proteins cannot be inherited.

ALKIMOS: So acquired traits can't be passed on?

DEMOCRITUS: I don't believe so. Yet, proteins are the most basic components of a cell. Proteins are soma on the atomic level.

EXTON: Indeed, the central dogma can be seen as a new molecular formulation of Weissman's germplasm. DNA is passed on along the germ-line, but proteins, or soma, are not. With every generation, they are newly created under the direction of genes. Think of it like this:

PROTEIN   PROTEIN   PROTEIN   PROTEIN   PROTEIN

DNA ⟶ DNA ⟶ DNA ⟶ DNA ⟶ DNA

Insulin can only change during the evolutionary process if the gene that governs it changes.

ALKIMOS: What's insulin?

EXTON: Insulin is a hormone that regulates blood sugar levels. Back then it was easy to obtain since large quantities were purified for use in treating diabetics. Moreover it's a relatively small protein.

ALKIMOS: And what did Sanger do with it?

EXTON: Sanger developed a chemical method by which he could cleave an amino acid at one end of insulin, the so-called N-terminus. This method allowed him to determine the sequence of short sections of the protein. Sanger used protein-digesting enzymes called *proteases*, that cleave the protein chain only next to specific amino acids. This resulted in short peptide chain fragments that he could separate using paper chromatography, and then figure out their sequence. By observing the overlap of sequences of the various peptide fragments he managed to establish the entire sequence.

ALKIMOS: What do you mean by the overlap of sequences?

EXTON: Let's say that one protease cleaved the chain to create these peptides: PEP TIDE SE QUENCE, and another created these: PEPTI DESE QUE NCE. He didn't know the order of them, but based on the two and how they overlap, he could put together the entire sequence.

DEMOCRITUS: And how did this sequence look?

ALKIMOS: Oh, that's easy, master. PEPTIDESEQUENCE!

DEMOCRITUS: Excellent, but I was thinking of insulin.

EXTON: I don't know it off hand. But we can find it in this book. Here:

> Phe–Cal–Asp–Glu–His–Leu–Cys–Gly–Ser–Gis–Leu–Val–
> Glu–Ala–Leu–Tyr–Leu–Val–Cys–Gly–Glu–Arg–Gly–Glu–
> Atg–Gly–Phe–Phe–Tyr–Thr–Pro–Lys–Ala

Insulin consists of two chains, and this is one, the B chain. Sanger published this in 1951, and two years later he published the A sequence, and within five years he determined the sequence of insulin from five different species.

ALKIMOS: We heard Crick talk about it.

EXTON: The determination of the entire sequence of amino acids in insulin was a milestone in molecular biology. Sanger proved that proteins were peptide chains that contained amino acids in a permanent, genetically specific order.

ALKIMOS: But Exton, there's no pattern to the sequence!

EXTON: That's right, Alkimos. Most biochemists expected the sequence to consist of repeated patterns. So the amino acids would follow one another as sugars do in a monotone line in cellulose or other polysaccharides. Others speculated that amino acids were built into the polypeptide chain at random, but in proportion to their frequency. In both cases this would mean that chemical principles determined their structure.

DEMOCRITUS: But according to the laws of chemistry, proteins of this type, which have a monotone or static structure, could be stitched together by a few enzymes.

ALKIMOS: And what stitches together the enzymes that stitch together the proteins?

DEMOCRITUS: Never mind that, Alkimos. You see the theory is incorrect, so there's no point in trying to work out its details. It seems that the completely irregular sequence of proteins is laid down somewhere.

EXTON: That's right. Sanger's work with insulin made it clear that the exact order of amino acids is not governed by chemical principles—it must be encoded somewhere.

ALKIMOS: In genes?

EXTON: In the sequence of bases in DNA. This sequence determines the order of amino acids in proteins. Deciphering how this works was the next great challenge of molecular biologists.

DEMOCRITUS: That means uncovering the DNA code of proteins!

EXTON: The *genetic code.*

ALKIMOS: Master, let's get to it!

DEMOCRITUS: I'm afraid we don't know enough to do that.

EXTON: A lot of people wanted to uncover the genetic code using only theoretical tools. A theoretical physicist named George Gamow made one of the earliest attempts. The articles by Crick, Watson, and Sanger inspired him to devise a scheme in 1954 that showed how the sequence of DNA directly determined the sequence of amino acids in a protein chain.

ALKIMOS: How could the DNA code determine it directly if DNA is made up of only four bases, but there are a lot more amino acids? It would only make sense if there were just four amino acids. Then one base would decide the position of just one amino acid in the chain.

EXTON: Unfortunately, there are twenty amino acids. So you're right; it's impossible that one base could be associated with just one amino acid. But Gamow found a solution. He reasoned that if genes were made of DNA, and DNA was none other than two chains made up of four different nucleotides that run alongside each other and are held together by their base pairs, then "the hereditary properties of any given organism could be characterised by a long number written in a four-digital system." In the four-digital or base-4 numeral system, only the numbers 0, 1, 2, and 3 are used.

DEMOCRITUS: So a DNA number would look like this:

01320132220101322000133201332022202311.

ALKIMOS: 0 is adenine, 1 if thymine, 2 is cytosine, and 3 is guanine.

EXTON: Yes, that's the idea. The number could have as many as a million digits. Gamow believed that if each amino acid were assigned a letter of the alphabet, then proteins "can be considered as long 'words' based on a twenty-letter alphabet."

ALKIMOS: So a protein text would look like this:

ACDEFHTRDSEEDDDDRTKILP

EXTON: Yes, it could. This figure shows the twenty amino acids occurring in proteins:

Glycine (Gly, G)   Alanine (Ala, A)   Serine (Ser, S)   Threonine (Thr, T)   Cystein (Cys, C)

Valine (Val, V)   Leucine (Leu, L)   Isoleucine (Ile, I)   Methionine (Met, M)   Proline (Pro, P)

Phenylalanine (Phe, F)   Tyrosine (Tyr, Y)   Tryptophan (Trp, W)   Aspartate (Asp, D)   Glutamate (Glu, E)

Asparagine (Asn, N)   Glutamine (Gln, Q)   Histidine (His, H)   Lysine (Lys, K)   Arginine (Arg, R)

It shows their chemical structures and the three-letter and one-letter abbreviations of their names. Learn them.

DEMOCRITUS: The question is how do we translate the base-4 DNA number into the twenty-lettered 'language of proteins'? It's clear that one DNA digit can't refer to one protein letter. But maybe two digits can?

ALKIMOS: A two-digit number?

DEMOCRITUS: Why not? Let's suppose that two bases that follow each other in a sequence determine one amino acid. But I'm afraid we'll still have trouble assigning a clear code to each of the twenty amino acids. There are four kinds of bases in DNA, so we can only pair them up sixteen different ways.

ALKIMOS: AA, AC, AG, AT, CA, CC, CG, CT, GA, GC, GG, GT, TA, TC, TG, and TT. Isn't that right, master?

DEMOCRITUS: That's right, but we need at least four more combinations.

EXTON: Gamow came up with a solution that resulted in exactly twenty combinations. He postulated that four bases near to each other created diamond-shaped pockets in the *major groove* of the DNA double helix spiral. The codes were not created by the sequence of four bases on one strand, but by two complementary pairs and just one member of the base pair that falls either above or below in the double helix. According to Gamow, the diamond formed by the four bases creates a hole in which exactly one amino acid fits, like a key in a lock. Exactly twenty different diamond holes are created in this way, as you can see here in Gamow's diagram:

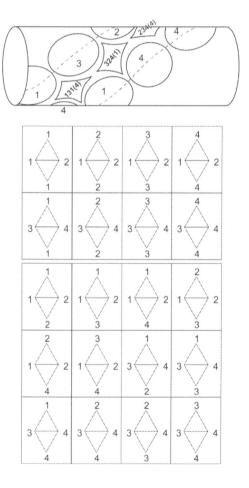

DEMOCRITUS: Exactly the number of amino acids!

ALKIMOS: But how does it work out to exactly twenty?

EXTON: Let's take a closer look. Each hole is defined by only three nucleotides, since two nucleotides located across the axis of the cylinder are related by base-pairing. In the diagram 1 pairs with 2 and 3 with 4. The first eight holes are simple, with a 1–2 or 3–4 pair flanked by two identical nucleotides. In the remaining twelve combinations the pairs are flanked by different nucleotides. Assuming that the right-handed and left-handed versions have the same surface, we cannot create more combinations. That's twenty altogether.

DEMOCRITUS: Amazing. It really is twenty!

EXTON: Yes.

DEMOCRITUS: And which amino acid corresponds to which diamond?

EXTON: Gamow of course had no idea.

ALKIMOS:

> *The dowry of mortals engraved in Gamow's diamond code,*
> *A puzzle that generates the strange living texts of enzymes.*

EXTON: In order to interpret the strange protein texts, scientists also generated theories about the code. Crick, who had a far better understanding of stereochemistry, immediately spotted the flaws in Gamow's theory. He considered it impossible that amino acids of such varying structures could fit exactly into the DNA pockets proposed by Gamow. But that wasn't his biggest problem with Gamow's theory.

ALKIMOS: Then what?

EXTON: At the time it was nearly certain that DNA did not regulate protein synthesis directly but used an RNA molecule as an intermediary, and the proteins formed not in the cell nucleus, but in the cytoplasm.

DEMOCRITUS: I see, and the DNA is in the nucleus.

ALKIMOS: Exton, what do you mean that RNA acts as an intermediary? Are you suggesting it's not one gene–one enzyme, but one gene–one RNA–one enzyme?

EXTON: A lot of observations certainly seemed to support the idea that RNA — the other nucleic acid form found in cells—played an important role in protein synthesis. In the early 1940s, Jean Brachet, a molecular biologist working in Brussels, developed a cytological method for staining RNA. He managed to directly and specifically stain RNA within the cell, and thereby was able to demonstrate that in cells that secrete lots of protein, the level of RNA is high. In cells at rest, such as muscle cells, the level is low. In other words, a high level of protein production is accompanied by a high level of RNA. The majority of RNA is found in the cytoplasm, while DNA is found in the cell nucleus. Brachet concluded that DNA was the hereditary material, while RNA played a role in protein synthesis.

DEMOCRITUS: That's a pretty bold conclusion given the data.

EXTON: Later Brachet showed that enucleated amoebas and algae are capable of protein synthesis.

DEMOCRITUS: That means the nucleus and DNA aren't necessary for protein synthesis!

ALKIMOS: Well then, why do we need genes?

DEMOCRITUS: I suppose for regulating the cytoplasm. Then it seems, the cytoplasm is capable of functioning on its own, at least for a while. If Brachet is right, the cytoplasmic RNA regulates protein synthesis, and the genes …

ALKIMOS: … regulate the synthesis of cytoplasmic RNA!

DEMOCRITUS: That's right. Since the genes are attached to the chromosomes, they can't leave the nucleus. They can, however, make a copy of themselves that can circulate freely and direct protein synthesis in the cytoplasm.

ALKIMOS: So RNA is a copy of a gene!

DEMOCRITUS: I believe so. RNA is also composed of nucleotides, as shown by Phoebus Levene. Why couldn't it form a chain that is constructed using a DNA template, according to base pairing?

ALKIMOS: But master, Exton said that in ribonucleic acid thymine is replaced with uracil …

DEMOCRITUS: Then we have to assume that like thymine, uracil is capable of creating hydrogen bonds with adenine.

ALKIMOS: To form a UA pair.

EXTON: Democritus, what brilliant reasoning!

DEMOCRITUS: Oh, it was nothing …

ALKIMOS: Exton, does RNA also form a double helix?

EXTON: Once the structure of DNA was deciphered, Watson experimented with RNA's structure using the same methods …

ALKIMOS: If RNA has a double helix, then we might be able to save Gamow's theory. The only difference is the amino acids would sit in the pockets formed by RNA rather than DNA!

EXTON: Ah, but that's not what Watson found. It seemed that RNA does not have a specific, repeating spatial structure. Later scientists discovered why: RNA does not form a double helix. Mostly it consists of just one thread, but it can create short double strands of complementary pairs by folding back on itself.

DEMOCRITUS: That makes it hard to save Gamow's diamonds. But perhaps the messenger RNA can give us some ideas of how to decipher the code. Go on, Exton!

EXTON: Others also had the idea that RNA molecules were messengers for genes, especially in light of studies done on the carbohydrate metabolism of bacteria.

ALKIMOS: What's that?

EXTON: Carbohydrate metabolism is the consumption of sugars. You see, bacteria love sugar. Their favourite is glucose, but if they can't get that, then lactose—the sugar in milk—will do.

ALKIMOS: I assume they're happiest if both are present, then they have a rip-roaring good time.

EXTON: No, they're pickier than that. If both sugars are present, they'll consume just glucose. They turn to lactose only when all the glucose is gone. Look at these graphs made by the French biochemist Jacques

Monod. The curves reveal the two-stepped growth pattern of the bacteria when more than one sugar is present.

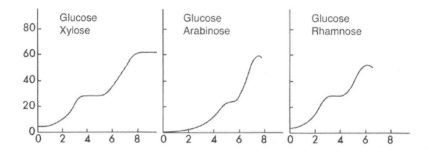

The culture grows as long as glucose is present. Then the bacteria have to shift their consumption to galactose or xylose or other less tasty sugars. The bacteria are reluctant at first, then resign themselves to eating the other sugar and so begin growing again.

ALKIMOS: I see. It's like a first and second course. But what does this have to do with RNA?

EXTON: At the end of the 1930s, Monod, who worked in Paris, began to investigate the carbohydrate metabolism of *Escherichia coli*, a bacterium found in the human intestine. He discovered that if the bacteria are fed lactose, this triggers the production of three proteins, including the enzyme *β-galactosidase*, which breaks down lactose. But none of these proteins are produced if the bacteria are grown on a medium containing glucose.

DEMOCRITUS: So the bacteria use their genes according to their needs. It's as if they adapt to their environment in a sort of Lamarckian fashion. Fantastic!

ALKIMOS: So they're able to switch on their genes?

DEMOCRITUS: And presumably also switch them off. But if you think about it, it's not all that surprising. Without such a mechanism genes would be in operation all the time, and a tremendous amount of protein would be produced unnecessarily. Not only bacteria, but all living things must be capable of using genes according to their needs.

ALKIMOS: But what happens when a gene is switched on?

EXTON: In 1957, Monod and his colleague François Jacob argued that a polynucleotide messenger must exist that copies the DNA—so that it consist of corresponding nucleotides—and then enters the cytoplasm where it quickly breaks down. They called this hypothetical molecule the messenger RNA, abbreviated to mRNA.

ALKIMOS: Ah, mRNA, the messenger of genes. But why is it so short-lived?

EXTON: Once the RNA has delivered its message to the proteins, there's no longer a need for it.

ALKIMOS:

> *Those despotic genes, such little time you're given,*
> *Handsome messenger! Just one round, you agile runner,*
> *Then gone: the polynucleotide, the bearer of veiled words.*

EXTON: Scientists managed to prove experimentally the existence of a short-lived RNA form. After phage infection, new RNA quickly appears in the affected bacteria and then very soon breaks down. At the same time, the cell is full of stable forms of RNA, and the quantity of these forms doesn't change during infection.

DEMOCRITUS: So there are stable and short-lived RNA forms.

EXTON: Before the short-lived mRNA decays it adheres to mysterious cytoplasmic particles. These particles are found in all cells, and there was some evidence that they are the site of protein synthesis. When, for example, these particles were extracted from red blood cells, a little haemoglobin was always found attached to them. Similarly, the particles extracted from the pancreas had insulin stuck to them. Yet, neither insulin nor haemoglobin is a permanent component of the particles. Instead it appeared that the proteins were made on these cytosplasmic particles; the proteins discovered on them were there because their production was in process.

DEMOCRITUS: So proteins are made on cytoplasmic particles according to instructions delivered by the mRNA?

EXTON: Exactly. The crucial experiment that revealed the importance of the particles was carried out in Paul Zamecnik's lab. Amino acids were marked with carbon-14 or ^{14}C, which is not a stable atom. When it decays it emits radiation that can be measured. ^{14}C is a *radioactive isotope* of carbon. The most common naturally occurring isotope of carbon is carbon-12, carbon-14 occurs in trace amounts.

DEMOCRITUS: I see, isotopes. A minor extension of my atomic theory.

EXTON: The researchers in Zamecnik's lab were interested in where protein synthesis takes place: in other words, where in the cell would they first find the radioactive amino acids? They injected rats with ^{14}C labelled amino acids, and then prepared an extract from their liver.

ALKIMOS: Quite a witch's brew.

EXTON: Radioactivity appeared first on the cytoplasmic particles, and only after was it found incorporated in the proteins.

DEMOCRITUS: So amino acids are built into proteins on these particles, and once the protein is complete, it detaches from the particle.

EXTON: That's right. They called these protein generating particles *ribosomes.*

ALKIMOS: The radioactive ribosomes of rat's liver. So I guess that's it for Gamow's diamond code …

EXTON: Don't worry, Alkimos. There are more elegant theories to come. Remember we still haven't cracked the genetic code.

DEMOCRITUS: But at least we know where the cells decode the information.

ALKIMOS: On the ribosomes.

EXTON: Yes, the ribosomes read the information encoded in the nucleotide sequence of mRNA and translate it into the amino acid sequence of proteins during the process of *translation.*

ALKIMOS: Very simple.

DEMOCRITUS: Son, I believe it deserves a couplet!

ALKIMOS:

> *From sequence comes sequence, from a gene a lanky, spindly protein,*
> *Its code cracked in the generous lap of a hearty ribosome.*

The Genetic Code

EXTON: Crick believed that because the genetic code was at least three billion years old, it might be impossible to reconstruct the series of events that led to its creation. The origin of the genetic code must be closely tied to the origin of life.

ALKIMOS: Three billion years old?

EXTON: Yes, life on Earth is more than three billion years old.

DEMOCRITUS: Incredible.

ALKIMOS: I can't even imagine that much time. The immortal gods of Olympus have been around that long?

EXTON: Possibly.

ALKIMOS: And the Titans?

EXTON: They've been around for even longer.

DEMOCRITUS: Why did the genetic code come into existence then?

EXTON: Crick argued that once the genetic code was established, it could not have changed, because even one small alteration would affect so many proteins that it would have certainly destroyed some of them.

ALKIMOS: So if a living creature experimented with changing the code, it would have died in the process.

DEMOCRITUS: So we can reject Gamow's code. Protein synthesis takes place on the ribosomes and is directed by the gene's messenger, the mRNA. Therefore, the double helix does not attract amino acids to itself. Instead the amino acids join up along the mRNA.

ALKIMOS: But if the mRNA doesn't form a double helix, then what kind of pockets can the amino acids sit in?

DEMOCRITUS: Maybe there are no pockets, no diamonds.

EXTON: Crick struggled with this question as well. He couldn't imagine how the DNA or RNA molecule could serve as a direct template for the large variety of side chains of amino acids. It appeared stereochemically impossible. But after a considerable amount of brainstorming, he came up with an unusual idea. He postulated that during protein synthesis, twenty small adaptor molecules made up of nucleic acid acted as transmitters between the amino acids and the mRNA.

ALKIMOS: So every amino acid has its own …

EXTON: Then Crick went on to imagine twenty special enzymes. According to his *adaptor hypothesis*, a special enzyme attaches a specific amino acid to the appropriate adaptor. The pair then diffuses to the mRNA template. In his mind the adaptor molecule was a small unit consisting of three nucleotides.

DEMOCRITUS: Brilliant!

EXTON: This small molecule connects to the appropriate mRNA template through complementary base pairing.

DEMOCRITUS: Right. It can only bind to the places where it can form the necessary hydrogen bonds.

EXTON: And with the adaptor bonded in the right place, it can transport the amino acid to the exact place it needs to be.

DEMOCRITUS: The ribosome then links the amino acid to the chain by a peptide bond; thus the protein chain increases by one amino acid.

ALKIMOS: And the next adaptor arrives with an amino acid on its back?

EXTON: That's the essence of Crick's idea.

DEMOCRITUS: It's ingenious!

EXTON: And it's correct. Crick's hypothetical adaptors were later proven to exist. However they contain about 90 nucleotides, not just three. But only three of their nucleotides bind to the mRNA by base pairing, which is the critical element of the theory. These molecules are called *transfer RNA* or tRNA.

ALKIMOS: A new kind of RNA!

EXTON: Of course the question still remained of how the sequence of bases in the nucleic acid corresponds to the sequence of amino acids in a protein.

DEMOCRITUS: It appears that three bases are needed to determine one amino acid. The sixty-four different possible combinations of bases are more than enough to code for twenty amino acids.

ALKIMOS: So there are sixty-four tRNA adaptors?

EXTON: In reality there are always less than sixty-four. But in any case, there are more base triplets, which Crick called *codons*, than amino acids. This means that more than one triplet could code for one amino acid. In Crick's words, the genetic code was *degenerate*.

ALKIMOS: It seems so haphazard. Why would there be more than one triplet for each amino acid?

EXTON: Indeed this degeneracy strikes us as imperfect. This is why Gamow's diamond code was so elegant: it provided us with the magic number twenty: twenty pockets for twenty amino acids. But you'll see Crick soon enough developed another amazing theory that possesses similar elegance: the comma-free code. But before I explain it, however, we need to clarify some things. The first crucial question was whether the codons overlapped?

ALKIMOS: What does that mean?

EXTON: When we read a genetic message, does a base operate in just one codon, or does it play a role in several overlapping codons simultaneously? This diagram illustrates the question:

```
GACGACGACGACGAC              RNA message
_____
GAC
 ACG
  CGA                        overlapping code
   GAC
    ACG
_____
GAC
 CGA
  ACG                        partially overlapping code
   GAC
    CGA
_____
GAC
  GAC
    GAC                      non-overlapping code
      GAC
        GAC
_____
```

DEMOCRITUS: So the genetic code in theory can contain overlapping, partially overlapping, and not overlapping codons.

ALKIMOS:

> *In my marrow, wet and gangling snails twist and turn,*
> *Full of codons, luring each other into a wicked dance.*

EXTON: As you say, Alkimos. Now, there's something we should notice. If the code is overlapping, that would put serious restrictions on the sequence of amino acids in a protein. In this case, the sequence GACC would determine two amino acids, in other words a dipeptide. Because there are four bases, this would mean that a maximum of 4 × 4 × 4 × 4, or two hundred and fifty-six different dipeptides could occur in proteins.

DEMOCRITUS: And in a non-overlapping code it would be 20 × 20 or four hundred possible dipeptides. An amino acid could be followed by any other amino acid, without restriction.

EXTON: An overlapping code therefore decreases the number of possible dipeptides in a protein.

DEMOCRITUS: But it's easy to verify the existence of an overlapping triplet code. All we need to see is how many dipeptide sequences Sanger discovered. If it's more than 256, then we can reject the concept of an overlapping code.

EXTON: But only the concept of an overlapping *triplet* code, or one consisting of three nucleotides.

DEMOCRITUS: Ah, you're so right, Exton. Overlapping codons consisting of four, five, or six nucleotides would still be possible.

EXTON: The difficulty is that Sanger and the biochemists who borrowed his method did not identify enough of the protein's sequences. But another daredevil of molecular biology, the South-African born Sydney Brenner, a good friend and close colleague of Crick's and another renegade of the time, proved in 1957 that an overlapping triplet code was not possible. He calculated that to code for all the known dipeptides at the time, a minimum of seventy different triplets would be necessary.

ALKIMOS: But there can only be sixty-four triplets!

EXTON: That's right. So we can exclude the existence of an overlapping triplet code. But there are other grounds for rejecting it. When we have a mutation in a gene that alters a protein, such as in sickle cell haemoglobin or in the various insulin sequences, it's only one amino acid that is altered.

DEMOCRITUS: Marvellous reasoning! If the code were overlapping, a mutant nucleotide would alter two neighbouring amino acids!

ALKIMOS: If we've rejected Gamow's code and the overlapping triplet code, then what do we have left?

EXTON: Partially overlapping codes, but they're not as elegant. We also have non-overlapping triplet codes. Here's how Crick described the critical problem posed by this idea: 'How is the base sequence, divided into codons? There is nothing in the backbone of the nucleic acid, which is perfectly regular, to show us how to group the bases into codons. If, for example, all the codons are triplets, then in addition to the correct reading of the message there are two incorrect readings which we shall obtain if we don't start the grouping into sets of three at the right place.'

DEMOCRITUS: If we start reading the triplets in the wrong place, then we'll completely misunderstand the message.

EXTON: This problem didn't exist with the completely overlapping triplets.

ALKIMOS: Why not?

EXTON: Let's look at our earlier example. If we have a non-overlapping triplet code of GACGACGACGACGAC, we can read it three different ways:

1st reading frame: GAC,GAC,GAC,GAC,GAC

2nd reading frame: G,ACG,ACG,ACG,ACG,AC

3rd reading frame: GA,CGA,CGA,CGA,CGA,C

DEMOCRITUS: So these are the three possible reading frames, and clearly the meaning of each is different. And probably only one of the reading frames makes sense. But how does the ribosome know where to start reading?

EXTON: Crick's comma-free code provides us with a brilliant solution.

DEMOCRITUS: We're all ears, Exton.

EXTON: Crick's idea was the following: imagine a code in which only some triplets make sense; in other words they code for an amino acid, and the rest are nonsense. Is it possible that there is just one reading frame that contains meaningful triplets, and the other two don't?

ALKIMOS: Then it would be clear which reading frame should be used.

EXTON: Yes, that's exactly what's so elegant about the comma-free code. Only one reading frame codes for amino acids. Here's how we can illustrate this:

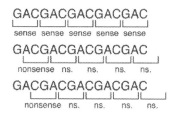

DEMOCRITUS: If I understand correctly, then only the CAG codon codes for an amino acid, and ACG and GCA don't.

ALKIMOS: How?

DEMOCRITUS: Look at the CAG CAG CAG sequence.

ALKIMOS: Oh, I see. If CAG is meaningful, then the other two reading frames are meaningless, so AGC and GCA are nonsense.

EXTON: That's exactly right.

DEMOCRITUS: So the question is: can this in theory be used to construct a code that works for all twenty amino acids?

EXTON: Crick showed that it could.

DEMOCRITUS: But again, if we apply this principle won't it severely limit the number of combinations?

EXTON: Crick also proved that the maximum numbers of amino acids that could be encoded is twenty.

ALKIMOS: The magical number twenty!

EXTON: And the proof was extraordinarily elegant. First, let's have a look at triplets made up of the same nucleotide.

ALKIMOS: AAA, GGG, CCC, and TTT.

EXTON: It's clear that these can't be used, because if two amino acids encoded by these combinations are next to each other, the reading frame will be ambiguous.

DEMOCRITUS: If we have AAAAAA, we don't know which three As are the codons.

EXTON: So we can exclude triplets made of the same nucleotides. The remaining sixty triplets can be divided up into twenty triplet groups. In each group are three combinations that overlap in a circular way, such as GAC, ACG, and CGA. It's easy to see that we could choose any one these, but no more than one.

DEMOCRITUS: I understand. If we assume that the code for the K amino acid is GAC, then the code for the KK dipeptide is GACGAC. According to the rules of Crick's theory, ACG and CGA must be nonsense.

ALKIMOS: So if just one remains from each of the twenty groups of three, then we have altogether twenty!

DEMOCRITUS: The next step is to figure out which twenty.

EXTON: This is one possible solution:

AGA	AGG	ACA	ACG	ACC
GCA	GCG	GCC	AUA	AUG
AUC	AUU	GUA	GUG	GUC
GUU	CUA	CUG	CUC	CUU

The other forty-four codes are nonsense. Whichever two 'sense' triplets we put side by side, if we shift the reading frame over, the codons will be nonsense.

ALKIMOS: Because the codons won't be among the twenty 'sense' codons.

EXTON: Give it a try! You'll see that it works.

DEMOCRITUS: Yes, it's clear. Only one reading frame codes for the amino acids, and the others are nonsense.

ALKIMOS: So have we solved it? Is this the genetic code?

DEMOCRITUS: Now slow down, Alkimos. We still don't know which triplet corresponds to which amino acid.

EXTON: What's more, in theory there are two-hundred and eighty-eight solutions to the comma-free code.

ALKIMOS: Two hundred and eighty-eight genetic codes? And we still don't know anything about the amino acids? Exton, please don't tell me this was just a pointless mental exercise!

DEMOCRITUS: An invigorating mental exercise!

ALKIMOS: Oh that Olympian, that Cloud-Gatherer, that ambrosia-eating …

EXTON: Let's not upset Alkimos. I must admit something that Crick also realised. The genetic code cannot be broken using theory. There are too many possible solutions. Whether we like it or not, genetics and theory have no choice but to yield to biochemistry. To decipher the genetic code, scientists needed to pull apart the cells; they needed to look at what was going on inside.

ALKIMOS: Oh no, we're breaking things apart again? I thought we were putting things back together?

EXTON: We are, we are. But we have to start with proteins.

ALKIMOS: And then we can get to the cells?

EXTON: Oh, we don't need cells right now. They aren't necessary for building a protein. Did you know that protein synthesis can take place *in vitro* if all the necessary components are present? Paul Zamecnik realised this in the early 1950s. He centrifuged rat liver extract at a slow speed. The resulting liquid was cell- and nucleus-free. But it did contain tiny microsomal particles—the ribosomes—and lots of protein. When he added ATP, the molecule that supplies a cell with energy, protein synthesis occurred. Zamecnik made a similar extract from *E. coli* bacteria.

ALKIMOS: A protein factory without cells? Genes cooked in a beaker!

EXTON: And artificial genes at that! It was a breakthrough in the deciphering of the genetic code when the American biochemist Marshall Nirenberg and his German postdoctoral fellow Heinrich Matthaei began to use synthetic RNA in Zamecnik's cell-free system of modified *E. coli.* By synthetic RNA I mean RNA produced in a test tube. When they added RNA made entirely of uracil, poly-U RNA, in the end they obtained only polyphenylalanine, meaning peptides made only of the amino acid phenylalanine.

ALKIMOS: I get it, so from UUUUUUUUUUUU they made FFFFF.

DEMOCRITUS: Son, we have the first codon of the code! UUU makes phenylalanine.

EXTON: In 1961, they made a discovery that proved beyond a doubt that a single-strand RNA molecule really can work as a messenger. The systematic deciphering of the genetic code had begun. Nirenberg and associates demonstrated that RNA made only of cytosine caused the formation of polyproline.

ALKIMOS: So CCC is the code for proline.

EXTON: You can see that this discovery immediately disproved Crick's comma-free code, since his theory stated that a triplet could not contain three of the same bases.

ALKIMOS: So we can reject 288 codes because of one experiment. All that thinking was for nothing.

EXTON: It certainly wasn't. Look at how many concepts were arrived at: a genetic code built of triplets, a non-overlapping system, even the notion that a code existed at all. And this was crucial. Without this fundamental idea, biochemists would have had nothing to work with, because they hadn't even understood there was a problem. The nomenclature, too, derives from all that thinking!

ALKIMOS: I guess I see that. So what do GGG and AAA code for?

EXTON: GGG for glycine and AAA for lysine. Up to this point it was simple. But what about, say, GUC?

ALKIMOS: It's easy, isn't it? Create a GUCGUCGUC polynucleotide, add it to the magical extract, and see what we get.

DEMOCRITUS: A simple experiment.

EXTON: But in 1961, no one could produce a polynucleotide with a specific sequence. They could make ones with random sequences, however, with the help of an enzyme called *polynucleotide phosphorylase*, discovered by Severo Ochoa in 1951. This enzyme can synthesise polynucleotides from mononucleotides. Using this enzyme, biochemists could synthesise for example RNA molecules containing C and A in a ratio of 5:1, but in a random sequence. But it's possible to calculate how frequently a certain triplet will appear in the polynucleotide. For example CCC is five times as likely as CCA, CAC, or ACC. If such an RNA is added to the extract, it directs the incorporation of several amino acids into the polypeptide chain in different proportions. By 1962, experiments like this had allowed scientists to establish the nucleotide composition, but not the sequence, of all codons.

DEMOCRITUS: Fantastic!

ALKIMOS: We're close now, master!

EXTON: To truly solve the problem, scientists needed to synthesise polynucleotides with a specific sequence. The person who did it was a chemist, Har Gobind Khorana. At the time, short DNA oligonucleotides with a specific sequence could be created. Khorana used these and the enzymes DNA-polymerase and RNA-polymerase. The first enzyme is capable of copying a DNA molecule, and the second can use a DNA molecule as a template to synthesise a complementary RNA molecule. With these enzymes and the short nucleotides Khorana produced short RNA molecules with a specific sequence.

ALKIMOS: Which he added to a cell-free extract.

EXTON: When he added the UGUGUG polymer, he obtained a valine–cysteine–valine–cysteine (VCVC) polymer.

DEMOCRITUS: In the UGUGUG polymer only the UGU and GUG codons are involved.

ALKIMOS: Which is the code for valine and which is the one for cysteine?

DEMOCRITUS: That's easy to determine, because we already know the composition of the amino acid codes.

EXTON: The code for cysteine contains two Us and a G.

ALKIMOS: So UGU is the code for cysteine!

DEMOCRITUS: Naturally GUG is the code for valine.

EXTON: Conducting similar experiments, Khorana managed to solve the entire genetic code. Nirenberg and associates had similar success using a different method. With the help of three-member nucleotides, they were able to bind tRNA to ribosomes. Using trinucleotides of a known sequence, they could determine the corresponding RNA and decide which amino acid linked to the RNA. This is how they obtained each letter of the code. Of the sixty-four triplets only UAA, UAG, and UGA don't code for any amino acids. These codons are called stop codons or termination codons and they signal the end of a protein chain. They also have been given names, UAA is *ochre*, UAG is *amber*, and UGA is *opal*. In 1966, the entire code was complete.

ALKIMOS: Let's see it.

EXTON: Here you are:

2nd → 1st ↓	U	C	A	G	3rd ↓
U	Phe Phe Leu Leu	Ser Ser Ser Ser	Tyr Tyr *Ochre* *Amber*	Cys Cys *Opal* Trp	U C A G
C	Leu Leu Leu Leu	Pro Pro Pro Pro	His His Gln Gln	Arg Arg Arg Arg	U C A G
A	Ile Ile Ile Met	Thr Thr Thr Thr	Asn Asn Lys Lys	Ser Ser Arg Arg	U C A G
G	Val Val Val Val	Ala Ala Ala Ala	Asp Asp Glu Glu	Gly Gly Gly Gly	U C A G

ALKIMOS: Master, the code!

DEMOCRITUS: Yes son, we now understand how the language of some of the cell's large polymers, nucleic acids, can be translated into the language of the protein's polymers. The genetic code is certainly one of the most fundamental aspects of biochemistry. It gave concrete meaning to abstract concepts of genetics. This code finally made it possible to link genotype to phenotype—at least on the level of DNA and proteins.

ALKIMOS: If we know the gene, we know the protein it's coding for!

DEMOCRITUS: But the proteins can't tell us the gene's structure, because the code is degenerate.

ALKIMOS: So if the protein contains valine, we won't know if the mRNA has GUU, GUC, GUA, or GUG.

DEMOCRITUS: The degeneracy of the code strengthens our belief in the central dogma—the idea that genetic information flows in only one direction.

EXTON: Let's summarise what we know about the structure of the generic code:

- The code's codons are composed of three bases next to each other, or triplets.
- A codon does not overlap with neighbouring codons.
- More than one triplet codes for every amino acid except tryptophan and methionine; thus the code is degenerate.
- Triplets that code for the same amino acid are generally similar.
- Three triplets do not code for any amino acid; these are stop codons; if a ribosome reaches a stop codon protein synthesis ceases.
- The message is read in triplets from a specific starting point.
- The gene sequence is colinear with the amino acid sequence; synthesis of the polypeptide chain begins at the N-terminus.
- The code—with only a few exceptions—is *universal*, which means it is generally valid for the entire living world.

DEMOCRITUS: Colinearity and universality!

EXTON: Experiment proof of colinearity is one of the best chapters in the genetics of phages. Sydney Brenner and associates are responsible for the proof. While studying phages, they found mutations that stopped protein synthesis prematurely. This happened when a mutation changed a 'sense' codon for a particular amino acid into a stop codon. The ribosome assumed it had reached the end of the message and stopped synthesis before the protein was completed. A stop codon can appear anywhere in the chain if a mutation has occurred; therefore truncated proteins of various sizes are associated with different mutations. These so-called *nonsense mutations* can be genetically mapped.

DEMOCRITUS: So Brenner and disciples had to show the order of the nonsense mutations on the genetic map. Then they had to show that in the same order the truncated proteins get shorter and shorter.

ALKIMOS: Why did they have to show that?

DEMOCRITUS: It's simple. If a nonsense mutation is at the beginning of the sequence, then, because of colinearity, the truncated protein chain would be short. If the mutation is at the end, then the truncated protein would naturally be longer.

EXTON: Yes, the farther the nonsense mutation is from the starting point of protein synthesis, the longer the truncated protein.

DEMOCRITUS: But how could they find the truncated proteins among the thousands of proteins in the phage-infected bacteria?

EXTON: The protein that forms the coating of the phage's head represents 70% of the proteins produced after infection. This made it easy to biochemically analyse this protein. Nonsense mutations were also isolated in the gene. By comparing the size of the truncated protein and the genetic mapping of the mutation, Brenner and associated proved that the gene and the protein it codes for are colinear.

DEMOCRITUS: Astounding! The language of the DNA polymer can be translated according to the genetic code. The DNA text consisting of three letters tells us the one letter of the protein text. The process doesn't work in reverse because of the degeneracy of the code. The code doesn't allow changes in the proteins to be retranslated back into the language of the DNA, which means these changes cannot be passed on.

EXTON: According to Jacques Monod: 'What molecular biology has done …is to prove beyond any doubt but in a totally new way the complete independence of the generic information from events occurring outside or even inside the cell—to prove by the very structure of the genetic code and the way it is transcribed that no information from outside, of any kind can, ever penetrate the inheritable message.'

ALKIMOS: Lamarck's blacksmith strengthened his arms for nothing. His son has to start over again from scratch.

EXTON: Let's put it a different way: a lot of muscle protein won't produce a lot of muscle genes. Here's how Monod continued his thought: 'with that and the understanding of the random physical basis of mutation that molecular biology has provided, the mechanism of Darwinism is at last securely founded.'

DEMOCRITUS: Because we understand how mutations occur and are passed on, and also how these changes affect the proteins that regulate and build somas.

ALKIMOS: Now I understand what the DNA sequence is good for. It's used to create an RNA sequence, which is used to make a protein sequence. But why does a protein need a sequence?

EXTON: To understand that we need to know more about proteins. We'll continue after coffee.

A Molecular Lung—*Max Ferdinand Perutz*

ALKIMOS: Crick also said that we need to know more about proteins.

DEMOCRITUS: The central dogma can help us.

ALKIMOS: It can?

DEMOCRITUS: Proteins can never pass on their sequence.

ALKIMOS: Okay, but what do proteins do with it then?

EXTON: We're going to find out. Direction Cambridge, 1968!

ALKIMOS: What's in Cambridge?

EXTON: We're going to visit Max Ferdinand Perutz. Let's go!

ALKIMOS: I'd rather not …

EXTON: Come on!

ALKIMOS: Aaaghh … here we go!

EXTON: Max Perutz was born in Vienna in 1914. He studied to be an organic chemist, but he found the subject boring and became more interested in biology. In Vienna, the field of biochemistry was lagging, so Perutz decided to go to Cambridge, England. In 1936, he began work in Desmond Bernal's X-ray crystallography lab. After toiling with some inorganic crystals, he turned his attention to the X-ray crystallographic study of haemoglobin.

DEMOCRITUS: Haemoglobin proteins can be crystallised?

EXTON: Yes, they can.

ALKIMOS: And do they create double spirals?

EXTON: Let's ask Perutz. We're here.

DEMOCRITUS: Ah, Cambridge—nowhere else have they discovered as much about the atoms of life as in this town!

ALKIMOS: Except for maybe Abdera.

DEMOCRITUS: Yes, perhaps.

ALKIMOS: Cambridge is the Abdera of the West!

EXTON: The Cavendish Laboratory is over there.

ALKIMOS: Couldn't we stop for a bite to eat?

EXTON: When we get back. We have a lot to do. Besides I don't think you'd find the food too exciting …

ALKIMOS: You can't be sure …

EXTON: Now come on, in we go. Up the stairs. There's Perutz's lab.

ALKIMOS: Is Crick here too? I think my master still has some things to discuss with him.

EXTON: I don't know, but be quiet. Excuse me, Dr. Perutz?

PERUTZ: Yes?

EXTON: Do you mind if we come in?

PERUTZ: Oh, not at all. Dr. Exton from Sussex, if I'm not mistaken.

EXTON: Yes, and these are my students.

PERUTZ: What a pleasure. Are you studying the structure of proteins too?

EXTON: Well, yes and no …

ALKIMOS: We're interested in the atoms of life and how they work. That's why we're here. Right, Exton?

EXTON: Yes.

ALKIMOS: We'd like to know why you're working with proteins and what you've discovered so far.

EXTON: But only if …

PERUTZ: Of course. So you ask, why proteins? Let's see, not too long before I moved to Cambridge to work with Bernal, it was discovered that enzymes catalyse every chemical reaction and every enzyme is a protein.

ALKIMOS: Yes, we've heard that.

PERUTZ: Then you've probably also heard that genes were thought to be proteins too. But we didn't know anything about the structure of proteins or the mechanisms by which they worked.

DEMOCRITUS: According to Crick, molecular biology is pointless unless we understand the secret of how proteins function.

PERUTZ: We thought that, too. Bernal had the idea that if we could figure out their structure, then we'd understand how they work.

DEMOCRITUS: So if we knew the position of every atom in an enzyme, then we'd understand how they can perform catalysis?

PERUTZ: Yes. For example, where do enzymes bind to the substrate?

ALKIMOS: Democritus, does he mean which atom of the material produced fits into which atom of the mould?

DEMOCRITUS: I believe so.

PERUTZ: Bernal's idea was bold and much more creative than any other idea postulated by enzymologists. In fact, twenty years later enzymologists still have their doubts about the crystallographers' approach. They were sceptical that resolving the structure of enzymes would lead to an understanding of the functional mechanisms of proteins.

DEMOCRITUS: But why couldn't Bernal be right? After all, discovering the structure of DNA shed light on the method by which complementary strands make copies.

ALKIMOS: A with T and G with C.

PERUTZ: Let me remind you that at the end of the 1930s, we didn't know anything about the role of nucleic acid—at least not here in Cambridge. The Secret of Life—in capital letters—lay in enzymes. We considered them the true components of a living cell. This is why we thought if we could determine their sequence, we could trace the workings of a living cell back to simple laws of physics and chemistry.

ALKIMOS: Well then, we should do it. What's the problem? We read a few things from the X-ray image and then we try to put together proteins—the way Crick and Watson did with DNA.

DEMOCRITUS: And Pauling with the alpha-helix.

PERUTZ: Yes, but the structure of a protein—let's say an enzyme molecule — is a thousand times more complex than the structure of DNA, which is relatively simple …

ALKIMOS: Simple?

PERUTZ: That's why it was possible to determine its structure through trial and error. A few X-ray crystallographic measurements were enough: the width of the double helix, the distance between the neighbouring bases in the sequence, and the height of the entire spiral. This data helped Crick and Watson to realise the spiral had a repetitive structure.

DEMOCRITUS: You mean the same structure always recurs, if we travel along the axis of the spiral.

PERUTZ: Yes, even when we have a length of several million base pairs. By taking into consideration the three key parameters, they could unravel the structure of DNA. Trial and Error.

ALKIMOS: Trial and error?

PERUTZ: Model building. But you couldn't determine the structure of a protein using that method, because a protein's structure isn't repetitive.

ALKIMOS: But the alpha-helix is just a spiral …

PERUTZ: But the alpha-helix's structure characterises only short sections of certain proteins. An entire protein molecule is much more complicated. Come with me. I'll show you a model of a molecule so you can understand the extraordinary structural complexity of a protein. To determine a structure like this we would need to read several thousand pieces of information from an X-ray diffraction image. Come!

ALKIMOS: Several thousand pieces of information? That would take years.

PERUTZ: I've been working on the structure of haemoglobin for thirty-one years …

ALKIMOS: Master, did you hear that?

DEMOCRITUS: Why did you decide to study haemoglobin?

ALKIMOS: Yes, why not DNA?

PERUTZ: Haemoglobin is easy to obtain and is one of the few proteins we've managed to crystallise. Nevertheless, when I started to work with it in 1937, what we knew about it was so little that—to quote the Cambridge physiologist Joseph Barcroft—it would fit on the back of a postage stamp.

ALKIMOS: And what have you discovered so far? How many postage stamp backs could you fill?

PERUTZ: How many postage stamps? Well, let's see: the results of my first X-ray diffraction study showed that the molecule is spherical and has an ordered, three-dimensional structure. This was quite surprising, because most people believed proteins to have a disorganised structure; they were imagined as a colloid of tangled threads.

DEMOCRITUS: The fact that some proteins could be crystallised was not evidence enough of an organised atomic structure?

PERUTZ: No, not for a lot of biochemists. The idea of an ordered protein structure was indeed a new one.

DEMOCRITUS: Sanger's work with insulin showed that the sequence of proteins is disordered. It doesn't show any periodicity. Is that true of their spatial structure too?

ALKIMOS: It doesn't have to be, does it? The sequence of DNA doesn't have any periodicity either; yet its structure is ordered.

DEMOCRITUS: You're right, son. I imagine we'll have to settle this question by experimenting.

PERUTZ: Gentlemen, here's my molecular model of horse haemoglobin.

DEMOCRITUS: Kronides! Look at all those atoms! Astounding!

ALKIMOS: Master …

DEMOCRITUS: Marvellous!

ALKIMOS: Compared to the alpha-helix … Pan himself couldn't play a flute of that many pipes all piled on top of each other!

DEMOCRITUS: How did you figure this out?

PERUTZ: Using X-ray crystallography. The X-ray illuminates the individual atoms. But unfortunately there are no eyes or lenses that can focus the diffracted rays.

EXTON: Here, let me show you an X-ray diffraction image of a protein:

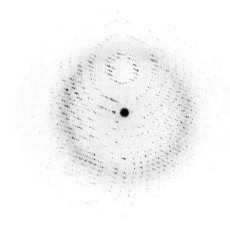

PERUTZ: Only a series of long, complicated calculations can help us determine the structure of the crystal based on the diffraction pattern.

ALKIMOS: So that's why it's taken so long.

PERUTZ: But I should mention that the mathematic and physical foundations of X-ray crystallography are relatively simple, or rather self-evident. That's why many physicists find it boring.

DEMOCRITUS: I suppose the method may be boring, but the result certainly isn't. I've never seen a more complex structure!

PERUTZ: Try to look at it from a different perspective. The sticks holding the amino acids complicate the model, but you could actually remove them.

ALKIMOS: But the amino acids would fall to the ground.

PERUTZ: Remove them only in your mind. As you can see, the model has two alpha-chains and two beta-chains. So the total is four protein chains.

ALKIMOS: Ah, that's why it's so complicated.

PERUTZ: The alpha-chain has 141 and the beta-chain 147 amino acids.

ALKIMOS: Insulin also had two chains of different lengths.

PERUTZ: Yes, but the two insulin chains didn't resemble each other. But in haemoglobin the alpha- and the beta-chains are very similar and the two sequences overlap if we leave a few gaps in the right places.

EXTON: Let me show this diagram to my students. They'll better understand the similarities.

PERUTZ: Of course.

EXTON: See how the sequence of the two chains can be made to overlap?

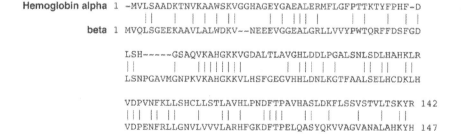

The bars indicate where the same amino acid can be found in the alpha- and beta-chains of horse haemoglobin. 43% of the residues are identical in the two subunits.

ALKIMOS: Crick talked about this! Related proteins are often different from one another in just one or two amino acids. But he was comparing horse to human haemoglobin, not horse to horse …

EXTON: Haemoglobin is made up of two alpha- and two beta-chains. The two chains are related to each other, so according to Darwin's theory of common descent they have a common ancestor.

DEMOCRITUS: The idea that related forms have a common ancestor doesn't apply only to the evolution of a new species. It seems that molecules too can be traced to one ancestor. One ancestral gene can give rise to two related genes!

EXTON: Through the process of *gene duplication*.

ALKIMOS: Gene duplication?

EXTON: Yes, and be careful. Gene duplication should not be confused with replication of the DNA chain. Replication occurs before every cell division. However, gene duplication—the duplication of a chromosome section—is a rare event and is worth talking about only on the evolutionary scale. Now, back to the structure of haemoglobin.

PERUTZ: We've already discussed the primary structure of haemoglobin, by which I mean the amino acid sequence.

ALKIMOS: So the primary structure is just the order of amino acids?

PERUTZ: That's right. An order that is determined by the DNA sequence of the gene that encodes haemoglobin.

ALKIMOS: Determined through the genetic code.

PERUTZ: A protein chain can be several hundred amino acids long. The chain folds into a complicated, three-dimensional ball, and the nature of this folding is determined, it seems, by the sequence of amino acids.

EXTON: Crick suspected this right away and brought it up when he discussed his sequence hypothesis, stating that the folding of the polypeptide chain is determined quite simply by the order of amino acids. The first to demonstrate this experimentally was Christian Anfinsen in 1962.

PERUTZ: Despite its complexity, we can, however, make sense of this three-dimensional ball. Its interior contains smaller, regular structures. These shorter sections of the protein chain are known as secondary structures. Examples of this are the alpha-helix and the beta-sheet.

EXTON: Beta-sheets are not found in haemoglobin, but they are in *keratin*, the protein in hair, and numerous other proteins.

ALKIMOS: What do they look like?

EXTON: The amino acids in beta-sheets or *pleaded sheets* do not form a spiral. Instead they're laid out in a planar structure. Pauling discovered this configuration of polypeptide chains. Here is a schematic drawing of a beta-sheet.

The structure is held together by hydrogen bonds, but these bonds are formed between protein chains rather than within a single chain as in an alpha-helix. But let's move on. Dr. Perutz, could you explain to us what is meant by a tertiary structure?

PERUTZ: With pleasure. It refers to the entire three-dimensional structure of one protein chain.

DEMOCRITUS: So the structure of each of the four individual chains in your model.

PERUTZ: Yes. The quaternary structure is therefore the structure of the entire tetramer. In other words, the position of the four chains relative to each other and their atomic structure together with the heme groups.

ALKIMOS: And the quinary structure?

PERUTZ: We don't generally talk about a quinary structure.

ALKIMOS: I thought, for example, how the model relates to the cupboard.

EXTON: Alkimos!

DEMOCRITUS: But why does it need four protein chains and thus such a complicated quaternary structure? What can it do with four chains that it can't with one?

PERUTZ: That's the hardest question. Why does it have this structure? How does it ensure the proper working of the protein? How does haemoglobin bind oxygen in the lungs and then release it in the tissues? I must admit that, despite our efforts, we haven't arrived at an answer.

ALKIMOS: So Bernal was wrong. Knowing the structure of haemoglobin hasn't helped us understand how it works?

PERUTZ: But we have an idea of how to proceed. We know that haemoglobin undergoes drastic structural changes when it takes up and then releases oxygen. We also know that when one oxygen molecule binds to one of the four chains it increases the ability of the others to take up oxygen.

EXTON: This is called *cooperativity*.

ALKIMOS: What does that mean?

EXTON: It means that if we steadily increase the oxygen concentration, the amount of oxygen bound by the haemoglobin does not increase linearly. Instead we get a sigmoid curve. In other words, the more oxygen that's present, the more readily the haemoglobin takes it up. The reverse is also true. When one chain releases a molecule, the others are able to do so more easily as a consequence.

PERUTZ: Haemoglobin is like a large excavator that completely fills its bucket in one round and then empties it in the next. Cooperativity is extraordinarily important because if the release of one oxygen molecule did not increase haemoglobin's ability to release the others, most of the oxygen would return unused to the lungs, and a person breathing normally would still suffocate.

ALKIMOS: So haemoglobin is like a large oxygen storing container?

PERUTZ: No, it's more like a molecular lung, like an organ in miniature.

EXTON: Dr. Perutz, allow me to show this image representing the atomic structure of horse haemoglobin to my students.

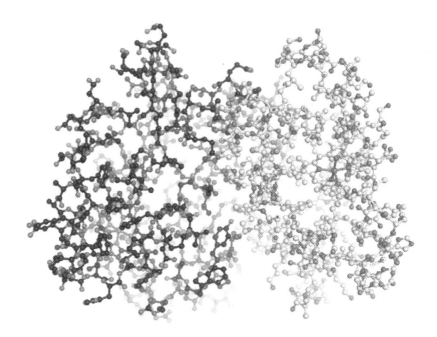

DEMOCRITUS: Fantastic! A molecular lung, formed by the specific arrangement of atoms.

PERUTZ: Can I see the diagram? Wow! How did you make such a nice illustration?

EXTON: We have a very good graphics department at Sussex.

ALKIMOS: Dr. Perutz, and can your stick model breathe?

PERUTZ: Excuse me?

ALKIMOS: I mean, can it change its structure, like real haemoglobin?

PERUTZ: No, I'm afraid not.

ALKIMOS: Well, what if we tried this. Master, can you hold the other end?

DEMOCRITUS: All right, I'll try.

ALKIMOS: Let's cooperatively push all four chains at once!

PERUTZ: What are you doing?

ALKIMOS: It's working. See, it moved.

PERUTZ: Let go of my model immediately! It took me months to build it!

ALKIMOS: But it's breathing! See!

DEMOCRITUS: The atoms have come to life! They've transformed into a lung. Look! It's respiring! Chronides!

ALKIMOS: What a noise!

PERUTZ: You're loosening the joints!

ALKIMOS: Master, what a curious chemical mechanism!

DEMOCRITUS: You're so right!

ALKIMOS: Master, let's swing it harder!

PERUTZ: Enough!

ALKIMOS: Now pull, master!

PERUTZ: Stop it!

EXTON: Alkimos, Democritus, do as Dr. Perutz says!

ALKIMOS: Oh, come on, Exton …

EXTON: We're leaving. Now!

DEMOCRITUS: I always knew it was the dance of the atoms that gives us life.

EXTON: I said now!

ALKIMOS: Dr. Perutz, thank you for the demonstration.

PERUTZ: My poor model. Look, what you've done—it's ruined!

EXTON: I'm starting the engine!

ALKIMOS: Aagggh … here we go!

Sugar-Consuming Bacteria—*Jacques Monod*

COOK: My friend, have you brought me something?

ALKIMOS: Brought you what?

COOK: You know—we talked about it. In return for a double portion of pudding.

ALKIMOS: Oh, that. Yes, yes, here.

COOK: What is that? What are all those balls?

ALKIMOS: It's a heme group.

COOK: Well, I'll be a chained titan! How'd you get it?

ALKIMOS: From a doddering Professor.

COOK: It's nice all right.

EXTON: Alkimos, where are you?

ALKIMOS: Over here, Exton.

EXTON: Come on. We're off to Paris.

COOK: Paris! The city of lights!

ALKIMOS: Paris? Already?

EXTON: Yes. Hey, what do you have there?

COOK: Oh, nothing special.

EXTON: Show it to me.

COOK: No, I'd rather …

EXTON: Alkimos, such ...grrr

ALKIMOS: It's just a souvenir.

EXTON: And what happens when Dr. Perutz notices …

ALKIMOS: He had three more. He won't miss it.

EXTON: Why Alkimos, such impudence, such audacity, such nefariousness, such … such… Lucky for you we have no time to deal with this!

ALKIMOS: What's the rush?

EXTON: We have a meeting in half an hour.

DEMOCRITUS: Half an hour?

EXTON: Yes, Democritus, I know what you're thinking. In a half an hour, and a few decades earlier.

ALKIMOS: But, can't we ...

EXTON: Everyone on board!

DEMOCRITUS: Come on, son! Direction Paris, 1961! We're going to visit Jacques Monod.

ALKIMOS: Jacques Monod ...

EXTON: Start the motors!

ALKIMOS: Aagghh .... Here we go!

DEMOCRITUS: Alkimos, stop fidgeting!

EXTON: We talked about Jacques Monod when we were discussing RNA and β-galactosidase.

ALKIMOS: Carbohydrate metabolism in bacteria.

EXTON: Yes, but now we're going to delve into more detail.

DEMOCRITUS: I have a few questions. Why are bacteria so choosy? And how can they use their enzymes so selectively? And how ...

EXTON: You'll have plenty of time to ask.

ALKIMOS: Yes, master, ask him as much as you can!

EXTON: If you want to better understand what Monod has to say, I need to explain to you a little about the sex life of bacteria.

ALKIMOS: Cloud-Gatherer Zeus, that sounds exciting!

DEMOCRITUS: We're all ears.

EXTON: Well, the essence of bacterial sex is ... Oh, it looks like we've arrived.

ALKIMOS: Oh, no, you have to go on! The essence ...

EXTON: Come on, everyone off.

DEMOCRITUS: Into such a dark alleyway ...

ALKIMOS: You call this the City of Lights? It can't hold a candle to Abdera. Can't even hold one to itself.

EXTON: We'll be out of here in a minute.

ALKIMOS: Oh, this is better …

EXTON: We'll find Monod in that restaurant.

ALKIMOS: I wonder what the *plat du jour* is …

EXTON: Let's go in.

ALKIMOS: When Monod gets here, can we order something?

EXTON: I wouldn't mind a bite myself. *Garçon, trois plat du jour, s'il vout plait.*

ALKIMOS: Now, tell us, what's the essence of bacterial sex?

EXTON: Oh right. Joshua Lederberg discovered the existence of bacterial sex in 1946. Basically what happens is that during *conjugation*, one bacterium—the donor—delivers some DNA through a tiny channel into the female—the recipient. What Lederberg discovered was that the bacterial chromosome passed into the recipient like spaghetti sucked through a straw. At the Pasteur Institute, François Jacob and Élie Wollman found that the injection of the DNA strand occurred at a constant, fixed speed. They also realised that the happy bacterial couple could be separated in a blender during the act. This made it possible to map the genes on the bacterial chromosome.

ALKIMOS: How?

EXTON: Let's say we have three mutations. During bacterial conjugation, a wild type bacterium can inject its DNA into a mutant cell. We can easily determine if the intact gene makes it across.

DEMOCRITUS: Because the mutant will again be capable of producing the necessary enzyme, or of growing on a nutrient-poor medium.

EXTON: That's right. The wild-type gene will complement the mutant gene.

ALKIMOS: But how will this produce a map?

EXTON: During conjugation, the donor chromosome at a certain point begins to inject itself into the recipient cell. If conjugation is interrupted at different points of time, we can determine when the gene in question made it into the recipient. The time of arrival depends on the gene's placement on the chromosome.

DEMOCRITUS: If we have several genes, we can determine the placement of each on the chromosome.

EXTON: And the end result is a genetic map of the bacterium's chromosome.

DEMOCRITUS: What an ingenious method!

EXTON: Monod used this to map the mutations in the β-galactosidase system. But we can ask him about this in person.

ALKIMOS: What's keeping him?

EXTON: He'll be here.

ALKIMOS: Should we order him a *plat du jour*, too?

EXTON: There he is.

ALKIMOS: What's that big box he's carrying?

EXTON: It looks like a cello case.

ALKIMOS: Cello?

EXTON: Yes, Monod is an excellent cellist.

ALKIMOS: What's that?

EXTON: I see we need to tackle the history of music next.

MONOD: *Bonjours, messieurs!*

EXTON: *Bonjour.*

ALKIMOS: *Plat du jour?*

EXTON: Professor, please join us. We'd like you to be our guest.

MONOD: *Merci, merci.*

EXTON: Excuse me, could we have a menu?

MONOD: Hmm, no, not the *plat du jour*. It's always a concoction of leftovers. Now let's see … *Garçon, aspics d'escargots et d'oeufs de caille, huîtres gratinées et cuisses grenouilles à l'ail.*

EXTON: Professor Monod, what we'd like to know is why you chose to study the regulation of gene expression?

MONOD: I was interested in *enzymatic adaptation*. Scientists have suspected the existence of this phenomenon for sixty years. What it means is a certain enzyme's production within a cell can be induced by adding a certain substrate for that enzyme.

DEMOCRITUS: So the cell adapts to the circumstances.

EXTON: For example, if an important nutrient appears, it starts producing the enzymes necessary to break it down.

MONOD: In the 1940s, scientists started to pay greater attention to this phenomenon. What fascinated them was that the production of a specific enzyme was caused by a specific molecule. It's obvious, I think, why geneticists and embryologists were so interested. An understanding of this could perhaps fill in the greatest gap between genetics and embryology by answering these questions: how can cells with identical genomes become differentiated during embryonic development? How do they become capable of synthesizing molecules with different patterns or configurations?

DEMOCRITUS: Yes, I see.

MONOD: From the start I thought the study of any phenomenon that involved the production of a specific enzyme by a specific external molecule would bring us closer to answering these questions. These phenomena are the synthesis of antibodies and naturally enzymatic adaptation.

DEMOCRITUS: If we understand these, then it might shed some light on embryonic induction.

ALKIMOS: That's right, Master. It might tell us why Spemann got a two-headed newt.

DEMOCRITUS: Indeed, since something like enzymatic adaptation happens during induction. The molecules synthesised in one region of the embryo impact the fate of another region. We need to study the enzymatic adaptation of the newt's head.

MONOD: Unfortunately, this phenomenon has thus far only been observed in microorganisms, so all our data refer to them. But it's such an effective method of gene regulation in these microorganisms that I'm convinced we'll discover it as a mechanism in multicellular organisms, too.

ALKIMOS: Then tell us more about enzymatic adaptation.

DEMOCRITUS: Yes, professor, please give us an in-depth look at how this phenomenon works.

MONOD: I'm sure you already know that the main job of genes is to determine the protein chains.

ALKIMOS: Yes, yes, we know it all—one gene, one enzyme, sequence hypothesis, central dogma, genetic code.

GARÇON: *Messieurs, trois plats du jours, voila!*

MONOD: Yes our picture has greatly expanded. We now know that the synthesis of a huge number of proteins—perhaps all—is regulated by specific molecules. In *E. coli* bacteria, for example, the enzyme *tryptophan synthase* is produced under the influence of trypotophan. This enzyme, as its name suggests, catalyses the last step in the synthesis of the amino acid tryptophan. If the bacteria grow in a medium that lacks tryptophan, the gene—whose location on the bacterium's gene map we know exactly—becomes active, meaning that under its direction tryptophan is produced. But if we add tryptophan to the nutrient medium, the production of this enzyme stops immediately.

ALKIMOS: Of course. Why should the bacterium make it if it's swimming in it?

MONOD: But tryptophan's ability to block production is specific. It inhibits production of all the enzymes needed in tryptophan synthesis, but not those needed for other processes. Moreover, aside from a few tryptophan analogues, which are similar to tryptophan, no other molecule is able to inhibit tryptophan synthesis.

EXTON: This blocking ability is known as *repression*. The presence of tryptophan represses synthesis of certain enzymes.

MONOD: That's correct. But let's have a look at the lactose system. In *E. coli*, production of the β-galactosidase enzyme is influenced by galactoside. β-galactosidase breaks down galactosides, such as lactose, allowing the bacterium to use these sugars.

DEMOCRITUS: We've heard that lactose can trigger β-galactosidase production.

MONOD: We found that as soon as we fed lactose to the bacterium, the speed of β-galactosidase production increases tenfold.

ALKIMOS: That's quite an effect.

MONOD: We called this effect *induction*.

ALKIMOS: Induction and repression.

MONOD: Induction is very specific. Lactose induces the production of only those enzymes needed for lactose metabolism—β-galactosidase, *galactoside-permease*, and *galactoside-transacetylase*. It had no effect on other enzymes. The reverse is also true: these three enzymes can only be triggered by galactosides. Other materials have no effect on them.

DEMOCRITUS: We know that enzymes bind to substrates and then change them, for example by cleaving them. But I don't understand how the enzyme–substrate bond is able to induce the production of the same enzyme from the corresponding gene.

MONOD: Initially we didn't either. But our series of experiments provided a very important clue. We examined several galactoside derivatives. This diagram shows a few of them:

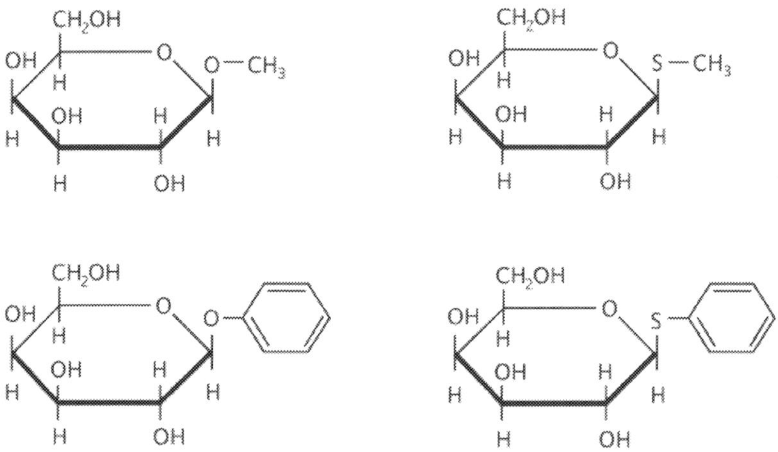

Interestingly, we found derivatives that proved to be great inducers, but β-galactosidase couldn't cleave them. We found others that were very good enzyme substrates, but weren't good inducers, such as phenyl-β-D-galactoside, in the lower left of the diagram. There's no connection, therefore, between being a substrate for a galactoside derivative and being a good inducer. It became obvious to us that the inducer did not act through the enzyme, but through another specific component of the cell. This is what we needed to identify.

DEMOCRITUS: Only genetics can help! If we discovered certain mutations that prevented induction, we might come upon this other, mysterious cellular component.

MONOD: We managed to identify two very important categories of mutations in the β-galactosidase system. One group of mutants behaved exactly as I'd predicted. Induction did not occur under any circumstances, so we called these *super-repressed* mutations. In the other group of mutants, on the other hand, the enzymes were expressed at a constant, high level. We referred to this group as *constitutive* mutants. Other researchers identified these two categories in other induction systems.

ALKIMOS: Then we're dealing with two genes.

MONOD: Surprisingly, mapping of the mutations showed that the two kinds of mutation occurred in the same gene. This gene was not β-galactosidase, because on the genetic map it's located far from the trio of β-galactosidase, permease, and transacetylase. We called this gene a *regulator*. Similar regulator genes were found in other inducible enzyme systems.

DEMOCRITUS: A gene that regulates other genes! Fantastic!

ALKIMOS: But how can you have two kinds of mutations in one gene?

DEMOCRITUS: That's simple, son. In the super-repressor mutants, the regulator gene that triggers enzyme production is damaged. In the constitutive mutants, the gene becomes too active, working even in the absence of an inducer.

MONOD: That's a reasonable interpretation, but it's not correct. Studies have shown that in several mutants that exhibit constitutive expression of an enzyme, the entire regulator gene is missing. Such mutations are known as *deletions* and can be clearly seen in the genetic map. Numerous other experiments also contradict your interpretation.

ALKIMOS: Master, if the regulator gene is missing, how can enzyme production be too active? It doesn't make sense.

DEMOCRITUS: Let's think about it … if the regulator gene is absent, then the enzyme production is constitutive, in other words continuous. But when the regulator gene is present, the enzymes aren't produced.

ALKIMOS: Until the cell senses an inducer. Then …

DEMOCRITUS: This suggests that without an inducer, the regulator gene suppresses enzyme expression. We also know that it's not the enzymes that sense the inducer, but another specific cellular component …

ALKIMOS: The regulator gene itself!

MONOD: We reached the same conclusion. A regulator gene is responsible for production of a *repressor*. The repressor inhibits expression of the three proteins until an inducer appears.

DEMOCRITUS: And the inducer somehow inhibits the repressors. It's a process of inhibiting the inhibitor.

ALKIMOS: Like a double negative! Two no's make a yes!

EXTON: The inducer is a positive regulator, because it negatively regulates a negative regulator.

ALKIMOS: Beautiful way of putting it!

GARÇON: *Monsier, aspic d'escargots et d'oeufs de caille, huîtres gratinées et cuisses de grenouille à l'ail, viola!*

ALKIMOS: I'd like some of that! Exton, this *plat du jour* is an abomination!

MONOD: Let me summarise the model I devised after two decades of work. The model distinguishes between two kinds of genes. The first are *structural genes*, whose mutations change the structure of the appropriate proteins. Mutations of this sort have been described in numerous adaptive enzyme systems. They affect the structure of just one protein—for example β-galactosidase—and on a genetic map they're found in one short section, always within one *cistron*.

EXTON: Excuse me Professor, let me explain to my students what a cistron is.

MONOD: Of course.

EXTON: A cistron is a section of the genetic material where two mutations in *trans* don't complement each other.

ALKIMOS: Umm, okay so you mean …

EXTON: It's the precise genetic definition of a gene.

DEMOCRITUS: And what does trans mean?

EXTON: It means that the two mutations are on two different DNA strands.

MONOD: That's right. Now let's return to structural genes. A very important characteristic of mutations that take place in these is that they have no influence on the regulator system. In these mutants, the regulator system continues to respond to the appropriate, specific stimulus, just as it does in wild types.

DEMOCRITUS: So a mutant β-galactosidase is just as inducible as the wild type.

MONOD: Yes, unless it is a deletion mutant. I should also note that enzymes that catalyse different steps in the same biosynthetic pathway are generally located next to each other in bacteria, creating a gene string.

EXTON: An example of this is the trio of β-galactosidase, permease and transacetylase. These three genes are located very close to each other and during *transcription*, they are copied onto an mRNA. The name of this sort of RNA is *polycistronic RNA*, if I'm not mistaken.

MONOD: Yes, that's quite right. But coming back to my theory... Genes of the second type are the *regulator genes*. Mutations of this type fall into two groups, as I mentioned. The essence of the model is this: structural genes are transcribed into short-lived messenger RNAs based on the complementarity of bases. The mRNA binds to the non-specific ribosomes in the cytoplasm. The protein chain is constructed on the ribosome based on the mRNA sequence and the genetic code. When it's complete it separates from the ribosome, which can begin reading a new mRNA. The synthesis of mRNA is a one-directional process that can only begin at a specific section of the DNA thread, known as the *operator* region. In many cases, one operator regulates the transcription of several structural genes lying next to each other. A group of these structural genes is known as an *operon*. The operon is none other than a unit of genetic regulation and transcription, as François Jacob and I defined it in 1961.

DEMOCRITUS: In the lactose system the structural genes β-galactosidase, permease, and ... what's the third?

MONOD: Transacetylase.

DEMOCRITUS: Yes, yes, these three form the operon.

EXTON: Together with the operator region.

MONOD: The coordination within operons is perfect. In the *Lac operon*, the three enzymes are always produced in the exact same quantities in response to the inducer.

ALKIMOS: I see. One unit of genetic regulation and transcription!

MONOD: In the genome, there are regulator genes as well as structural genes. The transcript made of the regulator gene—most likely an RNA transcript—binds to a specific operator—probably through base pairing—and blocks transcription of the entire operon and therefore the synthesis of the proteins which are coded by the operon's structural genes. The repressors are capable of binding specifically to certain small molecules beyond the operator region too. We call these *effectors*. In inducible systems, such as the lac system, only repressors unbound to effectors can bind to operators and express their blocking effect. In the presence of an effector—in this case an inducer—the repressors aren't able to work. In repressible systems the opposite is true. The repressor can only bind to the appropriate operator and block it if an effector is present. Thus, transcription of an operon comes to halt in the presence of an effector.

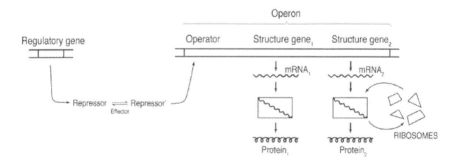

ALKIMOS: How simple!

DEMOCRITUS: Genes are repressible and inducible! The same bacterium can use different genes to suit different circumstances, and the whole time its genome remains unchanged.

ALKIMOS: Master, isn't that exactly the question we wanted to answer?

DEMOCRITUS: Absolutely! What else could explain why Spemann's newt had two heads? Certain genes can be turned on or off.

ALKIMOS: By the regulator gene!

DEMOCRITUS: Is this the secret of the development of multicellular organisms?

MONOD: Probably, but the truth is we really don't know.

EXTON: But we have a brilliant hypothesis.

ALKIMOS:

> *The coli sets off to suck and degrade, the gene turned on*
> *And off by sweet lac operon, after every snack.*

MONOD: That's very good!

DEMOCRITUS: And what about the development of multicellular organisms?

ALKIMOS:

> *How once the monster of Minos switched his own genes on,*
> *And a hideous pizzle sprouted between his wretched legs.*

EXTON: Professor, we should be going.

MONOD: But we haven't been served our *mousse au chocolat*.

EXTON: I'm afraid we're in a bit of a hurry. We need to pay a visit to someone.

MONOD: Ah yes, I understand.

EXTON: Come along, my students.

ALKIMOS: But Exton, I'd really like some mousse.

EXTON: We haven't got time!

DEMOCRITUS: *Au revoir*, Professor.

EXTON: All aboard!

ALKIMOS: Aaagghhh … here we go!

Part Seven

Genes in the Mortar

Epigenesis and Genetics—*Conrad Hal Waddington*

DEMOCRITUS: Dear Exton, tell us, is Monod's model correct?

EXTON: Yes, almost every word of it. There is only one mistake, at least in the way in which Monod and Jacob explained the model in 1961. The repressor is not an RNA molecule, but a protein.

ALKIMOS: So the repressor gene is also a structural gene? Since it codes for a protein?

EXTON: Indeed, the difference between a structural gene and a regulator gene becomes rather vague, but what's important about the model is its logic.

DEMOCRITUS: That one gene can influence the operation of another gene?

EXTON: Exactly. And this regulation is subject to another regulator.

ALKIMOS: The inducers or repressors. And likely this regulatory process is regulated by a third material, which ...

EXTON: Right. The model can be arbitrarily complex. So the pattern of switching on and off genes can also be as complicated as you like.

DEMOCRITUS: Just more regulator genes are needed and more external regulators, like hormones. In this way we can achieve any level of complexity.

EXTON: Most likely.

DEMOCRITUS: Amazing!

ALKIMOS: And if we can achieve any level of complexity, then it won't be difficult to make an elephant out of an *Escherichia coli*. We just need more genes and more regulators.

EXTON: The spectacular successes of molecular biology inspired many, not just you Alkimos, to oversimplify phenomena and make gross generalisations. However, even if we understand how a few *E. coli* genes are regulated, that does not mean we understand how, say, an animal embryo develops. It's true that within the framework of the one gene, one enzyme model, the genetic code and the central dogma, the simpler connections between genotype and phenotype become easier to grasp, such as the structure of proteins and certain metabolic steps. But in multicellular organisms the production of a phenotype based on a genotype during development is far more complicated.

DEMOCRITUS: What you mean is that in genetics there's no such thing as a one gene, one ear connection.

EXTON: That would be an extreme form of preformationism.

ALKIMOS: But what's wrong with that, Exton? Why can't one gene and one protein correspond to one trait? Perutz said our hair is constructed from one elongated protein. He also showed us what complicated forms proteins can come in. Why can't an ear be made up of thousands of ear-shaped proteins? The regulators only ensure that the ear or the wing genes are switched on in the right place at the right time.

DEMOCRITUS: Have you forgotten what Kühn taught us about fronts? And the considerable amount of evidence we have about epigenesis?

ALKIMOS: Master, I was just speculating … but to be honest, I still don't understand if it's epigenesis or preformation.

EXTON: Don't worry. I know which scientist can explain it to you: Conrad Hal Waddington. Hermes, prepare the Deus ex Machina. Direction Edinburgh, Scotland, 1957!

ALKIMOS: Scotland?

EXTON: All aboard!

DEMOCRITUS: Come on, son!

ALKIMOS: Aaagghh … here we go!

EXTON: Waddington was one of the most interesting figures in 20th-century embryology and genetics. He started out as an embryologist and was in close contact with Hans Spemann and other leading embryologists in the early 20th century. In the 1940s, he dealt with genetics. His main goal, in fact, was to bring the two fields together. Like D'Arcy Thompson, he was deeply theoretical and philosophical. But his strengths were in biology and experimentation.

ALKIMOS: I imagine the Scottish climate is favourable to philosophers.

EXTON: Probably more so than the southern Italian climate. Now, we'll hear the rest of the story from him. We're here.

DEMOCRITUS: It's dreadfully foggy.

ALKIMOS: Who's in the mood to philosophise in this rotten weather?

DEMOCRITUS: You said this was the right climate for it.

EXTON: We're not here to philosophise.

ALKIMOS: That's true. We're here to learn more about preformationism.

EXTON: Now let's go into the Institute of Genetics. Follow me.

DEMOCRITUS: Wait!

EXTON: Come on, this is it.

ALKIMOS: Is someone smoking a pipe?

EXTON: That would be Waddington.

WADDINGTON: Good evening!

EXTON: Professor Waddington, allow me to introduce myself. I'm Dr. Exton and these are my students, Mr. Alsmith and Dr. Dimatthews.

WADDINGTON: I'm pleased to meet you. Feel free to light up, if you'd like. I always enjoy a good smoke.

EXTON: Oh thank you, but we'd rather not.

WADDINGTON: Well, how can I help you?

ALKIMOS: We've been looking over the work of the great embryologists of the past and we'd like an answer.

WADDINGTON: Hmm. The biologists of the last century laid the foundations upon which we need to build. A quick overview would be useful to discover what is of true, lasting value. But the answers I'm afraid continue to elude us ...

ALKIMOS: Could you at least tell us, is it preformation or epigenesis?

WADDINGTON: Oh, that. Well, both of course.

ALKIMOS: Both? How's that possible?

WADDINGTON: Well, we know that the ovum contains preformed elements, such as the genes and certain regions of the cytoplasm. We also know that during development the preformed parts rely on epigenetic processes and connections to create the characteristics of the adult individual, which were not preformed in any way in the ovum.

ALKIMOS: Master, we never thought of that.

DEMOCRITUS: But we were close.

WADDINGTON: The science of epigenetics deals with the union of preformationism and epigenesis, or in other words genetics and embryology. As we know, genetics examines the inheritance of genes, while embryology studies the development of the embryo from a cellular, morphological, and comparative perspective.

ALKIMOS: von Baer's laws, Spemann's organiser...

WADDINGTON: That's right. In my view, embryology, organisers, germ layers, and the rest need to be expanded on through a genetic approach.

DEMOCRITUS: So the goal of epigenetics is to interpret the role of genes in development.

ALKIMOS: Master, epigenesis is regulated by preformed genes. How simple ...

WADDINGTON: I suppose ...

ALKIMOS: We've learned so much about genes! I'm sure we'll be able to understand epigenesis!

WADDINGTON: Indeed, it doesn't appear all that complicated. Most geneticists approach living systems with the same 'atomistic' metaphysics. They always formulate their questions like this: What does a given 'A' gene do and what decides if it's active or not? Let's review the work of several key researchers in this field. Goldschmidt, for example, concluded that genes determined the speed of a reaction. Muller, on the other hand, believed that genes control the quantity of a material produced, and can also produce an antigen, a material that has an opposing effect.

ALKIMOS: Right, we talked with Muller.

WADDINGTON: I beg your pardon? You know Muller?

EXTON: We know his work.

WADDINGTON: Ah, I see. Now, where was I? Yes, Garrod and later Beadle and Ephrussi determined that these materials were enzymes. Jacob and Monod showed how bacteria can regulate the expression of certain genes. It's clear that this kind of research was very productive, but in my opinion it doesn't solve the problems of epigenetics. After all, the differentiation of cells into nerve cells or muscle cells does not occur through the switching on and off of one gene. The transformation into a nerve cell, for example, is certainly a process that involves an enormous number of genes working together. Before Beadle and Ephrussi, I showed that the development of fruit fly wings was affected by at least forty different genes.

ALKIMOS: Forty wing genes? If there are forty genes, this means forty operons and regulators …

WADDINGTON: I managed to identify these forty genes by genetic analysis. The mutation of some genes results in notched wings, while others affect the size and venation of the wings, as you can see in this diagram.

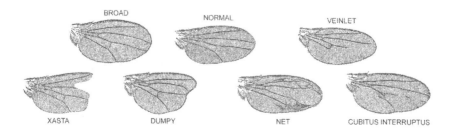

ALKIMOS: By studying the mutations, we can figure out which of the forty genes do what. One causes notching, the other venation ...

WADDINGTON: I'm afraid it's more difficult than that. The mutation of one gene might cause notching, but this doesn't explain how wings develop. Instead it presents a new problem.

ALKIMOS: A new problem? But —

WADDINGTON: Let's say that during a certain step in development, a gene regulates the flexibility or permeability of a membrane. We're never going to understand this using genetic analysis. Let's have a look at this diagram of newt gastrulation.

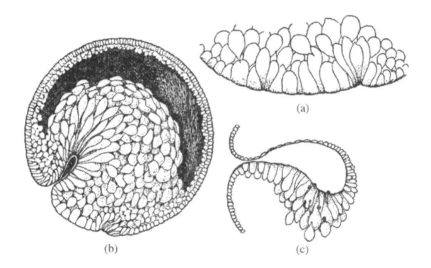

(a)

(b)   (c)

When the blastopore appears, we can notice flask-shaped cells. The change in behaviour of the cell membranes and the cytoplasm give rise to this cell form that triggers invagination. Of course, genetic analysis can help us identify those genes that play a role in this development process, and we can gain some insight into its complexity and the relationship of the genes to each other. But we'll learn very little about the steps—about the developmental mechanisms—that take us from the genes to cell formation and ultimately to a fully developed organism.

ALKIMOS: So, in addition to studying genes and embryos we have to investigate the flexibility and permeability of cell membranes?

WADDINGTON: And the living material in the cytoplasm. After all, genes don't act alone in developmental reactions; they join forces with the cytoplasm. At the same time, the cytoplasm can have an effect on the genes. And naturally several genes have an effect on the cytoplasm. In other words, there is no atomistic, one-to-one correspondence of the genotype to the phenotype. The entirety of the genotype is what determines the entirety of the phenotype by way of the cytoplasm, membranes, and developmental reactions.

DEMOCRITUS: Therefore, the epigenetic theory of development needs to link genes to the cytoplasm and cell membrane, and the cell membrane to tissues and organs. But developmental reactions and gene expression are so complicated! Our task is indeed a challenging one, and you've already said that genetic analysis alone won't be sufficient.

WADDINGTON: That's right. But a few conceptual anchors or metaphors can help us understand epigenetics. I've devised an abstract framework that can help guide our thinking.

ALKIMOS: Here comes the philosopher.

WADDINGTON: The developmental reactions or developmental steps occur along an epigenetic pathway or *chreod*, under the direction of the genotype.

ALKIMOS: One moment—*chre* meaning necessity and *hodos* meaning path or trajectory. Okay, please continue.

WADDINGTON: The developmental pathways of certain organs are protected from certain mutations or from the disturbing influence of external forces. Within certain limits, the system can readjust itself, like a good automobile that straightens itself after coming around a turn. I found the best way to illustrate this is with the metaphor of an *epigenetic landscape*.

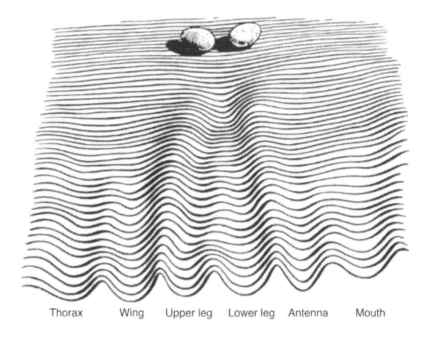

Thorax   Wing   Upper leg   Lower leg   Antenna   Mouth

DEMOCRITUS: If I understand, this diagram illustrates how cells, tissues, or organs develop from a rudimentary, undifferentiated state to a fully developed organism.

WADDINGTON: Yes. If the channels, or canals, are deep the pathway is very stable, and the cells have little chance of avoiding their fate. Even large disturbances are unable to set the cells on another path.

DEMOCRITUS: So the channels show the fate of certain cell types or body parts.

WADDINGTON: Correct. The diagram shows very clearly how the differences between the ear and the nose, for example, are not genotypic or phenotypic. They are the results of the two different developmental paths of two different tissues, of the genetic and non-genetic influences the two tissues are exposed to during development. The totality of these effects is what I call the tissue's epigenotype.

ALKIMOS: Master, does that mean that your nose and eyes are genotypically the same, but differ somewhat in their epigenotype?

DEMOCRITUS: It means the phenotype of a given organ is dependent on its genotype and epigenotype. It's clear that if we ever want to understand the connection between a genotype and a phenotype, we need to understand the epigenetics of development.

WADDINGTON: Epigenesis is the most important link between heredity, development and evolution.

ALKIMOS: Evolution?

EXTON: If you recall, in evolutionary biology, the embryo was useful only in comparative studies. Classical comparative embryologists used it to elucidate questions of phylogenetics.

WADDINGTON: That's true. But the union of genetics and embryology has opened up new vistas in evolutionary biology.

ALKIMOS: But why?

WADDINGTON: Look, the evolutionary synthesis was constructed from the notions of random mutations and natural selection. Reproduction and development were left out if it. But to understand evolution, it's not enough to understand genetic mutations and natural selection. You also need to understand how the genotype is transformed into a selectable phenotype. You need to understand epigenetics.

DEMOCRITUS: What you mean is that during evolution, genetic mutations lead to modifications in the epigenetic pathways and ultimately to variations in phenotype. Selection then acts on the phenotypic product of these pathways. In other words, understanding epigenetics is important so that we can understand how genes and their mutations lead to phenotypic change. For example, mutations can freely propagate within the deeper valleys of the epigenetic landscape until their effect is great enough to redirect the fate of the cell to another pathway. Epigenetics is extraordinarily important from an evolutionary perspective.

WADDINGTON: Yes, it is.

ALKIMOS: So to fully comprehend evolution, we also need to comprehend chreods.

DEMOCRITUS: And canalization.

ALKIMOS: And the stability of pathways.

DEMOCRITUS: If an epigenotype belongs to certain phenotypic elements, such as an ear or a nose, then selection ultimately affects epigenotypes.

To be more precise, selection affects the genotype by way of the phenotype and the epigenotype.

ALKIMOS: Aha, selection can make its way back along the pathway all the way to the starting point. Master, we could say that epigenesis travels downward on the pathway, and selection travels upward.

DEMOCRITUS: Yes, son. But in both directions the traffic is canalized.

ALKIMOS: Well, it depends on how deep the valleys are.

WADDINGTON: I'm glad to see the epigenetic landscape metaphor has been helpful.

ALKIMOS: Oh, it has!

EXTON: This entire conversation has been incredible enlightening. We've managed to put together several pieces of the puzzle.

ALKIMOS: That's right, Exton! I finally understand why its preformation *and* epigenesis!

EXTON: Finally. Now it's time to go.

WADDINGTON: Already?

EXTON: I'm afraid we must.

WADDINGTON: But we haven't even talked about genetic assimilation.

EXTON: I'm not sure we have the time to assimilate it into our conversation.

WADDINGTON: As you wish …

EXTON: Alkimos, Democritus, come on.

DEMOCRITUS: Canalized variation and evolution …

ALKIMOS: You should recommend us for membership in the Royal Society.

EXTON: Alkimos!

WADDINGTON: I'd be glad to.

DEMOCRITUS: I need to think about this some more.

EXTON: All aboard!

ALKIMOS: Aaagghh … here we go!

A Recipe for Making Mice—*Sydney Brenner*

ALKIMOS: We've done nothing but travel today.

EXTON: I know, today's been a little hectic, but I think our next visit will make up for it.

ALKIMOS: Why? Where are we going?

EXTON: You'll see. We're already here.

ALKIMOS: Where are we?

EXTON: Cambridge.

ALKIMOS: What are we doing here?

EXTON: We're here to see Sydney Brenner. He's expecting us.

ALKIMOS: Isn't he one of the renegades?

EXTON: Go ahead and ask him all you want about development.

ALKIMOS: Don't worry, we will.

EXTON: Waddington was an embryologist-turned-geneticist, while Brenner is a geneticist-turned-embryologist. This is apparent in their different ways of thinking, and it's also why I wanted to visit Brenner right after Waddington—so you can get a better sense of the difference in their approach. Waddington most likely considered Brenner's approach too atomistic.

ALKIMOS: And did Brenner think Waddington was too holistic?

EXTON: Probably.

ALKIMOS: Then master, I bet you'd choose Brenner.

DEMOCRITUS: We'll see, son.

EXTON: This quote says a lot about Brenner: 'I'm a strong believer that ignorance is important in science. If you know too much, you start seeing why things won't work. That's why it's important to change your field to collect more ignorance.' And Brenner indeed switched his area of research regularly. That was partly the secret of his success. He worked with plant pigments, the genetic code, phages, worms, and many other things.

ALKIMOS: I'll bet he didn't work with moth larva testes.

EXTON: No, I don't believe he did. Now come along.

ALKIMOS: But wait, we've already been here.

EXTON: Yes, and this time please promise not to touch anything. It took Perutz a week to reassemble his model.

ALKIMOS: At least he had an opportunity to think about cooperativity while he was doing it.

EXTON: Enough, Alkimos. We're here!

ALKIMOS: No stick models?

EXTON: Greetings, Dr. Brenner.

BRENNER: Ah, is it you? Finally. Come in!

ALKIMOS: I'm Exton's student, Alkimos, and this is my master, Democritus.

BRENNER: What amusing names you have.

ALKIMOS: Our aliases are Alchumbach and Demiescher.

BRENNER: Well that makes sense.

EXTON: Alkimos, enough jokes. We're here to discuss science.

BRENNER: What fields do you work in?

DEMOCRITUS: We're most interested in the role of the atoms of life in development and evolution.

BRENNER: You know that only three questions truly interest a biologist: how living things work, how they develop, and how they come into being during evolution. First of all, how do they work? How do they run around and around? How do they breathe? How do they eat? How do they die?

EXTON: *Physiology*, the study of how living organisms function, is a branch of science that searches for the answers to those questions. How do breathing, circulation, hormonal regulation, and plant photosynthesis work?

BRENNER: That's right. The second question is much more difficult. How does development work? We must answer this before we can address the question of evolution.

EXTON: I suggest we concentrate on this question.

BRENNER: Molecular biology and genetics have made it clear that living organisms have internal programmes. But this knowledge is insufficient for us to comprehend how a living thing comes into being.

DEMOCRITUS: So we need to shed light on the programme—the programme of development!

BRENNER: That's right. And in exhaustive detail.

DEMOCRITUS: On the level of atoms, of course.

BRENNER: You know that molecular biologists have always been concerned with mechanisms on the molecular level. They slowly taught biologists how to speak in the language of chemistry. Of course biochemists were already proficient in the language, and until the 1950s all they talked about was the origins and the flow of the materials and energy necessary for cell function. Molecular biologists, on the other hand, began talking about the flow of information. The discovery of the double helix suddenly made it just as possible to study information in biological systems as it was to study materials and energy.

EXTON: My students and I have already discussed how the main question for molecular biologists was the source of information that determines the sequence of amino acids and not the source of the energy required to create peptide bonds.

BRENNER: That's right. To hell with energy! Energy will take care of itself.

ALKIMOS: Biochemists will take care of it. And a fresh army of molecular biologists.

BRENNER: That's how I see it too. There are tons of details that need to be worked out in molecular biology. How are ribosomes constructed and how do they function? What proteins and RNA comprise them? What's their structure? The questions are endless. But you should know that the heroic age of molecular biology is over. The main questions have been answered. There won't be any more evangelists to spread the word.

ALKIMOS: But we want to be evangelists!

BRENNER: And so do I. That's why I think Crick and I made a good decision to turn our attention to another area of research: the development of complicated, multicellular organisms.

EXTON: Yes, let's return to the question of development …

BRENNER: Now, as I said, every animal contains its exact description. Its own developmental programme. To decipher it we need to conduct tons of experiments. We need to understand the language of the molecular and genetic description of organisms. And how an animal describes itself.

DEMOCRITUS: We need to learn a new language! One that has nothing in common with human languages. The language of the developmental programme.

BRENNER: At present we don't know anything about the nature of this language or the programme. But one thing is certain: it won't be enough to just be clear about general mechanisms.

ALKIMOS: You mean like the genetic code?

BRENNER: Yes. Or how ribosomes work, or the nature of gene replication. We've discovered a lot on the level of general mechanisms, but none of it tells us anything about developmental programmes. Let me give you an example. If you said to me, 'Here's a protein—how was it made and what's its genetic definition?' I'd be able to give you an extraordinarily detailed answer.

ALKIMOS: The code allows you to determine the gene sequence that codes for the protein …

EXTON: Although it's not clear because of the problem of redundancy…

ALKIMOS: We'd also be able to tell you how the protein was made. And with Perutz's help, we'd be able to determine the position of all the atoms.

BRENNER: Yes, but if you said to me, 'Here's an ear; here's an eye. How were they made? What's their genetic definition?' Well then …

ALKIMOS: That's exactly what we ask!

BRENNER: … I wouldn't be able to answer. I'd need to know the exact number of genes and their sequence, what each gene does and how they affect each other. I'd need to know the exact programme, moreover in the language of molecules. In theory, I'd need to know in such detail that I'd be able to write procedural instructions on how to make an ear or an eye.

ALKIMOS: What kind of procedural instructions?

BRENNER: The series of steps in the developmental programme.

ALKIMOS: Would this be the new gospel?

BRENNER: Today, the majority of molecular biologists would say the most important thing is to understand how genes switch on and off in multicellular organisms. What sorts of mechanisms are involved? But we could go further and ask: which genes switch on and off and when? And which genes are subordinate or superior to which? Which genes switch on at the same time as others and what is their effect? At this point, we shouldn't even deal with the exact mechanisms.

ALKIMOS: The mechanisms will take care of themselves.

BRENNER: That's right!

DEMOCRITUS: We should only deal with logical procedures.

BRENNER: Yes, since the programme is a series of logical procedures.

ALKIMOS: You like the sound of that, don't you, master?

BRENNER: To analyse the programme, we have to devise a theory of complex systems. Of course, in the meantime, we're going to run into the extremely complicated problem of organisational levels.

ALKIMOS: The problem of how proteins become cells, the cells tissue, the tissue organs, and the organs …

BRENNER: That's right. Sooner or later we're going to have to step beyond the notion of life processes being like simple mechanical clockwork. You see, there's no point in knowing which mutations arise on the level of a nucleotide. We're still not going to know anything until we understand how genetic changes relate to development, growth, and behaviour.

EXTON: And evolutionary ecology …

BRENNER: Many years ago John von Neumann wrote a very interesting essay. In it he posed the question of how to formulate a theory of pattern recognition. He concluded that such a theory is impossible. But he could provide a prescription for a device that actually performs pattern recognition. In other words, we're used to physicists think in laws, and molecular biologists thinking in mechanisms. The time has come for us to come up with a new approach in our discussion of development. An approach that could only be imparted using algorithms. Recipes. Instructions to be carried out.

ALKIMOS: I'll instruct my body to grow an ear here. Or rather my body instructs itself or …

BRENNER: If you ask me, 'what's a mouse? When does something become a mouse?' The only possible answer might be an algorithm, which you can use to build a mouse.

ALKIMOS: A recipe for making a mouse …

BRENNER: It's an algorithm which the mouse can use to build itself. It's important that we're not satisfied with merely a detailed description of a fully developed mouse.

DEMOCRITUS: A mouse is not just what it is, but how it's formed.

BRENNER: If we know the complete algorithm, the complete programme, and we're able to translate it into the language of a computer, only then will we be in possession of the description of a mouse, albeit in a strange language.

ALKIMOS:

> *The great secret of life, the program, lies deep within;*
> *The thirst for knowledge does all it can to pry it out.*

DEMOCRITUS: But what will such an algorithm look like?

BRENNER: Like this: Let's switch on the seventy-fifth and the two hundred and twenty-seventh gene. Let's turn off the hundred and eighth and slowly raise the level of proteins made by the twenty-ninth. Of course we'll have to decipher the molecular biology behind the entire programme. What is the two hundred and twenty-seventh gene and what exactly does it do? What other genes does it have a connection to? We'll need to do this in exhaustive detail.

ALKIMOS: It'll take a lot of scientists to do that!

BRENNER: And a lot of time. But I have a feeling in the future this is the direction molecular biology will take. Complicated logical, computational relationships, the programme and the algorithms of development need to be investigated. And all on a molecular level. If the two —molecular mechanisms and a logical description of the entire process—can be melded together, then we'll arrive at a new level of understanding of development.

DEMOCRITUS: Ah, how I wish we were there already.

ALKIMOS: But how do we start? How can we decipher this unknown language?

BRENNER: In my view, there are two ways we can approach the question. First, we can start with the organisms, by analysing mutations for example. That's the traditional approach.

DEMOCRITUS: That was the method of Morgan and associates, and of Kühn and Ephrussi.

BRENNER: And its success depends on the choice of animal and the degree to which we persist in getting at its secrets. The other approach is to use cell cultures. In the long term, I think this is the method we'll need to use. It's what will lead us to the answer.

ALKIMOS: Cell culture?

EXTON: The *in vitro* growing of cells. In the appropriate medium, the cells obtained from tissues can be kept alive and propagated.

ALKIMOS: Like bacteria in a test tube?

BRENNER: We need to try and build organs. If we can produce a retina in a cell culture, then we'll be able to say that we understand how an eye develops.

ALKIMOS: So we'll return to Paracelsus' homunculus? But instead of a squash or horse stomach, we'll use cell cultures?

EXTON: But we still don't have a recipe. We should start with the traditional one: genetics.

DEMOCRITUS: That's right! At first let's not create, let's just acquaint ourselves with things.

EXTON: Let the engineers create!

ALKIMOS: And the cooks!

DEMOCRITUS: And the poets! Come on, son!

ALKIMOS: Should I add a homunculus?

DEMOCRITUS: If you'd like.

ALKIMOS:

> Long ago a minute homunculus conceived in a violent storm,
> Tomorrow a fabricated eye winks, laughs, in the lab.

DEMOCRITUS: Tiresias could regain his sight.

EXTON: My dear students, we have a lot to do. It's time to go.

BRENNER: You're leaving so soon?

EXTON: We truly have a lot on our agenda.

DEMOCRITUS: Thank you for the conversation. Goodbye!

BRENNER: It was nothing. Bye!

ALKIMOS: Exton, can't we drop in on Crick? And Perutz?

EXTON: Not now. Come on.

ALKIMOS: All right, I'm coming. Are we finally going to decipher the programme?

DEMOCRITUS: Do you understand the task, Alkimos?

ALKIMOS: Of course, master. We have to discover the entire gene sequence, and also figure out which gene does what and when, and how they relate to the others. Easy as pie.

DEMOCRITUS: If we can do it, we can put together the developmental programme.

EXTON: All aboard!

ALKIMOS: After you, master.

DEMOCRITUS: Once we know the programmes in living things, we can compare them and perhaps understand their evolutionary origins.

ALKIMOS: And perhaps even more.

EXTON: I'm amazed neither of you broke anything.

ALKIMOS: Aaagghh… it's starting!

The Wiring of a Worm's Brain

ALKIMOS:

> *The groggy bed of my heart inspires dolorous songs.*
> *I sing the memories of those felicitous times of old.*
> *And wine the colour of honey resounds in my quieting words:*
> *How easily, how swiftly, the heavens put an end to joyous festivities.*

DEMOCRITUS: Marvellous verse, son.

ALKIMOS: You didn't want to wake up, master. I composed them while you were sleeping.

DEMOCRITUS: I have a harder and harder time getting up.

EXTON: Alkimos, why the despair?

ALKIMOS: I'm mourning the end of the heroic age of molecular biology.

EXTON: Is that so?

ALKIMOS: Yes, Brenner said it's over.

EXTON: Well that's true. In the early 1960s, molecular biology truly did pass from the heroic age into the classical period.

ALKIMOS: But I love the heroic age because anything is possible.

EXTON: When Morgan realised that as an embryologist, he'd exhausted the possibilities of classical embryology, he turned to genetics. Brenner called this 'Morgan and associates' deviation.' Morgan thought that with genetics he'd discover a new approach to developmental biology, which he believed had reached a dead end in the early 20th century. That's why he started experimenting with Drosophila, and he had great success with his investigation of fruit fly genetics. But in the 1930s, genetics too had reached the end of its heroic period and had entered its classical period. The creation of molecular biology in the 1950s—this was its heroic age—led once again to a change in approach, but when the classical period was reached, both Crick and Brenner decided it was time for a change.

ALKIMOS: A change? But they'd reached the level of atoms. We can't go any deeper than that!

EXTON: It's time to return to the true questions of biology, to the questions of Morgan and the early embryologists. How does tissue differentiation and development happen? How does the complicated genetic programme work in multicellular organisms?

DEMOCRITUS: Of course, we'd like to understand all of this on the atomic level.

ALKIMOS: Back to the embryos, epigenetics and the programmes!

DEMOCRITUS: Yes, son, that's what we have to do. We've learned what genes are and how they work. They operate the machinery of development through the synthesis of proteins. And, as Monod explained, the proteins can have an effect on the genes, turning them on and off. The essence of development is the switching on of the right genes in the right place.

ALKIMOS: So we've solved the problem.

EXTON: But your generalisation in itself isn't very helpful. In biology, the more general the theory, the emptier it is.

DEMOCRITUS: I agree with you, Exton. We need to delve into the details of the processes. We need to know which genes switch on when and what they do.

ALKIMOS: As Brenner said.

EXTON: I think you've realised by now it's not possible to discover everything or solve every problem at once. We need to select one problem within developmental biology.

ALKIMOS: Like the genes of the moth wing fronts!

DEMOCRITUS: Or the nature of cell differentiation.

ALKIMOS: Or the programme that guides the migration of a flounder's eye.

DEMOCRITUS: Or the canalization of the genes that regulate early cleavage in a frog egg.

EXTON: Sydney Brenner chose the development of the nervous system.

ALKIMOS: He jumped right into the thick of it. Why didn't he choose something simpler?

EXTON: Brenner said he chose it because it was absolutely clear to him that it could not be explained by one simple theory. His biggest concern was that in the end everything could be explained by β-galactosidase—by Jacob and Monod theories. There's a repressor that needs to be removed, and then that something can be induced, and that's all there is to it.

ALKIMOS: Just remove the repressor and the brain will sprout.

EXTON: If only it were so easy. Brenner searched for some time for the most appropriate animal to use in investigating the problem. He perused numerous works on zoology until he came across the ideal animal group.

ALKIMOS: Moth? Aphids? Salamanders? Perhaps flounder?

EXTON: No. Roundworms. *Nematodes.*

ALKIMOS: Like Boveri's *Ascaris?*

EXTON: Yes. The nervous system of the roundworm looked very promising. The German embryologist Richard Goldschmidt—we talked about him when we discussed crossing over—published very interesting and detailed observations in 1909 of the nervous system of the large *Ascaris lumbricoides* roundworm that lives in the guts of pigs. During dissection, he discovered that the animal's nervous system was very simple, consisting of a few hundred cells. Moreover, he was able to describe all of the connections between the nerve cells: a miniature nervous system with a specific and knowable wiring. Just for comparison's sake, the human brain contains approximately one billion nerve cells and the number of connections is 100 billion. The main type of cell in the nervous system of *Ascaris* is the *neuron*, or nerve cell, and this is true for every nervous system. Neurons are electrically excitable cells that grow long projections called *dendrites* and *axons*, as you can see in this illustration:

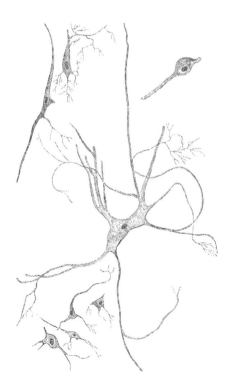

These projections extend toward other nerve cells, connecting to them at the synapses, structures reminiscent of the suction cups on the tentacles of an octopus. The nerve cells can also connect to muscle cells and other types of cells that they regulate. This is how nerve cells for example control the contraction of innervated muscle cells.

DEMOCRITUS: If I understand correctly, every neuron and every projection of the neuron in the roundworm is specific.

EXTON: Correct.

DEMOCRITUS: That must mean that its nervous system's wiring is found somewhere in the DNA, and is created when some kind of mysterious developmental code is read.

ALKIMOS: Let's decipher the code, master.

EXTON: That's what Brenner set out to do. He believed that the nervous system of an animal had a specific wiring and animals with identical genes have identical wiring. In theory it's possible to determine which genes are responsible for this particular wiring.

ALKIMOS: It looks like we're going to need mutants again. We can use the methods employed by Beadle and Tatum when they unravelled the metabolic wiring of black mould.

EXTON: Brenner also knew very well what his task was. He needed to choose a kind of nematode easily kept and replicated in a laboratory. Then he needed to create a wiring diagram of its entire nervous system, including all of its cells and connections. After that, he would have to isolate mutations that cause wiring mistakes. Then he'd be able to decipher the genetic programme that gave rise to the wiring. If he grasped how the nervous system was built, then perhaps he could answer the question of how it works.

DEMOCRITUS: Excellent plan. Let's try it! Let's get some worms!

ALKIMOS: We just need to build a few big pigsties, and then we can get started! Goldschmidt already mapped the wires of *Ascaris*.

EXTON: Now hold on, Alkimos. It wouldn't hurt to have a less complicated animal than *Ascaris*, one that does well in a lab environment and is considerably smaller. Brenner wanted to map the entire nervous system, including the neurons' cell membranes, which adhere to each other. Only an electron microscope can produce images with sufficient resolution, but because of its strong power of magnification, only small areas can be examined at a time.

DEMOCRITUS: So it's better to have a smaller worm.

EXTON: Brenner considered numerous worms before he hit upon one called *Caenorhabditis elegans*. The worm is altogether one millimetre long and can survive on nothing but *E. coli*. Its mode of reproduction is especially interesting and ideal for genetic experimentation. *C. elegans* is a self-fertilizing hermaphrodite.

ALKIMOS: Wow.

EXTON: That means the animal first produces sperm and then later it starts to ripen some eggs, which it fertilises with its own sperm.

ALKIMOS: It's as if Tiresias's male form impregnated his female form … or the left side a gynander mated with its right side … or …

EXTON: Sort of. What's crucial is that the isolated mutant offspring of a hermaphrodite are going to segregate according to Mendelian ratios. The cultures occasionally produce a male, and these can be crossed with certain lines, thereby moving the genes around. What's more, it's easy to produce mutants using the chemical mutagen *ethyl methanesulfonate* (EMS).

DEMOCRITUS: So everything's ready to carry out the grand plan!

ALKIMOS: Let's get to it! Put the worm under the electron microscope and make the wiring diagram. Then douse the wee beastie in EMS and let the mutants hatch!

EXTON: John Sulston, who worked with Brenner, created the detailed map of the nervous system, and what a tremendous amount of work it was. Brenner and associates also started the mutagenesis experiments. The first mutant *C. elegans* was isolated in 1967 in Cambridge. The mutant was given the name E1, because it was the first EMS induced mutant. Several hundred mutants followed.

DEMOCRITUS: And what about the wiring of the mutants?

EXTON: Unfortunately it wasn't too simple. Naturally it was impossible to map the entire nervous system of every mutant. They used a procedure of mutagenesis screening to identify the genes affecting the worm's brain function based on changes in behaviour rather than morphological alterations of the nervous system.

ALKIMOS: Behavioural mutants?

DEMOCRITUS: What a brilliant idea! If the wiring or the function of the nervous system is damaged by a mutation, it will manifest itself in the animal's behaviour.

ALKIMOS: You think so, master?

DEMOCRITUS: Yes, it's perfectly clear. Let's say a nerve wire doesn't reach the muscles of a worm because the gene that regulates its growth is mutant. In that case, the muscle in question won't contract, and the worm won't move.

ALKIMOS: But I don't understand how a nerve knows how it should be wired.

EXTON: Roger Sperry proposed a theory to explain how the wiring happens. He called it *chemoaffinity hypothesis*. According to his theory, every single nerve cell carries a unique *receptor* protein on its surface. This *receptor* molecule can connect to an appropriate *ligand* protein on the membrane of the target cell. The specific connections of nerve cells are guaranteed by molecules that bind to each other specifically.

DEMOCRITUS: So if a gene for a certain nerve cell receptor is ruined by a mutation, the nerve cell won't build normal connections.

ALKIMOS: And the worm's tail will flap strangely.

EXTON: That's the main idea. The problem with mutagenesis, however, is picking out the one or two mutant animals among the several hundred or thousand that are non-mutant. Let me put it a different way: creating a mutant is not in itself difficult, but finding it is. Using the right screening method of the animals is therefore critical. Brenner and associates managed to identify lots of interesting mutants based on changes in the way the worms moved or if they experienced total paralysis. One group crawled in an odd manner and were called uncoordinated mutants (*unc*-1, *unc*-2, etc.).

ALKIMOS:

> *Distorted movement, the worm's tail in the thick mud thrashes,*
> *Breathless nerves that have wired the stunted segments of muscles.*
> *If its vocal cord could vibrate, release a roaring sound;*
> *How it would curse scientists and all their great ingenuity.*

DEMOCRITUS: Yes, but I suppose it was Brenner and associates' great ingenuity that helped them determine the sequence of the given genes, which proteins they coded for, and what exactly the proteins did in the nerve wiring or in the muscles.

EXTON: They were successful in just a small number of cases.

DEMOCRITUS: But why?

EXTON: Brenner and his associates wanted to take their study to the molecular level; they wanted to know what certain genes produced. In other words, which protein was missing from certain mutant worms? They decided to start with proteins that are found in great quantities in the animal and whose biochemical nature was already known. Muscle proteins were the most obvious choice. A large part of the animal's body is made up of muscle tissue. The muscle proteins and the way in which the muscles function had been thoroughly studied by biochemists. In one series of experiments, Brenner and associates tried to identify muscle protein mutations. During genetic screens they found paralyzed mutants whose muscles were deformed. To discover which muscle protein was lacking in the mutant, they analysed them biochemically. This involved propagating the mutant worms in bacterial cultures grown in enormous containers, and then extracting the animals' muscle proteins. It took great effort, but they managed to find a few mutations affecting muscle proteins. The *unc-54* gene, for example, codes for the heavy chain of a muscle protein called myosin, and *unc-15* codes for paramyosin.

ALKIMOS: And the several hundred other mutants? What happened to them? Which genes were damaged in them? What was the use of creating so many mutants if they only managed to analyse two? And what does it tell us about the nervous system?

DEMOCRITUS: I've just had a terrible thought. If it's this much work just to extract the proteins in the greatest quantity in the animal, then how will it ever be possible to find those proteins which occur on the surface of perhaps just one nerve cell?

ALKIMOS: Like a wiring receptor?

DEMOCRITUS: Yes.

EXTON: Brenner was troubled by these questions too. To arrive at the answers required several hundred daring scientists and several decades, and the development of a revolutionary new set of technical tools.

ALKIMOS: A revolutionary technique to fish out wiring receptor genes.

EXTON: That's right. In general terms, scientists needed to learn how to isolate certain genes and replicate them in large quantities, read their sequences, change their forms, attach them to other genes, and then implant them back into living organisms. Only after these methods were developed were scientists able to examine in greater detail the workings of genes. These techniques are known as *gene manipulation techniques*, and they have fundamentally changed the way in which we conduct biological research. The number of possibilities in developmental biology has never been so enormous. Mutants and the method Brenner dreamed of and developed are now the indispensable tools of developmental biology.

DEMOCRITUS: Dear Exton, we'd love to hear all about this.

EXTON: It's my pleasure. We can discuss it over breakfast.

ALKIMOS: Now you're talking! Where's Hermes?

DEMOCRITUS: Hermes, come here immediately!

ALKIMOS: For the love of Zeus, get us our breakfast!

HERMES: Yes, of course, coming right up.

Recombinant DNA

EXTON: Come on, let's sit down.

ALKIMOS: With pleasure. Let's have that pudding!

HERMES: Pudding and tea. That's it.

ALKIMOS: What do you mean, that's it?

HERMES: It's the cook's day off.

ALKIMOS: Why that loitersack! That loll-poop! That ...

DEMOCRITUS: Alkimos, watch your tongue!

EXTON: Have no fear! Our subject for today is as rich and satisfying as a three-course meal. There's nothing like nourishing the mind, don't you agree, Alkimos?

ALKIMOS: Okay, but I'd really like some …

DEMOCRITUS: Oh fine, take mine.

EXTON: We're going to tackle genes. We're going to meet them head on and finally see them in their entirety.

ALKIMOS: Oh …

DEMOCRITUS: The age of atoms!

ALKIMOS: All right, I'm listening. Will it help explain the wiring of Brenner's worm brains?

DEMOCRITUS: Ah yes, Exton! Please, disclose how genes can be isolated, replicated, sequenced, and altered.

ALKIMOS: Yes, tell us, Exton! There's the DNA in the middle of the cell, but how can you get it out?

EXTON: That isn't all that difficult. Friderich Meischer managed to obtain fairly clean DNA from the trout sperm. The DNA of phages and other viruses is also easy to clean. All you have to do is remove any protein adhering to it.

ALKIMOS: Right, Avery explained that.

EXTON: But scientists had to learn the art of manipulating DNA molecules from bacteria. It turned out that bacteria mince any foreign DNA that's gotten into their cytoplasm as a way of protecting themselves against phages.

ALKIMOS: And I'm sure the phages fight back.

EXTON: They certainly do. They're in a constant evolutionary race with bacteria. They will do whatever it takes to infect the bacteria and the bacteria will do whatever it takes to thwart them. This has been going for a good three billion years. During this time, the bacteria managed to develop very sophisticated DNA-mincing enzymes.

ALKIMOS: Hm, finely ground phage …

EXTON: The enzymes break up the DNA at specific sequences of four, six or eight base pairs. These molecular scissors are known as *restriction endonucleases.* They're restrictive because they don't just randomly cut wherever they please like DNase, but do so only in these very specific places. The bacteria naturally protect their own DNA valiantly against these enzymes.

DEMOCRITUS: How?

EXTON: In addition to DNA-cutting enzymes, cells are also equipped with DNA-modifying enzymes. Each one can protect against the restriction endonuclease by altering—or *methylating*—the DNA at the spot where the endonuclease would sever it. Naturally, the alteration prevents the endonuclease from doing its job.

DEMOCRITUS: So the cell's own DNA is protected.

EXTON: But the foreign DNA—if it's not methylated at the right spot — is quickly dismembered.

DEMOCRITUS: What an ingenious system!

ALKIMOS: It is, but why do we need to know this?

DEMOCRITUS: Don't you see? If we could isolate this system, we could use it to carve up DNA.

EXTON: The first restriction endonuclease was identified at the end of the 1960s. Since then, several hundred forms have been discovered in various bacteria. These forms all chop up DNA in different sequences. Their names are all derived from the name of the bacterium they are isolated from. For example, PstI was obtained from the *Providentia stuarti* bacterium and EcoRI from *E. coli*. PstI only slices at CTGCAG sequences and EcoRI only at GAATTC.

DEMOCRITUS: Now I call that biological specificity!

EXTON: If we add an enzyme like this to a homogenous DNA extract—let's say an extract from a virus such as SV40—the DNA will be cut up into specific fragments that can then be isolated using agarose gel electrophoresis.

ALKIMOS: That's exactly what I wanted to say …

EXTON: We've already discussed the essence of electrophoresis. If the digested DNA is placed in agarose gel—or to be more precise into tiny wells in the gel—and then an electric field is applied, the negatively charged DNA fragments begin to move in the gel towards the positive electrode. The larger the fragment the more slowly it moves through the gel. If we stain the DNA, it becomes visible in the form of definite bands. Of course one band is not just one DNA molecule, but several thousand identical fragments. These bands of identical DNA molecules can easily be extracted from the gel.

DEMOCRITUS: That's how we can obtain specific sections of clean viral DNA, essentially the fragments of genes.

EXTON: Exactly. We can make a very accurate restriction map of DNA. If we use several different enzymes, we can infer the location of the restriction sites on the DNA based on the size of the fragments. In 1971, Daniel Nathans made the first restriction map, a circular diagram of the SV40 genome.

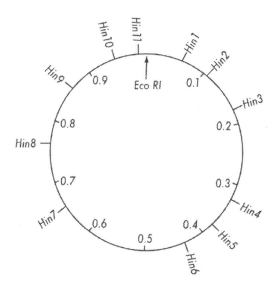

ALKIMOS: So we learned the art of chopping up DNA from bacteria!

EXTON: Bacteria are capable of doing a lot more than that with their DNA. They have to replicate their DNA, which requires uncoiling and separating the strands. They also need to correct mistakes in the DNA and rejoin strands that have been cut. They do this all with enzymes. The discovery of these enzymes and how they functioned opened up a host of new possibilities. Between 1965 and 1985 a series of new methods were developed to manipulate DNA and RNA *in vitro* that revolutionised the study of genetics.

ALKIMOS: We like revolutions, don't we, master?

DEMOCRITUS: Especially if we're talking about science.

EXTON: We've already discussed how the amino acid sequence of proteins was determined.

ALKIMOS: Right, Sanger and his insulin.

EXTON: The next step was to determine the sequence of nucleotides in nucleic acids.

DEMOCRITUS: Is that possible? If we can achieve that, there would be no need to read the sequence of proteins, since we would be able to deduce it from the gene sequence and the genetic code.

EXTON: You're correct.

ALKIMOS: So gene sequences would allow us to kill two birds with one stone.

EXTON: The first nucleic acid sequence was determined by Robert Holley and his colleagues in 1964. They worked with small segments of transfer RNA, consisting of about 75–80 nucleotides, instead of DNA. Using enzymes, they cut the RNA up into pieces, similar to the method used by Sanger when sequencing proteins. By removing various nucleotides they were able to piece together the entire sequence of the alanine tRNA found in yeast.

ALKIMOS: Can you show us?

EXTON: Here, have a look:

p G-G-G-C-G-U-G-U-G-G-C-G-C-G-U-A-G-U-C-G-G-U-A-G-C-G-C-G-C-U-C-C-C-U-U-I-G-C-I¥G-G-G-A-G-A-G-U-C-U-C-C-G-G-T¥-C-G-A-U-U-C-C-G-G-A-C-U-C-G-U-C-C-A-C-C-A_{OH}

When the bases pair up, it assumes a rather distinct shape, as you can see in Holley's diagram:

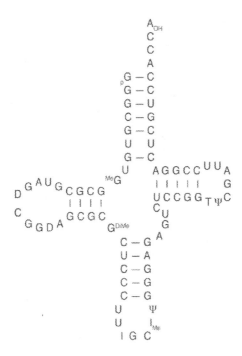

RNA's coiling is far more complicated than DNAs. It's not surprising that Watson failed to build a model of RNA structure. You see, the structure of single-stranded RNA is not repetitive. Every unique sequence has a unique structure.

ALKIMOS: Like enzymes.

EXTON: Yes. In fact, because of their distinct spatial structures, many RNA molecules can act as enzymes, known as *ribozymes*. The rRNA of ribosomes, for example, work as a ribozyme. One thing is certain: ribozymes had a crucial role in the early development of life, since they operate as both information-bearing genes and enzymes with a catalytic effect.

DEMOCRITUS: So the two most essential properties of the atoms of life come together in ribozymes. Ribozyme-based life must have preceded life constructed from DNA and specialised proteins. Astounding!

EXTON: But let's return to sequencing. The tRNA Holley mapped isn't an enzyme and it doesn't code for proteins. The first sequence of an RNA that codes for proteins was not determined until 1975. As the technology developed, the sequence of the single-stranded RNA of the bacteriophage MS2 was slowly mapped out in the laboratory of Walter Fiers. Light was shed for the first time on a nucleic acid sequence that codes for a protein. The codons coding for certain amino acids and the STOP codons that halt synthesis of the three phage protein chains were also revealed.

DEMOCRITUS: See, son? Here are the codons, so we can immediately determine the order of amino acids in the protein.

ALKIMOS: My Zeus, that's amazing! But I thought genes were built of DNA.

EXTON: MS2 is an exception; it's an RNA phage. DNA sequencing would have to wait until more restriction enzymes were discovered in the 1970s. Then it became possible to isolate large quantities of completely identical DNA segments. Because at that time no one was able to sequence DNA directly, scientists initially tried to sequence the complementary RNA transcribed from DNA with an RNA polymerase enzyme. This meant sequencing proceeded at a snail's pace, until Fred Sanger stepped into the limelight again.

ALKIMOS: The great master of sequencing.

DEMOCRITUS: What had Sanger come up with?

EXTON: Instead of chopping up DNA chemically, with enzymes, Sanger synthesised DNA using DNA-polymerase. To sequence DNA he needed a pure, homogenous sample of DNA ...

ALKIMOS: And he could obtain that using restriction endonucleases.

EXTON: ... and also DNA-polymerase and a short DNA fragment (about 15 to 20 nucleotides) complementary to a section of DNA—this is called a *primer*. The strands of DNA are separated using heat, and then the primer attaches itself to the appropriate DNA segment. DNA-polymerase can then start synthesis of the new chain. The four deoxynucleotide triphosphate monomers (dNTP) are of course present in the reaction compound. The polymerase incorporates each of these dNTPs, one after the other, in accordance with Watson–Crick complementarity. The enzyme forms phosphodiester bonds between the last nucleotide in the chain and the next arriving monomer. These bonds link the phosphate group of the new base to the 2′ oxygen of the deoxyribose ring of the preceding base. Sanger synthesised bases that lacked the 2′ oxygen, known as 2′3′ dideoxynucleotide triphosphates (ddNTP).

DEMOCRITUS: If I recall, the difference between DNA and RNA nucleotides, is that deoxyribonucleotides lack the 3′ oxygen.

EXTON: Exactly. ddNTPs lack both 2′ and 3′ oxygen. If ddNTPs are present in a polymerase reaction alongside dNTPs, then the polymerase unavoidably incorporates the ddNTPs into the chain too, since it can't tell the difference between the two kinds of triphosphates. After a ddNTP, however, the polymerase can't add on the next nucleotide because the oxygen necessary for bonding is absent. The chain can't grow any longer.

ALKIMOS: So the enzyme's tooth breaks off when it bites into the ddNTP. But how does this allow us to read the sequence? We'd have to ask the polymerase!

EXTON: Sanger conducted four reactions at once. In each one there were all four dNTPs but only one of the four of the ddNTPs (ddATP, ddCTP, ddGTP, and ddTTP). These were in smaller quantities than the dNTPs. The chains grew longer, but occasionally they stopped because of a ddNTP. In each reaction, the various DNA molecules had different lengths, because ddNTP can be added to the chain at any point. So, Sanger had four reactions, and in each one different segments. But in the ddATP test tube all of them stopped at the A bases. In the ddGTP, they all stopped at the G bases, and so on. The four samples are analysed on an acrylamide gel and the sequence can be determined.

ALKIMOS: After the DNA has been stained?

EXTON: In this case no. Sanger labelled the primer with a radioactive tag, so the gel bans were radioactive and could therefore be detected on film.

ALKIMOS: Very clever.

EXTON: Walter Gilbert and Allan Maxam developed a similarly ingenious method in 1977, which made it possible to quickly establish the SV40 virus' sequence of 5243 base pairs and the G4 phage's sequence of 5577 base pairs. With further refining of the DNA sequencing technique, the number of known sequences grew exponentially: hundreds, thousands, millions of genes, then dozens, hundreds of entire genomes. These sequences offered scientists a peek into the deepest secrets of evolution. But we'll talk about that later. For now let's remain in the heroic age of sequencing and DNA cutting.

ALKIMOS: Millions of genes? How is that possible?

EXTON: If a technique works well, it's not hard to do it on an industrial level. Of course it's not quantity that matters. The essence of recombinant DNA techniques is obtaining specific genes, replicating them, sequencing them, cutting them up, reassembling them, implanting them, or even eliminating them. This is how we come to understand how they work.

DEMOCRITUS: Dear Exton, I think we understand more or less how the DNA is cut up and sequenced. But how can they be reassembled or replicated?

ALKIMOS: How about the old-fashioned way?

EXTON: I haven't yet mentioned one of the important properties of restriction endonucleases: many of them do not cut straight through the DNA, which means they fail to leave behind a flat surface, known as a *blunt end*. Instead they leave unequal surface. By this, I mean the two DNA strands both end in a stretch of two to four unpaired bases. These two ends are naturally complementary, and will easily adhere to each other. Ends of this sort are known as *sticky ends*. For example, the sticky ends of EcoRI look like this:

———— G	AATTC————
———— CTTAA	G————

These sticky ends can again be fused together by the enzyme *DNA-ligase*.

ALKIMOS: So there are fusing enzymes too!

EXTON: Digestion by two distinct restriction enzymes results in two different types of sticky ends, which can't be ligated to each other. Ends created by identical enzymes, however, can be ligated, even if the DNA pieces have completely different origins.

DEMOCRITUS: After cleavage, any kind of DNA can be joined to any other kind of DNA?

EXTON: Of course. Remember, DNA is nothing but a molecule.

DEMOCRITUS: Incredible! That means we can create any kind of combination, even those that don't naturally occur.

EXTON: Right.

DEMOCRITUS: I don't dare to think about it. The possibilities are mind-boggling!

EXTON: The extracted DNA can be inserted into other DNA that can be propagated in bacteria. Bacterial *plasmids*, representing circular replicating pieces of DNA, are frequently used in this way. Plasmids are easily obtained from bacterial cultures and can be inserted without difficulty into bacterial cells. Within the cells, the plasmid will begin to replicate independent of the bacterial chromosome. If a plasmid is digested with a restriction enzyme such as EcoRI, its circular DNA opens up. Let's suppose that a sequence contains only one GAATTC site. EcoRI will cleave it only once. If we add a DNA segment—let's say a fragment of the SV40 virus—cut with EcoRI to the linearised plasmid then we can ligate two DNA molecules.

ALKIMOS: Because the sticky ends will adhere to each other.

EXTON: And the plasmid will again form a circle, but it will already contain the SV40 segment. The plasmid is then inserted into the bacteria.

ALKIMOS: How?

EXTON: Using a very simple procedure called *transformation*.

ALKIMOS: Avery did that!

EXTON: Yes, he did. Just as Avery's transforming principle could be introduced into a cell, a plasmid can be too. With a little salt and a little heat the *E. coli* bacteria, for example, will readily take up a plasmid. If the plasmid contains a gene that confers resistance to antibiotics and we spread the bacteria on a plate supplemented with that antibiotic, the transformed bacteria can easily be located, because only those will grow and form colonies.

DEMOCRITUS: And in those colonies, you'll find only bacteria whose cells contain plasmids with the foreign DNA. The foreign DNA has been replicated!

EXTON: It's called cloning. From one transformed bacteria we create an entire colony in a petri dish.

ALKIMOS: And what happens to the untransformed bacteria?

EXTON: They can't grow because they don't carry the antibiotic resistance gene.

ALKIMOS: So if a cell doesn't want to take up a fabricated DNA, then it dies. Clever trick!

EXTON: If there weren't antibiotics in the medium, then all the bacteria would grow and before long the plate would be covered in a thick bacterial lawn.

ALKIMOS: And then we could introduce phages!

EXTON: But let's follow the colonies bearing the antibiotic-resistant plasmid. We can grow the colony in a test tube. We put antibiotics in the solution so we don't lose the foreign plasmids.

ALKIMOS: Oh, that's all we need—to lose the foreign plasmid after all our hard work! Serve up that antibiotic.

EXTON: Once the culture is good and thick, we collect the cells in a centrifuge, then tear them apart. The plasmid DNA is then cleaned. The bacteria's chromosomes are much larger than the plasmids, so it's easy to get rid of them. The end result is a pure preparation of several million absolutely identical plasmids.

ALKIMOS: Each with the inserted gene right in the centre.

EXTON: Yes. We can then sequence the gene, replicate it, introduce a mutation, patch it together with other genes, and so on.

DEMOCRITUS: And that's how we clone a gene. Before our very eyes we have the physical entity of Mendel's invisible factors. Amazing!

ALKIMOS: Master, let's bring Mendel back a petri dish!

Striped Embryos—*Christiane Nüsslein-Volhard*

EXTON: The development of recombinant DNA techniques brought about fundamental changes in every branch of biology. The range of research possibilities expanded as never before. Perhaps the most dramatic impact was felt in developmental biology. One could say it sprang to life again. Finally the problems of classical embryology could be addressed on a molecular level.

ALKIMOS: The genes of development!

DEMOCRITUS: The dream of Morgan, Ephrussi, and Kühn.

EXTON: The identification of the first developmental genes finally fused genetics, developmental biology and molecular biology into one unified science.

ALKIMOS: Go on!

EXTON: If we want to discuss this in more depth, we need to travel again.

ALKIMOS: Oh, well …

DEMOCRITUS: Where, Exton?

EXTON: Tübingen, 1988. Come on, let's go!

DEMOCRITUS: Who are we going to meet?

EXTON: Christiane Nüsslein-Volhard.

ALKIMOS: Christiane? A woman scientist?

EXTON: She's the top of her field. But you are right Alkimos, in the 1980s women were terribly underrepresented in scientific circles, and they still are today.

ALKIMOS: But why?

EXTON: Probably because of the male ego. Nüsslein-Volhard also had many problems with that throughout her career. Now, let's go!

ALKIMOS: Aaaaggghhh! Here we go!

DEMOCRITUS: Please stop shrieking!

ALKIMOS: I'll never get used to the shaking.

DEMOCRITUS: Sit still.

EXTON: Alkimos, this will make you feel better: we're returning to one of your favourite model animals …

ALKIMOS: The sea urchin?

EXTON: No, the fruit fly.

ALKIMOS: Oh, well, it's not my …

DEMOCRITUS: So Morgan's work wasn't in vain.

EXTON: Of course not. Although, it's true that he couldn't answer the questions of development using genetics.

ALKIMOS: Maybe he just didn't have enough time …

EXTON: But others could, and it was thanks to the efforts of the early students of Drosophila and almost a century's worth of genetic experience. You'll see what I'm talking about in a minute. Here we are.

DEMOCRITUS: This is Tübingen?

EXTON: Yes, we're on Spemannstrasse.

ALKIMOS: After Hans Spemann?

EXTON: Of course. And here we are.

ALKIMOS: Max-Planck-Institut für Entwicklungsbiologie.

EXTON: Let's go in.

ALKIMOS: Abteilung Genetik. What a strange smell!

EXTON: That's the odour of fruit fly food.

DEMOCRITUS: Morgan's lab smelled worse. There they used mashed bananas to feed their fruit flies.

EXTON: This is the lab.

NÜSSLEIN-VOLHARD: Welcome.

EXTON: Good afternoon. I'm Exton, and these are my students.

ALKIMOS: Alchumbach.

DEMOCRITUS: Demiescher.

NÜSSLEIN-VOLHARD: Christiane Nüsslein-Volhard.

EXTON: We're here to ask you about your research on the genes of early embryogenesis and morphogens.

NÜSSLEIN-VOLHARD: I'd be delighted to discuss it.

DEMOCRITUS: We're very interested in the subject. Ever since the debate over preformation versus epigenesis, we've been trying to understand how development is possible. How can one fertilised egg cell develop into a sentient organism composed of a huge variety of cells? What's the role of proteins and genes in all of this?

NÜSSLEIN-VOLHARD: That's exactly what interests me, too.

EXTON: How did you start your research in this area?

NÜSSLEIN-VOLHARD: I wrote my doctoral dissertation on bacterial transcription, but I already had a nascent interest in the question of pattern formation. What fascinated me at first were experiments with hydras. As I'm sure you know, even small pieces of a hydra can regenerate into a complete animal.

ALKIMOS: Oh yes, we were very impressed by Abraham Trembley's experiments chopping up hydra, weren't we, master?

NÜSSLEIN-VOLHARD: This fantastic ability of *Hydra* makes it an ideal animal in which to study embryonic development. Based on his experiments with hydras, Lewis Wolpert posited his theory of *positional information* in 1969. According to this theory, during development, cells in a developmental field have positional values that correspond to their location. The cells interpret this value, and this interpretation determines the later state of the cell.

EXTON: Let me explain this to my students in more detail.

NÜSSLEIN-VOLHARD: Go right ahead.

EXTON: According to Wolpert's theory, the location of a cell in a tissue influences differentiation. The cells are capable of interpreting positional information within the coordinate system of a developmental field.

DEMOCRITUS: That explains why Spemann's organiser triggered the formation of another axis after it was transplanted: it defined another coordinate system. It wasn't just a question of local induction, but the relocating of the positional reference point of the entire embryo.

EXTON: Wolpert wasn't satisfied with Spemann's theory of induction either. But let's translate positional information into the language of molecules. It's simplest if we imagine a morphogen whose effect on the cell is dependent on its concentration. If the morphogen concentration is high, certain genes switch on. If the concentration is medium or low, then different genes are activated. If a morphogen like this forms a concentration gradient in a developing tissue, this leads to the creation of different cell types within a developmental field. Concentration gradients can arise, for example, if a morphogen is produced in a specific place and diffuses from there. In this case the dispersion of the morphogen determines the coordinate system. The morphogen concentration is the information that's received by the cells. Let me put it differently: the spread of the morphogen can determine the arrangement of information in the given area.

NÜSSLEIN-VOLHARD: Wolpert's idea is very attractive because it explains how tissue differentiation can occur in response to the difference in quantity of just one kind of molecule. Of course, Wolpert's theory was not immediately accepted, mainly because no one had ever identified morphogens of this type. Scientists experimented with different tissue extracts; they examined the effects of these extracts on pieces of embryos. But these experiments, which were burdened with a tremendous amount of methodological and theoretical difficulties, all failed.

DEMOCRITUS: The study of the wiring of a worm's brain by Brenner involved similar challenges. How is it possible to find biologically very active materials, whether these are cell-surface receptors or morphogens, when these perhaps come in very small quantities?

ALKIMOS: Then we need a new approach!

DEMOCRITUS: We need genetics.

NÜSSLEIN-VOLHARD: Yes, I too quickly recognised the advantages of a genetic approach. When I finished my doctoral dissertation, I began research on a new subject. I poured through the scientific literature in search of an organism which could help me bring together embryology and genetics. It didn't take me long to come upon Drosophila. In the early 1970s, several promising articles were published on the embryology of the fruit fly. For example, Illmensee and Mahowald demonstrated that if you inject cytoplasm from the *posterior* region, or the back end, of the fruit fly ovum into the *anterior* end, it will invoke development of pole cells typical of the back end in the front end.

EXTON: Excuse me, but I think I need to help my students. The pole cells are the first cells in a multinucleate *syncytial* fruit fly embryo to form cells in the posterior end of the embryo. They will eventually become the gametes of the adult individual.

DEMOCRITUS: The formation of the pole cells is therefore none other than the early separation of Weismann's germplasm.

EXTON: Exactly. Transplantation experiments revealed that the posterior cytoplasm contains something that's capable of triggering the formation of pole cells, even in the anterior region.

ALKIMOS: Is this an induction material?

NÜSSLEIN-VOLHARD: No, this has been traditionally called a *cytoplasmic determinant*. The posterior cytoplasm is a visibly different type of cytoplasm including special granules. It does not induce fate in the surrounding tissue as happens during induction.

EXTON: That's right. Illmensee's experiment was a beautiful demonstration of a localised determinant in mosaic development.

NÜSSLEIN-VOLHARD: They also described a mutant—although this was in another fruit fly species—in which the pole cells didn't form. This mutation is called *grandchildless*. Fruit flies without pole cells will develop into perfectly intact adult flies except they'll have no gametes; in other words, they're sterile.

ALKIMOS: So this mutation affects the cytoplasmic determinant in the posterior of the ovum.

NÜSSLEIN-VOLHARD: It affects the information content of the ovum—one of the materials that enter the ovum during its maturation in the mother.

ALKIMOS: Like a maternal care package.

NÜSSLEIN-VOLHARD: It's known as the maternal effect. The homozygote mutant offspring of heterozygote parents develops normally, forming pole cells and gametes. The homozygote females, however, lay eggs that lack something and no pole cells form in the ovum. There's a problem with the maternal effect. A homozygous mutant female will have children, but not grandchildren; she is thus *grandchildless*. It seemed obvious that other such maternal-effect mutations—those which influence the information contained in the ovum—could be identified in a fruit fly. *grandchildless* represents only one specific type of maternal effect mutation, a mutation where the pole plasm is missing. But there could be other types of maternal mutants, mutants in which unhealthy embryos are developing. A mutant embryo that lacks such a maternal component could then in principle be rescued by injecting a cytoplasm extract from a wild-type embryo.

DEMOCRITUS: So a deficiency that results from a mutation in a maternal-effect gene could be remedied with an injection from a healthy embryo. That's a fascinating idea!

NÜSSLEIN-VOLHARD: Yes, in this way we could even identify the component in question. This would be a much more sophisticated experimental setup than earlier ones. In 1972, Alan Garen and Walter Gehring published an experiment of this type. They managed to rescue an embryo with the *deep orange* maternal-effect mutation by injecting it with wild-type cytoplasm.

EXTON: As you see, the experimental manipulation of embryos can also be carried out in fruit flies. My students and I have already discussed the advantages of fruit flies in research. It's easy to collect a large number of embryos of identical age. During their development, they secrete a resistant cuticle, which mirrors the body structure of the embryo. By performing a preparation of the cuticle the segments are easily recognizable in the microscopy, as well as the dorsal and ventral side, and the anterior and posterior of the embryo.

NÜSSLEIN-VOLHARD: Here, have a look at this drawing.

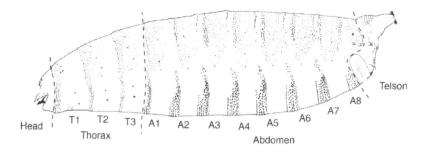

The mutations that influence the development of the embryo can be identified based merely on the morphology of the larval cuticle.

Eric Wieschaus and I relied on this when we undertook our first large-scale mutant screening in Heidelberg in 1979.

ALKIMOS: Following right in the footsteps of Morgan.

NÜSSLEIN-VOLHARD: In a way, except we were searching for the mutations that affect embryonic development, not those that influenced the phenotype of an adult fly. No one had yet performed a systematic screening like this of fruit flies, although mutants affecting the larval pattern had fortuitously been isolated. Naturally it wouldn't have been possible without the knowledge accumulated over eighty years and the genetic tools we now had available to us. As you know, Drosophila has only four chromosomes. This allowed us to develop highly refined genetic tools that would not have been available in other organisms. We now have so-called *balancer chromosomes* at our disposal that prevent recombination because they contain many inversions. These balancers, combined with numerous markers (such as *white*, *yellow*, and *Curly*), make it possible to select the desired genotype, follow the mutations and maintain the mutant strains.

DEMOCRITUS: How does a mutation screen work?

NÜSSLEIN-VOLHARD: We caused mutations in males using EMS, and then, through a number of crossings, produced homozygote animals. In the first screen we were looking for zygotic mutants. For technical reasons we only screened for maternal-effect genes later.

EXTON: Just a note for my students: zygotic mutants are those that show a phenotype in the homozygous embryos produced by heterozygous parents. In other words, a malfunctioning zygotic gene—that's a gene that originates from the genome of the fertilized ovum—causes the phenotype. In contrast, mutations in the maternal-effect genes cause an alteration in the phenotype not of the homozygous mother but in her offspring. So what's crucial in the zygote are the products synthesised from the gene of the mother and not the expression of the zygote's own genes.

ALKIMOS: This is getting complicated …

NÜSSLEIN-VOLHARD: Therefore, during zygotic screens, we were looking for mutations that caused the death of the embryo in the homozygous form. In fruit flies this means the larvae don't hatch. These kinds of mutations are called zygotic embryonic lethal mutations. Naturally we screened for the maternal-effect lethal mutations by examining whether larvae hatched from eggs laid by homozygous mothers. Of course the malfunctioning of many genes could cause the death of the embryo.

EXTON: These would include many so called *housekeeping genes*, such as translation factors, ribosomal proteins, cell cycle proteins.

NÜSSLEIN-VOLHARD: Indeed. But we weren't interested in these genes. We had to screen for those genes that specifically affected development. In order to properly screen for developmental genes we had to examine the cuticles of the mutants. Many lethal mutations have no effect on the cuticle pattern, so we weren't interested in these. What we were interested in were those that caused specific morphological changes that could easily be discerned in the cuticle pattern of larvae that never hatched. These accounted for approximately 4% of the lethal mutations.

DEMOCRITUS: They were the developmental mutations!

ALKIMOS: But what changes did they cause?

NÜSSLEIN-VOLHARD: The head and tail axis and the back and belly axis are both clearly visible on the embryonic cuticle. Numerous characteristic structures of the cuticle along these axes indicate the exact state and polarity of the larva. If you follow the length of the long, *anterior–posterior* axis you can easily identify the various segments of the larva. The back-belly, or *dorsal–ventral*, axis also displays specific variations in the construction of the cuticle. During the mutant screen we conducted in Heidelberg, we looked for mutations that caused significant differences in the cuticle pattern. We identified altogether 120 genes. Later screens we did in conjunction with other labs revealed a total of 30 maternal effect genes. Our first obvious and surprising finding was that in the case of the maternal mutants there were far fewer phenotypic categories than genes.

DEMOCRITUS: So several different mutations could result in the same phenotype.

NÜSSLEIN-VOLHARD: At least similar phenotypes. A significant number of mutations caused, for example, the so-called *dorsalised* phenotype. The back, or dorsal, cuticle structure spread to the belly, or ventral, region, and the belly cuticle structure was absent. One of these mutants is known as *dorsal*. *Toll* and *cactus* mutations, however, caused a *ventralised* phenotype.

ALKIMOS: The back became the belly …

NÜSSLEIN-VOLHARD: We discovered extraordinary zygotic mutations along the head–tail axis, too. We organised them into three groups: *gap* genes, *pair rule* genes, and *segment polarity* genes. Mutations in some of the *gap* genes, such as the *hunchback* gene, resulted in the absence of large individual sections of the embryo.

ALKIMOS: Ahah, so a literal *gap*, or hole, was visible in a larger stretch of segments.

NÜSSLEIN-VOLHARD: Embryos with mutations in the *pair rule* genes lacked every other segment. Examples of these genes are the *even-skipped* and *fushi tarazu*. Mutations in the segment polarity genes, such as the *engrailed* gene, caused distinct changes in every single segment.

DEMOCRITUS: So these are the genes responsible for the embryo's segments.

NÜSSLEIN-VOLHARD: Correct. These phenotypes suggested that the process of segmentation entails at least three levels of spatial organisation. The genes that regulate segmentation at first mark a wide stretch within the embryo. The embryo is then divided into seven sections, and then these are further divided until fourteen segments are created. After cloning these genes, during molecular and functional analysis, a tremendous number of extraordinary details were revealed. We can now say that we have a general understanding of the molecular background of the segmentation process.

ALKIMOS: The recipe for segmented embryos!

DEMOCRITUS: Fantastic! And what are those extraordinary details?

NÜSSLEIN-VOLHARD: It's quite difficult to explain, but I'll try and give you an overview. We discovered that a lot of the genes that regulate segmentation code for *transcription factors* that regulate the pattern of expression of identical or lower level genes.

EXTON: Let me add something. We've already spoken about the various techniques of cloning genes, but I need to mention an important method: *in situ hybridisation.* With this method we can establish in which part of the embryo, in which cells or tissues, a gene is expressed.

DEMOCRITUS: Brenner said we needed to discover which gene is switched on and where.

EXTON: And we can discover that with *in situ* hybridisation. If we've cloned a gene—in other words extracted and amplified it—then we can continue to make copies of it, DNA and RNA copies. The latter are made with the enzyme RNA-polymerase. The RNA prepared *in vitro* can be labelled with radioactive or chemically modified nucleotides. A tagged copy of the gene, known as a probe, can be added to the appropriately fixed and treated embryos. If the probe is antisense, or corresponds to the negative strand of the DNA, then it will hybridise with the positive strand of mRNA by base pairing. Of course only where the mRNA is present in the embryo—in other words where the gene in question is expressed. The non-hybridised probe is washed out. The tagging of the probe—either using radioactivity or dying techniques—allows us to establish where the gene is active.

NÜSSLEIN-VOLHARD: The segmentation genes give us beautiful expression patterns. Here, have a look. In the first picture you can see the *hunchback* gene, in the second the *runt* pair-rule gene, and in the third the *engrailed* segment polarity gene.

DEMOCRITUS: Let me see … Astounding!

ALKIMOS: Striped embryos!

NÜSSLEIN-VOLHARD: You can have these prints if you'd like.

DEMOCRITUS: Thank you!

EXTON: We started our conversation with morphogens. Before we leave, could we speak a little bit more about them? I know that based on the phenotype of certain mutants and thanks to the meticulous cytoplasm-injection experiments, you've arrived at a model that shows us the fruit fly embryo with two axes specified by two morphogens perpendicular to each other.

ALKIMOS: Yes, you must have found the morphogens' mutants, too.

NÜSSLEIN-VOLHARD: From the start, our models assumed that morphogens played a role in the pattern formation along the axes. Let me try to briefly outline the main points of our research on the *bicoid* gene without going into too much detail. We found it when screening for maternal-effect developmental genes. The phenotype of the *bicoid* mutation is distinct: the anterior parts, including the *acron*, the head and the thorax of the embryo are missing. In there place the *telson*—a region typically found in the embryo's posterior region—could be seen. The embryos thus have a structure of telson–abdomen–telson. We could show using injection experiments that anterior cytoplasm rescues the *bicoid* phenotype, whereas cytoplasm from other regions of wild type embryos does not. Furthermore, transplantation of anterior cytoplasm into posterior regions can make heads there. After Marcus Noll cloned the *bicoid* gene in his lab, it was discovered that the mRNA was found snugly at the anterior region of the egg. The mRNA is thus localised at the most anterior end; yet in the mutant both the head and the thorax are absent.

DEMOCRITUS: So the question is how can *bicoid* mRNA at the anterior end and the protein synthesised from it have an effect on the development of both the head and the thorax?

NÜSSLEIN-VOLHARD: And sometimes on the development of the very beginning of the abdomen, too. The protein is a morphogen that's produced in the anterior part, which then spreads by diffusion from the localised mRNA and creates a concentration gradient, could explain the phenotype. That's what we needed to demonstrate. Wolfgang Driever began to work on the problem and managed to obtain a very good antibody against the Bicoid protein. By staining the embryos with the antibody, we were able to prove the existence of a gradient. The level of Bicoid protein was highest at the head and then decreased exponentially as it progressed toward the tail, creating the gradient. The protein spread through about two-thirds of the ovum before its concentration grew so small it was no longer detectable.

ALKIMOS: A real morphogen! A concentration gradient along the head–tail axis!

NÜSSLEIN-VOLHARD: We managed to demonstrate the existence of a Bicoid protein gradient. But it still didn't mean that the protein is a true morphogen, that in different concentrations it leads to the creation of different cell types. We had to show this experimentally by changing Bicoid levels and then examining changes in the embryonic phenotypes.

ALKIMOS: How did you change the levels of Bicoid in an embryo? Did you inject it in different amounts?

NÜSSLEIN-VOLHARD: No, we relied on genetics. Bicoid levels drop by half in the offspring of heterozygous *bicoid* mutant mothers.

DEMOCRITUS: I assume because the mother has just one intact gene, so only half as many mRNA molecules are produced, from which only half as many proteins are synthesised.

ALKIMOS: But how can you increase the level?

NÜSSLEIN-VOLHARD: Alan Spradling and Gerald Rubin developed the P-element germplasm-transformation method. If the *bicoid* gene …

EXTON: One moment, please, let me explain this to my students. The P-element is a peculiar genetic element, a *jumping gene* or *transposon*. It can jump from one place to another in the genome with the help of the *transposase enzyme* that it codes for. The transformation technique takes advantage of this capability. If we clone a foreign gene into the centre of the P-element, the transposase will help the modified element jump into the genome of a fruit fly. The end result is a transformed strain that carries the desired gene or gene fragment.

ALKIMOS: I understand! You transplanted the *bicoid* gene into the P-element and then inserted it into the fruit fly. The animal then had two genes—the original and the inserted one.

NÜSSLEIN-VOLHARD: That's exactly what we did. These experiments proved that the level of the Bicoid protein and the fate of cells in certain parts of the embryo are closely tied. At higher protein levels, the structure more closely resembled a head; at lower levels, it appeared more tail–like.

DEMOCRITUS: One protein can determine the pattern of an entire axis. Incredible!

NÜSSLEIN-VOLHARD: Naturally, the *bicoid* doesn't act alone, but it is the crucial gene in determining the head–tail axis. The Bicoid protein is a transcription factor and it regulates the expression of other genes in the embryo.

DEMOCRITUS: Which genes and how?

EXTON: Demiescher, I'm afraid we don't have time.

DEMOCRITUS: But it's so interesting!

NÜSSLEIN-VOLHARD: Do you really need to go so soon?

EXTON: Oh yes, and right away I'm afraid.

NÜSSLEIN-VOLHARD: Well, I wanted to invite you for dinner. Anyway, in that case let me give you some reprints of our studies. They'll give you all the details.

DEMOCRITUS: Oh, I'm much indebted …

ALKIMOS: Master, can I have a look?

NÜSSLEIN-VOLHARD: It's been a pleasure.

EXTON: For us, too. Thank you for a fascinating discussion. Now, my students, the Deus ex Machina awaits us.

NÜSSLEIN-VOLHARD: I've never heard of that mutation. Could I …

EXTON: Come on, squeeze in, quick!

ALKIMOS: But I still wanted to ask about the male-egos.

NÜSSLEIN-VOLHARD: Then you should really stay for dinner…

EXTON: Come on, Alkimos!

DEMOCRITUS: Wait a minute. I'm coming!

ALKIMOS: Aaagghhh … Here we go!

Our Worm Ancestors—*Detlev Arendt*

ALKIMOS: Master …

DEMOCRITUS: Yes, son?

ALKIMOS: Have you read the articles?

DEMOCRITUS: I looked them over.

ALKIMOS: And?

DEMOCRITUS: It's incredible how far we've come. We couldn't have dreamed of this in Morgan's time. We've unravelled the function and molecular makeup of genes. We know how to interpret the molecular interplay of axes and segments. The molecular code of life now lies exposed before us.

EXTON: The molecular and functional analysis of genes has indeed progressed by leaps and bounds. Large-scale genetic screens for mutants have helped geneticists to genetically scour entire genomes of animals, in particular the fruit fly, *C. elegans* and more recently the zebra fish, or *Danio rerio*. In a lot of cases the gene sequence is also known. At first we considered genes to be abstract hereditary units; today we know their sequence, their location on the chromosome, the function of proteins synthesised from them (and often even their spatial structure) and their connection to other genes. Yes, it's mind-boggling the progress we've made. Now it's time to return to the question of evolution and re-examine it in light of our new understanding of development on the molecular level.

ALKIMOS: The evolution of genetic programmes!

EXTON: *Evolutionary developmental biology* or *evo-devo*. The intersection of genes, development and evolution.

ALKIMOS: What a marriage!

DEMOCRITUS: Waddington told us that evolution involves the changing of developmental pathways. With the help of genetics and molecular biology we began to uncover the genes and programmes behind these developmental pathways.

ALKIMOS: With the help of Brenner's and Nüsslein-Volhard's mutant screens and cloning of genes.

EXTON: That's right. If we compare the developmental programmes of various organisms at the genetic level we can gain insight into the evolution of development.

ALKIMOS: So we're not going to compare the embryos as von Baer did, but their genes. Amazing.

EXTON: Yes, their genes, morphogens, gene networks, and specified cell types. Do you remember D'Arcy Thompson's distorted fish? If, for example, we could compare the morphogens that determine the back-belly axis in these fish of various shapes to the gene expression patterns regulated by the morphogens, we'd learn a lot about the changes in body shape.

ALKIMOS: And the morphogens of the moth wing? Kühn's stripes could take on new meaning in an evolutionary perspective!

EXTON: Indeed they could. Evo-devo has a long history, going back to 19th century embryology. There are two major complementary research approaches. One has been called the neo-Darwinian approach by historian Ron Amundson in his must-read book *The Changing Role of the Embryo in Evolutionary Thought*. Practitioners of this approach try to find subtle genetic and developmental differences underlying phenotypic changes between closely related species. The other approach Amundson called the *structuralist approach*. Its goal is to go back deep in time and try to reconstruct what happened during evolution. Adherents of this approach are more interested in deep homologies between distantly related organisms, in the comparison of entire developmental cascades, or organ systems. It traces back to the eminent English anatomist Richard Owen who in 1849 reconstructed the vertebrate *archetype,* a conceptual form that best represents all vertebrates, as you can see in this picture.

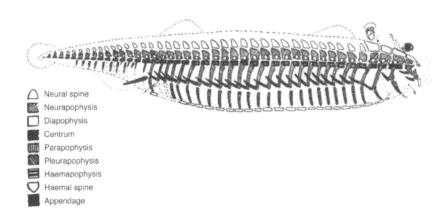

△ Neural spine
▨ Neurapophysis
☐ Diapophysis
■ Centrum
▦ Parapophysis
◨ Pleurapophysis
▤ Haemapophysis
▽ Haemal spine
■ Appendage

According to Darwin's theory of common descent this would correspond to the last common ancestor of vertebrates.

DEMOCRITUS: A hypothetical ancestor that we may never find in the fossil record.

EXTON: But we can try to reconstruct it. To revisit the past, we need to travel again, to 2014, and call on Detlev Arendt, one of the leading representatives of the structuralist approach in evo-devo.

ALKIMOS: So it's time to board?

EXTON: Yes, it is. We're going to Heidelberg!

ALKIMOS: Aaagghhh … here we go!

EXTON: Detlev Arendt works at the European Molecular Biology Laboratory in Heidelberg. This institute is one of Europe's scholarly citadels atop a magic mountain. Whoever works within its walls quickly loses a sense of time. Scientists work often far from their families, their homelands, and live only for science. It's as if time stops and only science progresses.

ALKIMOS: It's like a Deus ex Machina.

EXTON: You could say that. Come on, we've arrived.

ALKIMOS: Heidelberger Turnverein …

DEMOCRITUS: I don't see a mountain or an institute …

EXTON: Oh, we're not meeting at the institute. We're meeting in Detlev Arendt's garden. It's safer—we don't want to lose our sense of time …

ALKIMOS: Exton, we think in centuries as it is. What sense of time are we going to lose?

EXTON: Come one, we mustn't be late!

ARENDT: Over here!

EXTON: There you are! Detlev, *grüß dich*!

ARENDT: *Willkommen in meinem Garten*!

ALKIMOS: Cherries!

ARENDT: That's the neighbour's tree.

EXTON: Alkimos, did you hear?

ARENDT: Please, have a seat.

EXTON: We've come to ask you about the evolution of development and your gene-level comparisons of developmental processes. What are the connections between embryos, genes, and evolution?

ARENDT: Well, what are you most interested in?

ALKIMOS: Master, I think you should tell him.

DEMOCRITUS: We've talked a lot about genes and proteins. We've also heard about genetics, about Jacob and Monod, gene regulation, gene replication, and molecular research. We have basic understanding of the work of Boveri, Hans Spemann, Alfred Kühn, and Christiane Nüsslein-Volhard. We've talked about evolution, Darwin and natural selection, Dobzhansky and evolutionary synthesis, and Ernst Mayr and the origin of species. We now would like to understand how to approach the problem of the evolution of development. We heard about evo-devo and the two fundamentally different approaches to look at evolution, the neo-Darwinian approach and the structuralist approach.

ARENDT: Yes, we can call the two approaches like that. What you call the neo-Darwinian approach I like to call the population genetics approach. This group of evo-devo researchers is focusing on closely related organisms, and try to genetically map mutations that are responsible for subtle phenotypic changes such as spots on the wings of a fly. But I am more interested in reconstructing morphologies deep in evolutionary time and their transformations that happened through several million years. These changes of course entailed thousands and thousands of mutations in both structural genes and regulatory elements in the genome that cannot possibly be all reconstructed. But we can reconstruct the main sequence of events and see how form and function evolved in different groups of animals.

ALKIMOS: The structuralist approach.

ARENDT: I could use an analogy to compare these two approaches. The population genetic approach is like reading the daily newspaper to see what happened in the world during the previous day. The structuralist approach instead tries to understand events such as the Punic Wars. Obviously, we cannot understand that by reading the daily newspaper. We have to dig deep for archaeological remains or try to read other kinds of historical signals.

DEMOCRITUS: But how can we go back so deep in time to understand these major evolutionary patterns? Perhaps we need to study fossilised embryos?

ARENDT: Fossils, the petrified remains of organisms that lived long ago, can indeed help us tremendously in understanding past events; in fact, for a long time studying fossils was the only way to understand the past. But to understand changes in development and cellular differentiation, we must turn to other tools. Since fossilised embryos are extraordinarily rare, it's best if we compare these processes in animals alive today.

DEMOCRITUS: But what can the comparison of present-day organisms tell us about evolution?

ARENDT: The founder of evolutionary developmental biology was Ernst Haeckel. An enthusiastic follower of Darwin, he was one of the first embryologist to formulate a connection between development and evolution.

EXTON: The famous and controversial *biogenetic law.*

ARENDT: According to Haeckel the developing embryo *recapitulates* the stages of evolution; in other words, it goes through all the evolutionary stages as it passes from embryo to adult. To put it a different way, each phase of embryonic development corresponds to an evolutionary one. The progress of evolution is thus partially revealed in the development of the embryo. Therefore, if any living thing changes, a new phase is added to its development.

EXTON: Haeckel's law is different from von Baer's laws that stated that the general characters appear earlier in development than the group-specific characters.

ARENDT: Yes, and now we know that both of these views are over simplified. However, contemporary thinkers tend to dismiss Haeckel and resort to fashionable 'Haeckel-bashing' while uncritically favouring von Baer's explanations. However, von Baer's laws cannot explain gill arches and notochords in mammalian embryos. These structures are not generalised forms of adult organs, but transient embryonic structures that reveal something about the ancestors of these animals.

EXTON: We could say that both Haeckel's and von Baer's theories are right that we can learn a lot about evolution by studying embryos. But we have to keep in mind the limitations of these law.

ARENDT: This is correct. For example, recapitulation is only partial and often blurred, and no embryonic stage directly corresponds to the adult stage of an ancestor. The gill slits in early human embryos do not functionally correspond to the gill slits in adult fish. But their presence tells us that at one point in evolution our ancestors very likely had functional gill slits. We also now know that any stage of development could change under the appropriate selective pressure, and not only the final stages by the process of terminal addition or the modification of the adult forms, as postulated by Haeckel and Darwin. For example, as animals increasingly moved over land, their bodies had to make the adjustments to their new surroundings. Among the many new requirements were hard shelled eggs and ova that prevented the embryo from drying up.

DEMOCRITUS: So a new step was not added to the end, if I understand correctly, but at the very beginning with a change in the formation of the egg.

ARENDT: Exactly. Likewise, if selection favours shorter life spans, significant changes can occur in the development stages. For example, Drosophila development involves numerous shortcuts. The ovum starts out elongated rather than spherical and is full with the products of the maternal effect genes that regulate the head–tail and back–belly axes.

ALKIMOS: So the presence of *bicoid* and other maternal effect genes shorten the development period?

ARENDT: Yes. In fact, bicoid is only patterning anterior structures in some dipterans and it does not represent the ancestral head morphogene in insects. Also in many other aspects the fruit fly doesn't faithfully recapitulate ancient developmental states. The embryo is like a sketch of the developed animal showing the basic axes and rudimentary organs. It's enough to think of the imaginal discs, which already during the larval stages begin to develop into adult structures.

DEMOCRITUS: Fruit fly development reminds us in many ways of preformation.

EXTON: Oh yes, a serious argument in favour of preformationism during the great preformation-epigenesis debate were observations of insects in the pupal stage. In fact the imaginal discs, the rudiments of an adult organism, could be observed not only in the pupal stage but in the larval stage.

ARENDT: As you can appreciate, it may be difficult to understand ancient developmental processes by studying highly derived fruit flies. The problem is the choice of organism. The development of other species, especially many marine animals, is much more morphogenetic. What I mean is that the adult structures gradually appear, and there are no signs of them in the embryo.

DEMOCRITUS: If the development of certain sea animals is morphogenic, this means it's not abbreviated as it is in some land animals.

ARENDT: The developing embryos of marine creatures swim in the sea like plankton. In their earliest forms, these spherical-shaped larvae in no way resemble the adult creatures. They are unsegmented—in fact the segments of some marine annelids don't form until two weeks into their development. They have rudimentary organs, simple mouth and anal openings, and no circulatory system or long intestine. In these animals, the development of organs is much slower and more gradual. After a planktonic phase of several days, weeks, or even months, these larvae settle on the ocean floor and begin their benthic, adult life. This is called the *biphasic life cycle*. Such a life cycle with a benthic adult and a planktonic larval stage is characteristic of the majority of animal phyla.

DEMOCRITUS: That means the development of some sea animals more faithfully recapitulates evolution.

ARENDT: Yes, we can say that. Comparative larval biology has long been considered key to understanding the evolution and relationship of animal body plans. For example, the discovery by Alexander Kowalewsky of the tunicate tadpole larva revealed that ascidians are chordates. Similarly, the discovery of the nauplius larva by John Vaughan Thompson demonstrated that barnacles belong to the crustaceans.

ALKIMOS: Then we need to study the embryos of Poseidon!

ARENDT: If we want to understand the early history of animal life on the planet, then definitely yes. Life started in the oceans. It's obvious that the most ancient forms of development should be sought in the seas. That's why the development of sea animals can reveal so much about the course of evolution. In these animals, the ancient modes of development have changed little. In contrast, land animals encountered numerous environmental challenges. First of all, water was still necessary for development; thus egg shells evolved to decrease evaporation. As animals increasingly subjected themselves to life on land, they had to make ever more changes in their early development.

EXTON: So it makes sense to study sea animals. You and your colleagues have been doing that for years, examining the development of *Platynereis dumerilii*, a polychaete or bristle worm. This animal belongs to the phylum of annelid worms or *Annelida*. Why did you choose this species?

ARENDT: It has one big advantage: it can be easily propagated in the laboratory. We couldn't do this with most other marine species. We would have to regularly collect specimens from the sea.

EXTON: And finding a seashore in Heidelberg isn't too easy.

ARENDT: That's right. But even far from the sea, we can maintain a healthy *Platynereis* culture, thanks to the protocols worked out fifty years ago by Albrecht Fischer. By fertilizing the eggs of just one female, we can attain several thousand transparent embryos that develop simultaneously. This is another huge advantage when studying development. Another is that the animal's body structure and development is very archaic. Similar wormlike creatures lived in the Cambrian Period more than 500 million years ago. These ancestors of the *Platynereis* were also sea dwellers and lived in a fairly similar way.

ALKIMOS: So it's a family tradition. My father, grandfather, and great-grandfather were all fishermen, but I left the sea and took a stab at philosophy. I'm not recapitulating anything.

DEMOCRITUS: You're not a polychaete, son.

ARENDT: According to the present theory, the entire Earth's surface may have turned to ice about 600 million years ago. By then the first animal life forms had appeared and the icy blanket around the world would have presented a considerable challenge to them. Shortly after the ice melted, bilaterally symmetrical animals began to evolve and diversify rapidly.

ALKIMOS: The tree of life branched out.

ARENDT: This was the *Cambrian explosion*, about 535 million years ago. Today, bilaterally symmetrical animals can all trace their origins to this period. One big question is what the common ancestor of these animals looked like. This animal somehow survived the ice age and became the starting point for a huge wave of diversification.

DEMOCRITUS: This creature is the common ancestor of insects, worms, fish and humans?

ARENDT: Yes. And the ancestor of snails, polyps and sea urchins, too. All bilaterally symmetrical animals, or *Bilateria*, have the same ancestor, the *urbilaterian*.

DEMOCRITUS: I thought sea urchins have pentamerous radial symmetry.

ARENDT: Indeed, but that represents a later evolutionary derivation.

ALKIMOS: Then the Bilateria doesn't include Abraham Trembley's cylindrical hydras.

ARENDT: Indeed, these animals have radial symmetry ancestrally.

DEMOCRITUS: How can we reconstruct a common ancestor? We're obviously never going to find a fossil.

ARENDT: Morphological and molecular similarities form the basis of the reconstruction. What we know for sure is that the common ancestor was much more complex than we once believed. For example, the common ancestor had a full *Hox cluster* and complex patterning along the antero-posterior axis.

EXTON: Let me explain this to my students. Hox genes encode a class of transcription factors that specify regional identity along the antero-posterior axis of bilaterians. Mutations in Hox genes cause *homeotic transformations* that entails the change of the identity of one body region towards another region. The first such mutant, *bithorax*, was isolated by Calvin Bridges in Morgan's laboratory in 1915. In *bithorax* mutants the third thoracic segment is transformed towards the second thoracic segment. The Hox genes are arranged in genomic clusters and are expressed in a spatially *colinear* fashion—anterior genes in the cluster are expressed towards the anterior end of the body, whereas posterior genes are expressed towards the posterior end. The discovery in 1989 that the ordering of Hox genes within clusters has been conserved between Drosophila and the mouse was the first striking example of the deep evolutionary conservation of gene networks that play a crucial role in body patterning in distantly related animals. The neo-Darwinians were completely puzzled by this.

ARENDT: Indeed, the discovery of the Hox genes and other genes with similar conservation of function completely changed our thinking about animal evolution. Why, thirty years ago, most scientists thought the common ancestor of the human and the fruit fly was nothing but a spherical ball of cells. Now we know, thanks to molecular comparisons, that this isn't so. The common ancestor also certainly contained molecules such as receptors and hormones that regulated the functioning of the nervous system, indicating that a molecularly complex nervous system with diverse cell types was already present. Based on morphological comparisons we know it was a wormlike animal, with a brain at the front end and two sets of nerve bundles that ran in parallel down the ventral side. We know that it had a circulatory system and heart, a mouth and anus, and a gut.

ALKIMOS: How can you know this common ancestor had all these parts?

ARENDT: The existence of a gene in two lines of descent clearly means this gene was present in the common ancestor of the two lines. Naturally, we're not going to find two identical genes in distantly related species; we're always going to find discrepancies. But if the similarity in sequences is great enough then we can be sure the two genes, called *homologues*, descended from the same gene in a common ancestor and are not the results of independent evolution. If the gene's function has also been conserved—meaning those genes that have descended from a single ancestor carry out the same function in the various species—then we can be quite sure the ancestral gene also performed this function. That's how we know this wormlike ancestor had for example *serotonin receptors* and cells sensitive to serotonin neuromodulators.

DEMOCRITUS: So we can examine something on many levels: the level of genes, cells, tissues, and organs. I understand that two gene sequences can be similar if they both derive from the same ancestor. But is that true for organs too? Why couldn't two different animal groups independently develop a heart, eyes, and a gut?

ARENDT: That's an excellent question. Indeed, morphological comparisons are limited in determining whether organs or characteristic body structures are homologous. Segmented animals can be found among arthropods and deuterostomes, but also among the Lophotrochozoa, a large group that includes our sea worm, snails, and other animals. It's an open question whether segmentation in these different groups emerged independently or whether this characteristic was present in our common ancestor. Only molecular comparisons can answer the question.

ALKIMOS: So if we find homologous segmentation genes everywhere, then we can assert that segmentation itself is homologous?

ARENDT: Something like that. The basis of these studies is the comparison of genes' expression patterns and functions.

ALKIMOS: *In situ* hybridisation and the analysis of mutants.

ARENDT: Exactly. If we find related genes that display similar expression patterns in various groups, for example along the segments, then our hypothesis of homology based on morphology gains strength.

EXTON: Some people would object and add that we need to test homology by looking at the distribution of a character across phylogeny, using a *cladistics approach*.

ARENDT: The cladistics approach, mapping the presence or absence of a character on a tree and then inferring the most parsimonious scenario of gains and losses is a powerful approach if we have a tree with several nested nodes and a dense sampling of organism. This approach, however is too simplistic and can be misleading if we want to analyse deep divergences, that are often bifurcations, and we don't have many nested branches. For example, within the deuterostomes, we only have a clade of chordates and a clade of the *ambulacrarians*, that includes organisms such as sea urchins with highly derived morphologies. A cladistics analysis in such cases is very difficult, due to the extreme derivation of characters and the few branching events. People also rarely appreciate how frequently character loss can occur in evolution, invalidating a simple parsimony approach. In these cases we have to compare the complex molecular underpinnings of characters, such as organs or cell types and infer homologies based on that.

DEMOCRITUS: In this case, I presume the more similarities we find, the more certain we can be that two characters are homologous.

ARENDT: Yes. If we find many similarities, we can decide in favour of homology since it's unlikely that the particular combination of given genes that determine a specific cell morphology was arrived at independently in several different organisms. Our method boils down to making connections on the level of genes and cell morphology. We develop an understanding of the evolution of cell types, and thereby an understanding of the evolution of development, organs, body shapes, and animal groups. And in the process, we examine the evolutionary history of a huge numbers of genes. At the moment we're involved in a large-scale *in situ* hybridisation screen of larval *Platynereis*. Our goal is to determine the expression patterns of several hundred genes.

ALKIMOS: Several hundred genes?

ARENDT: Yes, and at cellular resolution. This will allow us to finally establish the molecular fingerprint of every organ and cell type. In another approach we use *single-cell transcriptome sequencing*. We dissociate the larvae and isolate single cells. Then we sequence many thousands of cDNAs from every cell. Given the large number of genes for which we know the spatial expression based on the *in situ* hybridisation data, we can then map back these dissociated cells to a spatial reference. We will soon arrive at a model of the larva where we know the position of every cell and we know pretty much every gene that is expressed in each cell. We can then compare our data to the results of similar approaches in other animals such as the fruit fly and zebrafish. This will tell us a tremendous amount about the origin of cell types, their distribution and function.

DEMOCRITUS: This is a fascinating approach. So what have these studies revealed about the evolution of cell types and organ systems?

ARENDT: For example, our work clarified the history of photoreceptor cells in animals, providing new insights into the evolution of our eyes. In the case of photoreceptors or light-sensing cells, morphologists have distinguished between two main cell types: ciliary and rhabdomeric photoreceptors. We did a molecular analysis of these cells in the worm's brain and established a molecular fingerprint—a unique combination of certain transcription factors and the effector genes they regulate—which was characteristic of one or the other cell type. This fingerprint is conserved in the photoreceptors of *Platynereis* and people. Therefore, these photoreceptors were present in our last common ancestor with the worm, and through the course of evolutionary change, they transformed into the light sensitive cells of our eyes, with certain essential components preserved in both groups.

ALKIMOS: So you shed light—haha!—on the origins of photoreceptors in the human eye!

ARENDT: And on the origins of many other types of cells.

ALKIMOS:

> *Our eyes emerged from the brain of a worm, strange orbs!*
> *How they cling to shrivelled jelly, how they yearn for the distant sea.*

DEMOCRITUS: That's why our tears are salty.

ALKIMOS: I miss the sea, master.

DEMOCRITUS: Now we can see how molecular biology extended the horizon of comparative anatomy and evolutionary biology. The history of genes is tied to the history of cell types and embryos. During the course of development, the genome thus creates cell types and other manifestations of phenotype. Selection modifies these over millions of years. We've been the witnesses of true synthesis.

EXTON: Detlev, we need to be going. Thank you for this intriguing conversation.

ARENDT: Please come again.

DEMOCRITUS: All our earlier efforts have brought us to this marvellous synthesis! We have before us a unified picture of genes, embryos and evolution. This is what we've longed for. It's time to go home. How I've missed the sea!

ALKIMOS: Come on master, let's pick some cherries.

EXTON: Alkimos!

NACHBAR: *Was machen sie denn da?*

EXTON: Alkimos, that's the neighbour's tree. Look! He's furious!

NACHBAR: *Das ist aber … So eine Unverschämtheit!*

ALKIMOS: Give my best regards to the worms!

ARENDT: I will.

EXTON: Alkimos, come on!

ALKIMOS: Aaagghhh … here we go!

The Age of Genomics—*Eugene Koonin*

ALKIMOS: Master, did you like the worms?

DEMOCRITUS: Yes, they reminded me of the sea.

ALKIMOS: We're all descended from Oceanus …

DEMOCRITUS: Yes, our worm ancestors lived in the sea. We can reconstruct them with a comparative study of development and genes. If we compare every branch of the tree of life we'll eventually come upon the common ancestor of all living things.

ALKIMOS: The ancestral bacteria.

DEMOCRITUS: Well, if we start with ancient bacteria, we'll wind up with endless variety. The tree will become thick with branches.

EXTON: *Idem et diversum.* Sameness and difference. The most apt description of the tree of life, the billion years of evolutionary events. The astounding conservation of the genetic code, fundamental biochemical reactions, DNA replication, transcription, and protein synthesis—this is the *idem* aspect, the legacy of our common ancestors. At the same time the incredible diversity of forms, colours and strategies, and on the molecular level, too—this is the *diversum* aspect, the dynamic diversity of evolution.

DEMOCRITUS: The study of this duality is perhaps the most exciting scholarly undertaking.

EXTON: After the publication of *On the Origin of Species*, Darwin's theory that every living thing descends from one common ancestor was immediately accepted in wide circles. All at once it was possible to explain observations that were otherwise incomprehensible without the assumption of a common origin. Studies such as von Baer's in comparative anatomy, or Linnaeus's hierarchical classification of organisms based on shared characteristics, or observations of the bio-geographic distribution of species all make sense if we accept common origins. The idea of common origins never appeared in evolutionary theories before Darwin. Lamarck believed species developed separately along parallel lines of descent. Darwin, however, argued that new species arose from older ones, after segregation. An emergence of a new species meant a new branch on the evolutionary tree.

DEMOCRITUS: The similarity of living things on the level of form, developmental pathways, cell types and genes is the proof of common origins and evolution.

EXTON: That's why a study of related species gives us some insight into the process of evolution. If we examine the cell types and development of present-day organisms using a comparative approach, we can infer the characteristics of extinct ancestors, as Arendt explained to us. But we can learn a lot also by comparing genomes. The development of increasingly affordable sequencing techniques has allowed us to determine the complete genome sequences of numerous animals. Gene sequences and a deluge of other data have led to a new branch of science: *bioinformatics*. Students of bioinformatics don't look at embryos under microscopes; they don't prepare mouse skeletons or sequence genes. Their job is to interpret and analyse the data to obtain a new understanding. Their work methods are closer to those of programmers than zoologists.

ALKIMOS: Why? What does a bioinformatics scientist do exactly?

EXTON: You'll see. Direction Bethesda, Maryland, 2015! We're going to visit Eugene Koonin.

ALKIMOS: We're going to travel again? The engine hasn't even cooled yet.

EXTON: All board!

ALKIMOS: Aaaaggghh … here we go!

EXTON: Koonin works in the National Center for Biotechnology Information in Bethesda, Maryland, on the east coast of the United States. The main task of the institute is to maintain and develop various biological databases. As recombinant DNA techniques have become more effective and widely used, the quantity of DNA sequences has grown exponentially. This has led to the creation of separate databases. The NCBI sequence database contains all of the public DNA sequences. Anyone can search it, with the help of a computer of course.

ALKIMOS: So genes have moved from the test tube to the computer.

DEMOCRITUS: I'd like to see that, son.

EXTON: Come on, we're here.

ALKIMOS: What a strange place.

EXTON: We're on the rooftop of the NCBI building. We need to go down those stairs.

DEMOCRITUS: Wait for me!

ALKIMOS: Hurry, master. Let's find those fruit flies! Or the computer they're trapped in!

EXTON: This is the floor.

ALKIMOS: I don't see any vials.

EXTON: This office here.

ALKIMOS: Master, don't get left behind!

EXTON: Eugene, greetings!

KOONIN: Von Echstand, how nice to see you!

EXTON: We're lucky to have found our way.

KOONIN: Yes, it's easy to get lost around here.

EXTON: These are my students.

KOONIN: Pleased to meet you. Have a seat.

EXTON: Thank you. As I said in my letter, we're here to discuss your work.

KOONIN: So you're interested in computational biology?

DEMOCRITUS: We're interested in everything related to the role of genes in living things.

ALKIMOS: I'm interested in where the labs are. And the microscopes.

KOONIN: Oh, you won't find any of that here.

ALKIMOS: Then I don't understand. What do you do here if you don't extract DNA or cross fruit flies or observe frogspawn?

KOONIN: Don't worry. Most biologists still work in labs, although these days, lab workers spend increasingly less time handling mice or frogs and more time moving test tubes in and out of sophisticated instruments and machinery. And bioinformaticians, of course, have moved even farther away from living organisms. We spend our days in boring offices sitting in front of computers.

ALKIMOS: I'd rather deal with frogspawn. At least those are alive.

KOONIN: Yes, I'm afraid the object of our research is not a living thing but information. Primarily the information encoded in the long DNA chains of genomes. Nevertheless, the questions that interest us computational biologists are actually not that far from those that traditional experimental biologists investigate. How do genes and cells work? During the course of evolution, how did they transform into the fantastic molecular machinery we're familiar with today?

DEMOCRITUS: We're interested in those questions too. We just don't quite see how this huge amount of data will lead to answers.

KOONIN: Computational biologists try to consolidate in some way the tremendous amount of data generated by experimental biologists. We store genome sequences and the large quantity of other information in databases. Then we try to glean biological knowledge from it. Analysis of the genome sequences has been a huge help in these endeavours.

DEMOCRITUS: Of course. Comparing the sequence of genes will give us some insight into evolutionary events.

ALKIMOS: Crick mentioned this when he talked about the haemoglobin of various animals.

KOONIN: That's right. The thousands of complete genome sequences of different organisms that we've completed so far—not counting viruses—have already revealed an incredible amount about evolution and the relationship between organisms.

ALKIMOS: Thousands of genome sequences? My Zeus!

KOONIN: These genomes include those of a few hundred eukaryotes, and the rest are prokaryotes. Among the eukaryotes are yeast, fruit flies, *C. elegans*, mice, rats, mustard grass, rice and, of course, humans. And if you think about how many genomes are awaiting sequencing, you'll understand why we need an institute such as this that deals with the processing of this information.

ALKIMOS: But why do we need so many gene sequences?

KOONIN: Because by comparing sequences we can eventually discover the general laws of evolution and come to understand specific events. We need to sequence as many genes as possible, so we can reconstruct past events with even greater accuracy and certainty. Maybe this is nothing but a scholarly exercise, but still, it would be a grave error to think all this is useless from a practical perspective, for example from a medical perspective.

DEMOCRITUS: But why would a doctor need to know about the past, about evolution?

KOONIN: Because living systems formed during the incredibly efficient process of Darwinian evolution. Any interpretation that fails to formulate the general principles and reconstruct the details of evolution is likely to be shallow at best, most likely misleading. This includes an interpretation of how the human body works.

ALKIMOS: And most likely the soul, too …

KOONIN: According to the greatest evolutionary geneticist of the 20th century, Theodosius Dobzhansky, 'nothing in biology makes sense except in the light of evolution.' I don't think this is an expression of 'evolutionist chauvinism,' and it's not even a metaphor, but a simple, self-evident truth.

ALKIMOS: We can understand living things through the lens of evolution. To unravel evolution we need the help of genomes. Okay, then let's sequence!

KOONIN: I am convinced that we need to sequence a lot more genomes than we have so far. When taking samples we need to take care that every branch is represented. Naturally, it is not yet feasible to sequence the genomes of every existing species or even the majority of them, which is why we need to make our choices judiciously.

ALKIMOS: I recommend the aphid. We need to discover the genes responsible for pathogenesis. Or the hydra, so we can understand the genes that regulate regeneration, or the flounder so we understand the genes that regulate the migration of the eye, or …

KOONIN: The flounder—now that's an idea!

ALKIMOS: And the genes that control moth wing fronts, and the fruit fly …

KOONIN: The fruit fly's genome was complete in 2000, *Hydra* and the pea aphid were sequenced in 2010.

ALKIMOS: Well, how about the newt? In honour of Hans Spemann …

EXTON: Thank you, Alkimos, that's enough.

KOONIN: Oh Eschtand, it's all right. I'm delighted that I've managed to inspire your colleague. You know that people perform their best research when driven by curiosity?

EXTON: Yes, of course. That's true of learning, too.

DEMOCRITUS: What did genome sequences teach us about the general principles of evolution and individual events? That's what we're truly interested in.

KOONIN: Gene sequences have revealed striking similarities between diverse groups of organisms. Certain elements—especially the housekeeping genes responsible for basic cellular function—display astounding conservation in all or most organisms. We have many genes that are clearly related to bacterial genes. This conservation opens a door to the past, to events that took place on our planet 3.5–4 billion years ago. Of course the picture, at least so far, is incomplete, and certain traces have disappeared forever.

ALKIMOS: Master, what do you say to all this?

DEMOCRITUS: Incredible! The unity and common origin of everything is revealed in the sequence of genes.

KOONIN: If something works, evolution is unlikely to change it. Many proteins have been around for three billion years and since then, their form has barely changed. If a form is proven to work in a bacterium, it will in a human too.

DEMOCRITUS: Does that mean that if a gene is found in a bacterium or fruit fly and in a similar form in a human, then the knowledge obtained about that gene by examining the bacterium or fruit fly is applicable to a human too?

KOONIN: Yes, often this is the case.

DEMOCRITUS: And it's easier to experiment on a bacterium or fruit fly than on a human.

KOONIN: It certainly is. Although sequencing human genes is easy too. You know that it doesn't involve experimenting directly on a person.

ALKIMOS: But a fruit fly still isn't a person, so even if many of the genes are the same, there's a lot that's different.

KOONIN: That's correct. But when we compare genome sequences, we take into account the differences. We learn even more this way. Using the methods of evolutionary biology to sequence genomes, we can then reconstruct, based on the similarities, the genomes of our ancestors who lived several million or several hundred million years ago. If we pay attention to the differences, we can also understand why our ancestors developed in a different way than those of the fruit fly.

ALKIMOS: We can find out what happened to our wing genes …

EXTON: I think we understand the importance of the genome sequence databases. Could you tell us what other databases you have?

KOONIN: Yes, we have separate ones for storing the spatial structures of proteins.

ALKIMOS: Perutz would be happy about that.

KOONIN: And others that store information on the interaction of proteins, and ones for gene expression patterns.

EXTON: So a database of which gene is expressed in which tissue and under what circumstances.

DEMOCRITUS: How exciting!

KOONIN: In other databases we've collected the effects of different mutations on phenotypes.

DEMOCRITUS: That means that sooner or later we're going to know when and where all genes are expressed.

ALKIMOS: And what they do!

DEMOCRITUS: We'll know the exact molecular programme. We'll know how an ear or a mouse is made.

ALKIMOS: The recipe for making a mouse. We can compute Sydney Brenner's mouse in the language of math as D'Arcy Thompson wanted!

DEMOCRITUS: Can we really consolidate all this data? Will we really arrive at a true understanding of living systems?

KOONIN: A uniform analysis of the huge variety of data is the goal of a new branch of science frequently called *systems biology*. Such an invigorating objective …

ALKIMOS: Master! The system in the language of atoms! It's what you always hoped to achieve!

DEMOCRITUS: You're right. The atoms of life have slowly assembled to form mice, shells, and cypresses.

ALKIMOS: We can compute a mouse and evolution, too! Soon we're going to know everything about living things!

KOONIN: To suggest we know everything about life is rather presumptuous. It's like the zoologists in the mid-19th century who proclaimed they knew everything about mammals and birds …

ALKIMOS: The progress of science is breathtaking!

DEMOCRITUS: Son, we've been waiting for this moment!

KOONIN: … it's true they'd discovered almost all the species, but their level of intellectual exchange was rather, well, insufficient by today's standards …

EXTON: I think we need to go.

KOONIN: … yet we're not so different. You see we know about almost all the molecules that make up living systems, but we haven't yet acquired a true understanding …

ALKIMOS: The atoms of life, encoded in the genome, swoosh through life!

DEMOCRITUS: Son, we've achieved out quest.

EXTON: Eugene, thank you so much for talking with us.

KOONIN: My pleasure, but …

DEMOCRITUS: Back to the life-giving sea!

EXTON: On to the roof!

ALKIMOS: Thank you, thank you. We finally understand everything!

KOONIN: I think you've misunderstood. I never said that we ...

EXTON: Come on, take you seats!

ALKIMOS: Aaagghh ... here we go!

Part Eight

Beyond Genes

Postcard to Thrace

COOK: Are they really being sent home?

HERMES: Yes, the old codger at least. Your pal is staying on though. I guess Exton and the codger think he still needs to learn a few things.

COOK: Hm. Well, if it's moderation, they're wasting their time.

HERMES: No, I don't think that's what they're worried about.

COOK: He keeps going on about Alexandrian women—how he can hardly wait …

HERMES: That's not going to happen. They have another trip planned for him.

COOK: And when is the geezer leaving?

HERMES: Tomorrow.

COOK: So soon? Right back to Thrace?

HERMES: Yes, to Abdera.

COOK: Lucky him.

HERMES: Why?

COOK: There's nothing better than Thracian wine.

HERMES: And nothing more beautiful than … Wait a minute!

COOK: What?

HERMES: I need him to take something for me.

COOK: What?

HERMES: A postcard.

COOK: To Abdera?

HERMES: Yes.

COOK: But, you've never been there, have you?

HERMES: Hah! I'm a messenger. People of my profession have been everywhere.

The Cedar Forest of Abdera

ALKIMOS: Master, are you awake?

DEMOCRITUS: I had a strange dream. I was in Abdera, by the bay. I was just preparing to launch my boat when who should appear, but Pallas Athena, in the form of Leucippus. But I knew by the way she rose from the sea and by her gleaming, owl eyes it was her. She stepped toward me and said, 'Democritus, your time has almost run out, and you haven't finished your job.' I told her I knew that, but we'd discovered so many interesting things. I'd been distracted. 'We still have time,' I assured her. 'You fool!' she cried, 'How do you know how much time you have left? Only the gods know that. Go on, don't waste another minute!'

ALKIMOS: Master, you're shaking!

DEMOCRITUS: I'm exhausted. I don't have the strength to continue.

ALKIMOS: Democritus, you don't mean —

DEMOCRITUS: Yes, Alkimos, I'm going home. I've acquired the knowledge I was seeking. The secret of the atoms of life, and the place of us human beings in life's great chain. In my remaining days I'd like to contemplate life in its entirety, as I used to in my youth. I must return to Abdera. There, I'll know that the cicada is governed by the atoms of life, that it develops and reproduces under the direction of genes, that its distant relatives are the cedars upon which it feasts insatiably. All of this belongs to the chain, to the great tree of life. I now understand this. But I am still awestruck by the thousands of fish and coral in the Bay of Abdera, by the desiccated foliage of the forest and brush, by the extraordinary scent of the crabs and shells lurking in the gravel. Why is the smell of the sea the only one capable of soothing my uneasy heart? I don't think I'll ever understand why this natural world came into existence, and why it fills me with such wonder. It's so far removed from the atoms of life; the level or organisation it requires is extraordinary. What solutions and bonds, what stimulators and inhibitors, inductors and repressors, love and death—what endless complexity! I must return so I can spend my old age tasting the fruit of the cedar forest. It's not my mind or ambition that requires nurturing. No, not anymore. Only my heart. Now it's time for me to understand the gods, and my own soul. Dear Exton, please, take me home!

EXTON: As you wish. In fact, I've been waiting for this moment. All the preparations have been made.

ALKIMOS: Excellent! I have to admit, I've grown weary of all this learning. Home sounds good to me, too.

DEMOCRITUS: No, you're staying.

ALKIMOS: What?

DEMOCRITUS: You no longer need to follow me.

ALKIMOS: Okay, but—

DEMOCRITUS: Exton, shall we depart?

ALKIMOS: Master, no! You can't leave me!

DEMOCRITUS: I must.

ALKIMOS: Please, let me read my latest poem. It's not quite done, but … well, it's about you …

DEMOCRITUS: The one that you started before our journey? You're still working on it?

ALKIMOS: Yes, master. Listen to the beginning! Who knows when we'll see each other again?

DEMOCRITUS: Oh Alkimos, relax. When your poem is finished we'll read it together.

ALKIMOS: In Abdera?

DEMOCRITUS: Of course.

ALKIMOS: Okay, but listen, master—just the first few lines!

> *It's you I quote, my pearly-eyed muse, my lyrical one.*
> *Come down from there, that rocky crag, sit beside me,*
> *Whisper the rhythm in my ear, your words of happiness and joy!*
> *My poetic impetus is strong as I've told you so oft' before,*
> *What music rumbles from my spotless lungs and resounds in my throat,*
> *Forming words that reflect a mind full of artificial notions,*
> *Which cause a throbbing pain in the hearts of those that hear them.*
> *And you may inquire, with your own heart pounding, which 'til now*
> *Was full of nothing but sweet ideas. My dearest friend,*
> *What words of happiness and joy will my pearly-eyed muse whisper?*
> *What murmurous, susserous, rapturous words will hum in my ears,*
> *Tinged with dolour? And look, how your desire's fulfilled, because*
> *I'll tell you straight away the reason I remain, linger;*
> *I'll tell you where the words of my song, my air of heroism, come from.*
> *Behold, for I've said it; a song of courage, of pluck, of valour,*
> *One not of youthful affection, sung in twilight hours.*
> *Who are the heroes? Where does their manful struggle unfold?*
> *That too will soon emerge, in a burst of sunlight, illuminating*
> *The bare tonsure of their precious noggins and the rich earth,*
> *Upon which their feet are placed, giving support to ample bodies.*

DEMOCRITUS: Bare tonsures?

ALKIMOS: Master, that part refers to Leucippus ... to your encounter with him.

DEMOCRITUS: I see. So it's a poem about Abdera. Excellent.

ALKIMOS: Yes.

DEMOCRITUS: Well, make sure you do justice to Leucippus.

ALKIMOS: Oh, I will, master.

EXTON: Democritus, time to go!

DEMOCRITUS: May Zeus be with you, Alkimos.

EXTON: Have a good journey home!

ALKIMOS: And with you, master!

DEMOCRITUS: Aaaagghh, here I go!

The Philosophy of Biology—*Ernst Mayr*

ALKIMOS: Master, I'm never going to see you again! Oh, Pallas Athena!

EXTON: Alkimos, be happy. Your master's restless heart has finally found peace.

ALKIMOS: No, Exton. He'll never be content.

EXTON: Well, at least he might rest.

ALKIMOS: Oh, nearly two thousand five hundred years have gone by! I should have returned with him.

EXTON: No, Alkimos, you need to go on.

ALKIMOS: But where? What am I without my master?

EXTON: A true master does not groom students; he grooms the next generation of masters.

ALKIMOS: The next generation?

EXTON: Yes. And Democritus was a true master—a true scholar and educator. That means that you're ready to forge ahead on your own, Alkimos.

ALKIMOS: I suppose. But haven't I learned everything about genes?

EXTON: Not at all. There's still a lot more to know, but lucky for you we don't have time. Instead, in honour of Democritus, we're going to turn our attention to philosophy. Direction, America! Bedford, Massachusetts, 2004! We're going to visit Ernst Mayr, the greatest evolutionary biologist of the 20th century and an apostle of the new philosophy of biology! He's just turned one hundred!

ALKIMOS: One hundred? That's not so old. In two weeks I'll be two thousand five hundred years old.

EXTON: Climb in, Alkimos.

ALKIMOS: Aaagghh, here we go!

EXTON: Sit down. We'll be there soon. We've already talked about Ernst Mayr. Do you remember? When we spoke about the biological species concept and evolutionary synthesis. Now we're going to discuss with him the foundations of the philosophy of biology.

ALKIMOS: What a shame Democritus can't be with us.

EXTON: Mayr was active throughout nearly the entire 20th century. In the 1920s, he led ornithological expeditions to New Guinea. In 1942, he published his book *Systematics and the Origin of Species*. Like Dobzhansky's *Genetics and the Origin of Species*, it's among the foundational works of the evolutionary synthesis. How species arose was a central question of his inexhaustible research. His books on the history and philosophy of science are also fundamental. Mayr is an exceptional champion of the Darwinian philosophy of science.

ALKIMOS: And even at one hundred he's still working?

EXTON: I guess we'll see. That's his house over there. Come on, I'll ring the doorbell.

MAYR: Yes, who's there?

EXTON: Dr. Exton.

MAYR: Wonderful! I was afraid it was another newspaper reporter. Ah, you've brought your student. Come in, come in. Today's my one hundredth birthday. Did you know that?

ALKIMOS: May Apollo bless you with another hundred years!

MAYR: I'd prefer the blessing of the Papuan gods.

ALKIMOS: Papuan?

EXTON: It's a tremendous honour to be here, Professor. We are great admirers of your work.

MAYR: Oh, Dr. Exton, really now …

EXTON: We loved your latest book.

MAYR: I've got three planned for next year.

EXTON: How do you have so much energy? Not many people write a book at age one hundred!

MAYR: Well, you know I attended a German *gymnasium*. They're quite rigorous at that time. We didn't just acquire knowledge—we learned to ask critical questions. We learned how to think. The concept of the German *gymnasium* was based on the guiding principles of education set forth by Wilhelm von Humboldt—he established the University of Berlin in 1810. His idea was that our culture rests on the pillars of Greek culture, and the ability to think deeply can only be truly developed by contemplating the fundamental questions of nature. I often say that my achievements are all thanks to the cultural heritage of the German *gymnasium*.

EXTON: So for one hundred years you've been contemplating the fundamental questions of nature?

ALKIMOS: No wonder you can't stop working! The legacy of Greek culture is enormous. That's why my master never lets up either!

EXTON: Professor, I believe another profound impact on your life and career were three expeditions to the tropics, between 1928 and 1930.

MAYR: Yes, that's right. I spent two years in New Guinea and nine months on the Solomon Islands in the Pacific, studying birds.

EXTON: Alfred Russell Wallace also made a trip to New Guinea, back in the 1850s. He was a contemporary of Darwin's who independently conceived the laws of evolution by natural selection.

MAYR: Yes, Wallace was also a visitor to the island. But unlike Wallace, I left the inhabited parts and delved deep into the jungle. It was seventy years after Wallace, yet my conditions were far worse. In New Guinea I was able to study directly the most important questions in evolutionary biology: the origins of adaptation and biodiversity. Why do living creatures have eyes and wings? What's the origin of the connection between the species? Why are there so many different kinds of organisms, from bacteria to giant redwoods, from birds of paradise to elephants? The oceanic islands give us an opportunity to observe evolutionary change from island to island.

EXTON: The most famous observations were made by Darwin. He examined the various finch species on the Galapagos Islands.

MAYR: In New Guinea, I was able to observe the changes from mountain to mountain, not island to island as Darwin observed for finches. We managed to collect and identify about the same number of bird species from every mountain chain, but the species on each chain were distinct.

An examination of this phenomenon helped us to solve the problem of the origin of species.

EXTON: My students and I have already discussed this. At the beginning of the 20ᵗʰ century, geneticists such as de Vries and Bateson were unable to unravel the problem of the origin of species. They imagined that species came into being by great leaps—saltations—as a result of single mutations.

MAYR: Geneticists studying Guinea pigs in cages never saw the evolutionary processes of nature, and therefore were unable to develop a fundamental understanding of the origin of species.

EXTON: Your book from 1942, *Systematics and the Origin of Species*, was the first to offer a clear discussion of the question.

MAYR: Yes, but this was only possible after adaptation and diversity had been studied in the natural setting. To truly understand species, we need to observe them in their own environment.

ALKIMOS: Into the wild!

EXTON: As you see, Professor, we're also born naturalists! My friend Alkimos is a great fish expert.

ALKIMOS: That I am!

EXTON: But Professor Mayr, if it's all right with you, I'd like to turn our conversation to a somewhat different topic. In fact, the reason we've come is to hear more about the development of your interest in philosophy and the fundamentals of the philosophy of biology.

MAYR: I've had a passion for the philosophy of science since my youth. At first I was interested in vitalism. I had a feeling that biology was quite unlike physics or chemistry—it could not be construed as merely another kind of physical science. At that time only the vitalists thought this way. When I went to New Guinea I brought only two books with me, one by Bergson and the other by Driesch. During my long stay I studied them thoroughly and discovered that the vitalist explanations offered by both authors were seriously flawed. That led me to quickly abandon vitalism. At the same time, I realised that the so-called philosophers of science also made claims that were problematic. You see, they based their philosophy on the laws of physics. Biology and physics, however, are two very different sciences, and biology cannot be forced into the straitjacket of physicalism. To put it simply, a philosophy of science based on the laws of physics is incapable of explaining biology.

EXTON: In your philosophical writings, you state that the tenets of the new philosophy of biology are based on Darwin's principles.

MAYR: Darwin radically altered the foundations of western thought. He called into question core beliefs that had never before faced scrutiny. In Darwin's time a devout Christian interpreted the Bible literally. Everything we see in the world was thought to have been created by God, and nothing had changed since Creation. Moreover, God had designed all living things to fit perfectly into their designated place in the world. But Darwin shook all three pillars of this system of beliefs. He stated that the world was undergoing constant change; that species did not arise at the time of Creation but evolved from one common ancestor; and that adaptation was not a result of God's design but rather natural selection. According to Darwin's theory, the development of the natural world did not require divine intervention or supernatural powers. Darwin replaced the notion of a living world created by God with a world shaped by the laws of Nature. This was revolutionary.

EXTON: But it wasn't only the notion of God's intervention in the development of human beings and the natural world that Darwin upended. Other important philosophical principles were also called into question. Could you talk about them?

MAYR: Of course. One was the principle of essentialism. This can be traced back to Plato and Pythagoras. According to essentialism the existing things in the world can be categorised into a limited number of classes. Every single class has an exact definition. The variety within a class is imperfect and negligible. Reality is none other than the absolutes behind every class, the defining essence of the class.

ALKIMOS: So reality is not the class but rather its definition?

MAYR: Yes.

EXTON: The perfect *eide* or *essence*.

ALKIMOS: Is it something like the *wild type* to geneticists?

EXTON: Yes, and just as meaningless in biological terms.

MAYR: In physicist circles and in the philosophy and scholarly thought dominated by physics, this principle is generally accepted. Chemical elements and molecules really are constant and sharply separated from each other. Darwin however showed us that this kind of essentialist or typological thinking is utterly ill-suited to describing the diversity of life. Organisms are not created according to absolute patterns; they are members of biological populations. Within the population every individual is different. Just think of us humans. There are six billion of us and not one is the same as another.

ALKIMOS: So the human species has no essence, no ideal form?

EXTON: Although everyone likes to think that he or she is the ideal …

ALKIMOS: All six billion of us!

MAYR: Darwin introduced population thinking, which was unknown in philosophy at the time. Within reproductive communities, or biopopulations, every individual is different. Reality is not the perfect pattern underlying every group, but the group itself, and the variation between all of its members. The group's statistical average, the pattern, is just an abstraction.

ALKIMOS: So essences don't exist, only diverse individuals.

MAYR: That's right, and this diversity of nature—first understood by Darwin—is the key to natural selection and evolution. In the physical sciences, however, this kind of diversity doesn't exist; population thinking therefore has no role in physics. Imagine a gas in which every molecule is different!

ALKIMOS: I think only Democritus could do that … Oh my master!

MAYR: Essentialism, however, is deeply embedded in our way of thinking. It must be uprooted if we are ever to succeed in the battle against racism and inequality. With the introduction of population thinking, Darwin made a fundamental contribution to modern western thought.

EXTON: Could you also address another basic tenet of 19th-century philosophy: *teleology*, this idea of end-directedness?

MAYR: Teleology can also be traced back to the Greeks. Its revival is associated with Emmanuel Kant. Kant attempted to devise a philosophy of biology based on the philosophy of Newtonian physics. It was a total failure. He finally concluded that biology was different from physics ...

EXTON: And he was absolutely right, as we've already established.

MAYR: ... and that a new element needed to be introduced to the philosophy of biology that doesn't exist in Newton's physical system. Kant believed he'd found this component in Aristotle's notion of the 'final cause.' To explain all biological phenomena that couldn't be explained by Newtonian physics, Kant thus invoked a 'final cosmic cause,' or pre-determined goal—this is teleology.

EXTON: To what degree did this cosmic final cause resemble the life force presumed by Bergson, Driesch and the other vitalists?

MAYR: Enough for us to see that it too is completely unnecessary. Darwin showed that all the phenomena in the living world can be understood without cosmic teleology. Natural selection gets rid of the most inefficient among the countless variations in each generation. It has the power to explain everything that until 1859 could only be explained with teleology.

EXTON: The teleological aspect of development—that it has a final goal, the fully developed individual—also has an explanation: the genetic program.

MAYR: That's correct. The existence or non-existence of the genetic program is the other incredibly important difference between biology and physics. In biology, everything can be categorised into two sets of causes. One set is the laws of physics, which no living thing can violate, to which all phenomena in the living world conform.

ALKIMOS: That's exactly what D'Arcy Thompson said!

MAYR: But there is another set of causes which impact life phenomena: the genetic program. Nothing happens in a living creature that has not been influenced in some way by the genetic program.

ALKIMOS: The genetic program has been the focus of our investigations!

EXTON: Yes, we've immersed ourselves in the disciplines that examine the nature of genes and the genetic program: genetics and molecular biology. Since we're on the subject, may I ask how, in your opinion, molecular biology has contributed to evolutionary biology?

MAYR: Molecular biology has demonstrated that the genetic code is universal and evolved only once. Therefore all life—including bacteria—have one common ancestor. This is extraordinarily important.

ALKIMOS: The universality of the genetic code.

MAYR: Molecular biology has also shown that information passes from the DNA to the proteins, but it cannot move in the reverse direction, from proteins to DNA.

ALKIMOS: The central dogma.

MAYR: This was the final nail in the coffin of Lamarckism. It became clear that the inheritance of acquired traits is impossible. The discovery by François Jacob and Jacques Monod that a group of genes regulate another group is also very important.

ALKIMOS: Repressors, activators, and transcription factors.

MAYR: The identification of these regulatory genes and the techniques of molecular biology revived the field of developmental biology, which had withered with the decline of Entwicklungsmechanik in the 1930s. Developmental biology is one of the most important fields in biology—if not *the* most important; it owes its advancement exclusively to molecular biology.

EXTON: Without molecular biology the study of genetic programs would be impossible. We'd be stuck with vitalism, teleology, and physicalism.

MAYR: Right. There's nothing in the physical world comparable to the genetic program.

EXTON: Except maybe computer programs.

MAYR: Perhaps.

ALKIMOS: Of course! We used a program to instruct the Deus ex Machina to take us back in time to visit the Professor, and we used another program to send Democritus back to Abdera!

EXTON: Enough mumble jumble, Alkimos.

MAYR: What's he saying? What's this about Democritus?

ALKIMOS: He's my master.

MAYR: How did you send him back?

ALKIMOS: With a time machine. We had to program it …

MAYR: You mean that odd contraption over there?

EXTON: Don't listen to him, Professor. He likes to joke. That's just an everyday …

MAYR: Could it take me back to New Guinea?

EXTON: To New Guinea?

MAYR: To the year 1928?

ALKIMOS: Easily!

EXTON: Impossible!

ALKIMOS: Why is it impossible? Exton, you know that …

EXTON: Alkimos, I said no mumble jumble! We have to be going.

MAYR: No please. Just one moment …

EXTON: Excuse us, we really need to …

ALKIMOS: Exton, have some compassion! It's probably a matter of the heart!

MAYR: You see, in 1928 …

ALKIMOS: See, Exton! Who wouldn't give up one hundred years of contemplation for a tryst in the jungle!

MAYR: Excuse me, but that is not at all what I'm referring …

EXTON: Come with me, Alkimos. Immediately!

ALKIMOS: Exton, how can you …

MAYR: But all I ask …

EXTON: On board, now!

ALKIMOS: Sorry Professor, I tried!

MAYR: But, but …

ALKIMOS: May the Papuan gods be with you!

EXTON: Let's get this engine started!

ALKIMOS: Agghh … here we go!

The Genetics of the Biosphere

EXTON: Alkimos, have a seat. It's dinnertime.

ALKIMOS: Excellent, I'm starved!

EXTON: We're going to try some highly unusual dishes.

ALKIMOS: I'm ready for it!

EXTON: Because tomorrow we're embarking on a long journey.

ALKIMOS: Oh ...

EXTON: And this evening we need to consider some very difficult questions.

ALKIMOS: Umm, I believe that's what we've been doing all along.

EXTON: Yes, but I'd like us to re-examine everything from a somewhat different perspective.

ALKIMOS: Okay, that doesn't sound too bad. From the perspective of poetry?

EXTON: No. From the perspective of the biosphere. We're going to take into consideration the cosmic scale of time, space, and diversity.

ALKIMOS: Cosmic unrest.

EXTON: Exactly. Metaphysical unrest—the study of the gods or our own origins—is no longer of interest to us. Instead we're going to examine the consequences of our existence.

ALKIMOS: So we have to broaden our horizons.

EXTON: You know, Alkimos, the well of the past is unfathomably deep. Life began 3.5–3.8 billion years ago. We don't know how—we can only speculate, develop theories. We know that organic, carbon-based, molecules can arise from inorganic matter. Urea (CH_2NH_3) was the first organic molecule scientists managed to synthesise, in 1828. Friederich Wöhler thus demonstrated that organic molecules could be created without a life force.

HERMES: Shall I serve dinner?

EXTON: Yes, please.

HERMES: Saffron risotto with asphodel sauce.

ALKIMOS: A large serving please!

EXTON: In 1953, Stanley L. Miller and Harold C. Urey conducted a famous experiment. They put water, ammonia, methane, and hydrogen in a test tube and then zapped the solution with an electric charge. They were trying to mimic the contents of the primordial ocean and the storms and lightening of the primordial atmosphere. When the solution was exposed to electricity, chemical processes were set in motion. The liquid slowly browned and the result of the electric charge was the creation of organic molecules from simple ones, and not just any organic molecules. Miller and Urey discovered that amino acids—the building blocks of proteins — had formed in the liquid.

ALKIMOS: Like Spallanzani's soup?

EXTON: Haeckel's soup. Darwin got the idea from Haeckel, and the world got the idea from Darwin. Life sprang from this soup. Spontaneous generation.

ALKIMOS: I knew it was possible when we visited Spallanzani. I should have taken a sample of his soup.

EXTON: What Urey and Miller showed is that there is nothing extraordinary about the building blocks of life and their origins. Amino acids, nucleic acids, sugars, lipids—they can all arise in a non-living environment. As brilliantly articulated later by the Hungarian chemist Tibor Gánti in his *chemoton theory*, life is a complicated autocatalytic system with three coupled subsystems, a boundary or cell membrane, a metabolic system, and a genetic system. This doesn't mean the origin of these three subsystems and how they came together is easy to decipher, but we have now very sound ideas about how this may have happened.

ALKIMOS: Exton, you said, for example, that RNA has forms that are both enzymes and genes simultaneously.

EXTON: The ribozymes.

ALKIMOS: Does this mean we've killed two birds with one stone?

EXTON: Yes, it does. Because it is conceivable that Darwinian evolution started when the first ribozyme capable of self-replication appeared.

ALKIMOS: An autocatalytic gene.

EXTON: Exactly.

HERMES: Gentlemen, wolf spider soup with tarragon dumplings.

ALKIMOS: Eww!

EXTON: I told you we'd have some unusual dishes tonight.

ALKIMOS: Okay, but you never said wolf spiders. Hmm, well, I am hungry ...

EXTON: I had no doubt, Alkimos ...

ALKIMOS: But Exton, back to what you were saying. So, living things that are capable of reproduction and are also regulated by some kind of program appeared one day?

EXTON: Yes. About 3.5 billion years ago. And thanks to reproduction, the regulatory programs, and mistakes in copying the programs, evolution was already underway.

ALKIMOS: That makes sense.

EXTON: We don't fully understand how the first self-replicating ribozyme may have appeared and what was the dynamics of such primordial self-replicating systems. But it is likely that early on such replicating ensembles of RNA molecules were encapsulated into simple vesicles to form the first primordial cells. This was followed by the evolution of the first ribosome-containing organisms and the advent of the DNA–RNA–protein world we know. For a long time, prokaryotes ruled the earth. Then about one and a half billion years ago eukaryotes, cells containing a nucleus, appeared. These cells were feeding on other cells by the process of *phagocytosis*, catalysing a massive ecological transformation of the planet. We've already talked about the Cambrian explosion that ensued and the transitioning of some creatures to dry land. The details aren't important at the moment—it's the scale we need to understand in terms of time, space, and diversity. Modern man appeared approximately 100,000 years ago. In the 3.5-billion-year history of life, that's just a blink of the eye.

ALKIMOS: And what is a human anyway?

EXTON: Just one of the many millions of species on the planet. But it's the only one capable of understanding its own origins.

ALKIMOS: Yes, that's what we're trying to do.

EXTON: But we rarely think about the scope of our task. We've been studying the genes and the development of fruit flies for nearly one hundred years. Thousands of researchers have worked and thought for millions of hours. More than ninety thousand scholarly articles have been written on Drosophila, and we still don't know how fruit flies work. Oh, we've discovered many exciting details, but the workings of the entire system — how the thousands of genes and their products, the membranes and the metabolites, all interact to create all the cells and the entire organism—is beyond our understanding.

ALKIMOS: Didn't we already establish that to truly understand it, we need to compute it? All of it—development, operation and evolution. Isn't that what Brenner suggested?

EXTON: In a way, but it's almost a hopelessly complicated task. And one model of a living organism is not enough. We have to return to the level of biological organisation and the incredibly complicated problem it presents.

ALKIMOS: Democritus suspected this.

EXTON: We've talked a lot about genes and molecules and have strived to understand how this level determines cell function and development. We've also discussed the evolution of the genetic program. But Mayr stressed the importance of levels of organisation beyond the individual. Evolution takes place on the level of populations. Genetic diversity, the indispensable raw material of evolution, can only be understood on this level. Individuals create populations—reproductive communities. These communities and their members can be considered to belong to one species if they reproduce with each other and are separated from other populations by mechanisms of reproductive isolation. Although these species are segregated in terms of reproduction and the mixing of genes, on a higher level—the level of ecological associations—they form wide-ranging connections with each other.

ALKIMOS: The cedar forests of Abdera are this kind of association?

EXTON: Exactly. An integrated unit of several hundred or thousand species — plants, animals, fungi, and microorganisms—with a predictable composition. Within this unit some species live in symbiosis; others have a host–parasite relationship. The food chains and the flow of nutrients, the network of what eats what, hold the system together. But the organisation of associations and the entire biosphere are stretched across a geophysical and geochemical framework that includes the water, wind, rock, and solar cycles. From 1799 to 1804, Alexander von Humboldt travelled around South America. He collected plants, made precise geological and meteorological measurements, and observed the geographical distribution of plants. By studying the climate and vegetation of mountainous areas and at various latitudes, he was able to understand how climate impacts plants. Humboldt was the first to recognise the zonation of earthly life. During his travels he strove to carefully survey the environment and interpret the phenomena he observed in light of the complex interaction of physical, geological, and biological processes. His studies and methodology had a profound influence on the next generation of naturalists. Darwin drew considerable inspiration from Humboldt's works. The focus of Humboldtian science was the entire Earth. He dissected the elaborate interconnectedness of biosphere, geosphere, hydrosphere, atmosphere, and lithosphere. His aim was to understand why the phenomena he observed existed.

ALKIMOS: If we can get from the why of genes to the why of interaction among spheres, then perhaps we can say that we understand something about how organisational levels build upon each other?

EXTON: It would be rather daunting to start the Humboldtian program at the level of genes. I fear we won't have time for it.

ALKIMOS: Why not?

EXTON: To thoroughly understand the working of the biosphere, we need to establish the role and the origins of millions of species. It would take centuries. And since there are no general laws in biology—because everything is unique, non-recurring and unrepeatable—the study of every single species would require a tremendous amount of time.

ALKIMOS: We've been studying fruit flies for a hundred years …

EXTON: However, a thorough examination of a few selected species—this would entail gene sequencing, molecular analysis, and a determination of their developmental programme—could be very helpful in understanding other species.

ALKIMOS: Given that all living creatures have one common ancestor.

EXTON: Yes, and because of this, they display a remarkable number of similarities, on the molecular level too. This allows us to make generalisations in many cases. Let's say we very carefully work out how an enzyme functions in a particular species; we can rest assured that this same enzyme works in a very similar fashion in other species. But of course if everything is conserved, then evolution would not exist. To understand evolution, we need to study those elements that are not constant, whose role has changed—whether it's a gene, an organ, or the link between two species in an association. Generalisations will not help us to understand unique characteristics.

ALKIMOS: But why don't we have enough time? We're not in a hurry. The present genetic researchers will be replaced by a new generation, and step by step they will unravel the secrets of the biosphere, from the level of genes to the level of biomes.

EXTON: Alkimos, perhaps *we* have time, but the planet does not. You see, the biosphere is not what it was in Humboldt's time. When Humboldt travelled the world, it was in a nearly pristine state and displayed astonishing diversity.

ALKIMOS: What's changed?

EXTON: People now occupy a far greater portion of the earth's surface and have cultivated it, manipulated it, and in many cases seriously damaged it. Human population has also increased at an unprecedented rate. More than seven billion people live on the planet today and the number continues to grow. At present, we humans use 40% of primary biological production— the total amount produced through plant photosynthesis—for our own purposes. We eat it, feed it to our animals, clear it away, or burn it.

ALKIMOS: That's impossible …

EXTON: And in the process, we cause the destruction of countless other species. So far scientists have described 1.7 million species, but an estimated 10 million species are thought to inhabit the Earth. So you see we've documented only a fraction of them, and we'll never discover the majority.

ALKIMOS: But Exton, I don't understand. Why are we destroying things so carelessly?

EXTON: There's an illusion deeply entrenched in Western thought that the world is constantly developing, progressing toward some kind of endpoint—that the passage of time has an aim. This thought is most explicitly realised in Christian theology. Time began with Creation, and it will end with the triumphant second coming of Christ. This notion presides over every impulse in the Western world. Increasing the gross domestic product, decreasing unemployment, expanding agricultural output. Bewitched by the illusion of progress, people fail to notice the devastating effects their efforts to achieve their often greedy goals have on the natural world. They never look back, only forward. They want everything to be bigger, faster and more plentiful. In their minds this is progress.

ALKIMOS: It sounds like cosmic teleology!

EXTON: But cosmic teleology doesn't exist. If we view the world and the biosphere on a cosmic scale, we can see that in the course of one or two human generations there is no progress in any direction. The entire one- to two-million-year history of the *Homo* genus is the blink of an eye in the 3.5-million-year history of the biosphere. Of course humanity can have goals; it can strive for something.

ALKIMOS: For example to understand the biosphere and humanity's place in it.

EXTON: Yes, but it seems that these goals are not set by Humboldt, Socrates or other similarly wise individuals.

ALKIMOS: Socrates was executed.

EXTON: Yes, he was executed because he had a different idea of progress.

ALKIMOS: He thought it involved contemplation and independent thought.

EXTON: Unfortunately, contemplative people do not shape the world, only themselves. Thus the shaping of the world is left to those who don't care to observe or think.

ALKIMOS: Right. Athenians.

EXTON: Indeed. Those who cut down entire cedar forests to build more and bigger ships.

ALKIMOS: But why?

EXTON: Because they thought they had a complete picture of the world— that they knew everything there was to know—and their way of thinking was the right way.

ALKIMOS: But Socrates wasn't so sure of that.

EXTON: We can consider such narrowness of thinking as a legacy of the evolution of the human mind. This kind of worldview was highly advantageous to our hunter–gatherer ancestors. But as soon as you couple it with humanity's technical prowess it becomes dangerous. Humans try to shape the world in accordance with their narrow understanding.

ALKIMOS: The world shaped humanity, and now humanity wants to shape the world. Somehow this seems inevitable.

EXTON: In 3.5 billion years, the tree of life has branched out beautifully. Every so often it's been trimmed back by a cataclysmic event, such as at the end of the Cretaceous period 65 million years ago, when a meteor struck the Earth.

ALKIMOS: Yes, I see.

EXTON: But its injured branches have always managed to sprout new life. And one day, an unusual fruit began to ripen on one of the branches: the genus *Homo*. Its most successful species was called *sapiens*; however it cast a dark shadow over the other branches. It wanted all the sunlight for itself.

ALKIMOS: I do enjoy a little sunlight on my face.

HERMES: Cedar sprouts in wine sauce with some Thracian toast.

ALKIMOS: Ah, yes please! And take this soup away. I couldn't eat it. I think whoever cooked the soup should eat it!

HERMES: All you needed to do was to pull the legs off. They're a bit hairy, I know, but ...

ALKIMOS: Never mind that! We're talking about dark shadows ...

EXTON: Ah, but something marvellous also lurked in the dark shadows, Alkimos. You see, from time to time there was a flash of light. A spark of human spirit.

ALKIMOS: And the gods? Which branch of the tree did they grow on?

EXTON: Many believe the gods give meaning to this tree, that they're its roots, that they give it life. But the truth is the complete opposite. Without the tree, without the most unusual branch produced by nature, we wouldn't have any gods.

ALKIMOS: So the tree has nurtured gods too? Perhaps they can save the tree from the darkness!

EXTON: The gods are nestled away on Mount Olympus, where the sun sets late. I fear the world will fall into total darkness before they take notice.

ALKIMOS: Then, to the top of Olympus! Let's bid farewell to the setting sun!

Biscuits Baked in Ash

COOK: Wake up, my friend!

ALKIMOS: Is that you?

COOK: I've made you a special breakfast.

ALKIMOS: Really?

COOK: Yes, because you're going on a long journey today.

ALKIMOS: Oh, yes … Exton did say something about that … How long, I wonder …

COOK: I have no idea.

ALKIMOS: Well can you pack us some food for the trip?

COOK: If you don't eat up everything now.

ALKIMOS: Well, maybe you should have prepared us a bountiful breakfast rather than a special one.

COOK: Here you are: tamarisk biscuits baked in cedar ashes.

ALKIMOS: Oh, not bad at all!

EXTON: Alkimos, come over here!

ALKIMOS: What for?

COOK: Don't forget your biscuits, my friend!

ALKIMOS: Thank you.

EXTON: Come on, it's time for you to go!

ALKIMOS: Me? What about you? Aren't you coming?

EXTON: No, not this time.

ALKIMOS: But, what will I do without you?

EXTON: Don't worry. They're expecting you.

ALKIMOS: Expecting me? Who?

EXTON: My master.

ALKIMOS: Your master? Then why aren't you coming with me?

EXTON: I can't.

ALKIMOS: But …

EXTON: I couldn't bear it.

ALKIMOS: To see your master?

EXTON: Oh no, I'd give anything to see him again.

ALKIMOS: Then what can't you bear?

EXTON: The future. I'd rather not know.

ALKIMOS: But you have no problem sending me? I don't understand.

EXTON: Alkimos, you wanted to see everything.

ALKIMOS: I suppose, but I didn't think I'd have to go alone.

EXTON: I'm programming the Deus ex Machina for two trips, so be careful. You mustn't get left behind.

ALKIMOS: I'll be careful.

EXTON: Farewell, Alkimos!

ALKIMOS: May Zeus be with you!

EXTON: And with you!

ALKIMOS: Agghh, here I go!!!

On the Island of Bensalem—*Pál Juhász-Nagy*

PAULUS MAGNUS: So you've arrived.

ALKIMOS: Are you talking to me?

PAULUS MAGNUS: I've been waiting for you for some time.

ALKIMOS: Who are you?

PAULUS MAGNUS: I'm Paulus Magnus. I'm a hermit. Or perhaps you could say I'm an islander. Shipwrecked.

ALKIMOS: Where am I?

PAULUS MAGNUS: The island of Bensalem.

ALKIMOS: Pardon?

PAULUS MAGNUS: Or Cuba, Hawaii, Crete, Borneo, the Easter Islands. Whatever you'd like.

ALKIMOS: I don't understand.

PAULUS MAGNUS: The world looks the same wherever you go. The main thing is we're high up, so we can look down.

ALKIMOS: On what?

PAULUS MAGNUS: On the Earth. Or Bensalem Island.

ALKIMOS: I'd like to have a look at Abdera.

PAULUS MAGNUS: I don't advise that. It'd cause you too much pain. Instead just have a look at what I show you. Observe!

ALKIMOS: All right.

PAULUS MAGNUS: Imagine that from the depth of the ocean emerges a shellfish capable of thought, and it opens its valves in the sunlight. Then imagine it can only remain for a few seconds in the immense, varied universe before it's swept back into the dark mysterious ocean currents. What else can the shellfish do but observe everything, observe the universe?

ALKIMOS: Are you a shellfish too?

PAULUS MAGNUS: During our short lives, while we see the sunlight, what else can we do but observe and think?

ALKIMOS: That's what motivated my master and me. Observation and the desire to solve the mysteries of Nature.

PAULUS MAGNUS: At the same time we need to be aware that Nature is immeasurably more than a thorny tangle of mysteries to be solved.

ALKIMOS: Yes, but this aspiration, this spirit is what drives people, what propels all of humanity forward. The desire to know.

PAULUS MAGNUS: No, you're wrong, Alkimos. What has dominated humanity's thinking—and thus determined our fate—is not the desire to observe and to know, and not the desire to be good and to be beautiful.

ALKIMOS: What?

PAULUS MAGNUS: No. Aggressive warfare and the unscrupulous lust for power have always gained the day.

ALKIMOS: Lust for power? What of Socrates? And Democritus? And the other great philosophers and scholars? And the sculptors and playwrights?

PAULUS MAGNUS: They're exceptions. An insignificant minority.

ALKIMOS: I see that the Persians were certainly dominated by this lust for power, but not the Greeks.

PAULUS MAGNUS: Don't fool yourself, Alkimos. What about the Trojan massacre?

ALKIMOS: But the gods were responsible for ...

PAULUS MAGNUS: No, Alkimos. You're here to look truth in the eye. People, *Homo insapiens*, have caused monstrous acts of destruction.

ALKIMOS: Why do you say *insapiens*?

PAULUS MAGNUS: In Linnaeus' lifetime there was a mistaken belief that humans were wise. Linnaeus was unable to see the truth, so he inaccurately labelled us '*sapiens.*' But this erroneous classification must be changed.

ALKIMOS: I still don't understand. What monstrous destruction are you referring to?

PAULUS MAGNUS: During the past one hundred years, ethnic, geographic, and biological treasures have been lost, entire cultures destroyed. Both biological and cultural diversity have been severely reduced. Legions of species have been wiped out, the majority without ever being examined by science. Water, soil, and air have all been wantonly polluted. The entire biosphere, all its cycles and regulatory processes, which went on undisturbed for millions of years, have now been irrevocably thrown off balance. And saddest of all, the warning signs were there, but they went unheeded—this is indeed the most telling example of the infinite stupidity and short-sightedness of humans. They took no steps to contain their unbridled greed, to temper their breathtaking arrogance and technocratic conceit.

ALKIMOS: But why?

PAULUS MAGNUS: People adopted this mentality during their struggles against nature, but sadly this militaristic mindset became entrenched. And now this same approach, which once meant survival, has driven humanity to the brink of annihilation.

ALKIMOS: But humanity will never be annihilated. It's impossible. Exton said humans are reproducing as never before.

PAULUS MAGNUS: And that was one of the main causes of the collapse. Overpopulation. The other is the wide array of technology created during the industrial revolution.

ALKIMOS: You mean like the Ford Model T?

PAULUS MAGNUS: You know, Alkimos, there is something I haven't quite figured out yet. Is it humanity itself that has sets its own limits and the world's course into the future? Or are we also prisoners of some kind of demonic automaticity? Have we become entrapped in some kind of *circulus vitiosus*, a vicious circle of which rampant urbanisation was just another aspect alongside vanishing biodiversity?

ALKIMOS: I'm not sure I follow you. How does biodiversity vanish? The world is enormous, almost endless. The Earth's inhabitants are tiny; their presence is negligible.

PAULUS MAGNUS: Your conception of the world is tragically flawed. I suppose you also believe the world's resources are inexhaustible? It's a childish, naïve notion sanctified in the Judeo-Christian tradition—this idea that nature was designed to serve humanity. This erroneous idea was the underpinning of the foolish global ideology known as consumerism. According to the logic of this ideology, the reason for increasing production is so that we may consume more. But no one has given any thought to the reckless squandering of resources this involves. This ideology has nurtured people's greed, leading to the consumption of nearly everything. People in their own craze for more and more have wrought monstrous destruction upon the Earth.

ALKIMOS: Horrible!

PAULUS MAGNUS: No, it's more than horrible. It's a cosmic sin. Human beings are nothing but abominable cosmic criminals.

ALKIMOS: We've committed sins against the gods.

PAULUS MAGNUS: We've systematically plundered the living world, ravaged the biosphere and demolished incomparable and irreplaceable natural wonders that were created over billions of years of evolution.

ALKIMOS: Cosmic sin.

PAULUS MAGNUS: But not just a sin. A mistake. 'Wise' humans have dug their own graves.

ALKIMOS: But why didn't scientists intercede? Why didn't they do anything?

PAULUS MAGNUS: They didn't have the power. But also, too few scientists seemed to understand the gravity of the situation, and of course, no one wanted to listen to those who did. I'm afraid the majority of scientists behaved as if they weren't living on this planet. I'll give you an example. Our global knowledge about the species that inhabit our biosphere was deplorably inadequate. And this ignorance was a particular problem because of the rapid destruction of the natural world. We were incapable of assessing what was being lost forever and how and at what pace. Therefore science was unable to produce reliable data at precisely the moment when it was most vital.

ALKIMOS: And what did scientists have to say about that?

PAULUS MAGNUS: If any were willing to deal with such questions, they often gave a pragmatic, simplistic answer worthy of only the most primitive utilitarian distortions of their troubled time. Generally they argued that unknown species might contain valuable material, useful in medicine for example. Or they pointed out how our ability to manipulate genes in the future might rest on the unique characteristics of genes not yet discovered.

ALKIMOS: So the only reason to preserve biodiversity was to benefit mankind? I see it's the height of foolishness. My master and I have never been driven by the principle of utilitarianism. Curiosity, the desire to know and understand the living world has been our motivation.

PAULUS MAGNUS: Then you certainly understand that the realm of better, broader truths lies far from the dominion of utilitarianism.

ALKIMOS: But I don't understand how the principle of utilitarianism managed to distort science.

PAULUS MAGNUS: The root of these distortions lay with the inability of most to understand what the greatest problems of humanity were. I don't think there was even one politician who did. But even too many biologists were preoccupied with the fight against cancer, Parkinson's disease, or infectious diseases. The institutions that support the sciences were equally short-sighted. The end result was the relegation of biology to a mere adjunct of medical science. While thousands of scientists studied certain model organisms, dozens of other species vanished on a daily basis, never to be known by science. Unfortunately, the 'elite' branches scorned taxonomy as an outdated, superfluous exercise. 'The age of the butterfly collector is over,' they said. But for us to understand the working of the entire system, we needed to record all of the basic components, by that I mean all the species, of the biosphere.

ALKIMOS: Well, why don't we? Let's revive the field of taxonomy!

PAULUS MAGNUS: It's too late. It was very sad to watch how the ancient forests of the Amazon River basin shrank with every passing year until they disappeared completely, and with hardly a cry of protest from biologists.

ALKIMOS: Let's go back in time! Come on Paulus. Let's board the Deus ex Machina! I'm sure I can reprogram it. We can warn people of the cataclysm in time!

PAULUS MAGNUS: It's pointless. No one listened to Cassandra.

> *For in my heart and soul I also know this well:*
> *The day will come when sacred Troy must die,*
> *Priam must die and all his people with him.*

ALKIMOS: No. I don't care what Homer wrote! Troy can be saved!

PAULUS MAGNUS: You know, Alkimos, regardless of what great poetic masterpieces have been written, what is most astonishing is that *Homo insapiens*, the Crazy Ape, the Unfinished Animal, to this day looks upon the world that it made ugly as if it had just emerged from its cave completely unable to find its place in the 'great order of nature.'

ALKIMOS: But I know my place in the great order of nature. On my boat in the bay of Abdera. I'm satisfied with just one or two fish a day.

PAULUS MAGNUS: As people destroyed the last ancient forests and the other 'ancient' entities, they also lost their points of reference. Now they will never know the laws of natural diversity, the secrets of nature, or themselves.

ALKIMOS: Paulus, we have to do something!

PAULUS MAGNUS: I'm sorry. I don't understand this machine of yours. But why don't you give it a try. It may take you somewhere.

ALKIMOS: Yes, I have to go. I have to try!

PAULUS MAGNUS: Then go on.

ALKIMOS: And what about you? What will you do?

PAULUS MAGNUS: I'm also going to try to find my place in the great order of nature. As József Eötvös said, 'The longer we live, the better we can see that the mind of which we are so proud was not given to us so that we may rule over Nature, but so that we may follow and obey it.'

ALKIMOS: You're going to follow Nature? What will you live on if there are no fish to catch, no birds to lay eggs, no sheep, no hay, no vegetation?

PAULUS MAGNUS: It's true; there are no such things anymore. But before the collapse, I began collecting a fantastic library here in my tower, in more or less wise anticipation of what was to come. That is what I live on.

ALKIMOS: You can't live on books.

PAULUS MAGNUS: No one else needs them. Pillaging hordes have come through but left these behind. They may have great bibliographical value, but because you can't eat them, they're worthless. But I have found a great use for them.

ALKIMOS: You read all day?

PAULUS MAGNUS: No, not anymore. I feed them to the termites living outside my tower.

ALKIMOS: The termites eat books?

PAULUS MAGNUS: Yes, it sustains them. And then I eat the termites.

ALKIMOS: Rare and valuable books?

PAULUS MAGNUS: Oh, not so valuable after all. There was no point in all that writing.

ALKIMOS: How can you say that?

PAULUS MAGNUS: Everyone wrote in the hopes of redemption.

ALKIMOS: Redemption?

PAULUS MAGNUS: Yes, that the wisdom of their words might bring redemption. But most people read very little, and if they did, they understood too few of the words on the page.

ALKIMOS: How sad.

PAULUS MAGNUS: Yes, it makes you want to cry, doesn't it?

ALKIMOS: Oh, Zeus, son of Cronus! Don't abandon us now! Oh, Pallas Athena! Save us miserable, contemptible humans!

PAULUS MAGNUS: Farewell, Alkimos.

ALKIMOS: I don't know where to go. Oh Chronides! My Master!

PAULUS MAGNUS: I need to go too. I have to check my termite traps.

ALKIMOS: Here, take this bag of food. At least …

PAULUS MAGNUS: Thank you. Now you should go.

ALKIMOS: May the Owl-Eyed Goddess be with you!

PAULUS MAGNUS: And with you!

ALKIMOS: Aagghhh … here I go!

Epilogue

On Mount Olympus

HERMES: Time to wake up! Come on! What happened to you?

ALKIMOS: Ohhh …

HERMES: Pardon?

ALKIMOS: No, no …

HERMES: Alkimos, open your eyes!

ALKIMOS: No, I can't, it's too horrible …

HERMES: Alkimos, stop yammering!

ALKIMOS: Where am I? Help!

HERMES: Relax, you're all right. Here, a little ambrosia liquor …

ALKIMOS: Ohhh …

HERMES: It's good, isn't?

ALKIMOS: I suppose …

HERMES: Pull yourself together. You can't make an appearance looking like that.

ALKIMOS: An appearance? What kind of appearance? Hermes, where am I? Where's my master? I need him!

HERMES: Alkimos, we're on Mount Olympus. You're to make an appearance before the Owl-Eyed Goddess.

ALKIMOS: Who? You don't mean ...

HERMES: Of course. Pallas Athena.

ALKIMOS: Oh my Zeus!

HERMES: There's no need to pray to the gods here—that's only a custom down below. Now come on! She's waiting.

ALKIMOS: Oh, I just don't ...

ATHENA: So you're here, Alkimos.

ALKIMOS: Oh, such thunder in my ears ...

ATHENA: Do stop shaking!

ALKIMOS: Oh, the hecatomb! I forgot to offer a hecatomb. I'll have to ...

ATHENA: That's not necessary, Alkimos. Now look at me.

ALKIMOS: Doctor Spallanzani?

ATHENA: Of course not. I just took on a familiar appearance to help you relax.

ALKIMOS: You did what?

ATHENA: I always have to do that. You mortals are so skittish.

ALKIMOS: We're afraid of death.

ATHENA: Ah yes, but such a waste of energy, don't you think?

ALKIMOS: Doctor Spallanzani ... or madam goddess ... or ... am I here to die? Or have I died already? Has the sun set?

ATHENA: What nonsense. I don't deal with death or destruction. I only deal with creation.

ALKIMOS: Destruction! Oh, Owl-Eyed One! You have to help us! We have no one to turn to and we've nearly destroyed our planet!

ATHENA: How can I help?

ALKIMOS: Do something! You're a goddess! You're Pallas Athena!

ATHENA: Oh, dear mortal, I've done so much already. But it was all in vain. Don't you see?

ALKIMOS: In vain?

ATHENA: Yes. In vain. You mortals are terribly thick and you have such an inflated opinion of yourselves. I'm afraid your fate is unavoidable—it's been determined by your very nature.

ALKIMOS: No, please!

ATHENA: Listen to me, Alkimos. Apollo and I tried everything. We gave you art — music, sculpture, poetry—to ennoble your spirit. We endowed you with intellect—the ability to learn and understand, to develop your mental powers. And what did you do with these gifts? You placed them in the service of greed and ambition, while you remained stupid and brutish. Ah, but you did have faith and hope—but such foolishness. You believed in the impossible—that no matter what abominations you came up with, there would be no consequences. But of course there were. You believed you were indestructible, but how pitifully you've destroyed yourselves. And all along you could have applied the same energy to protecting the beauty that already existed. All you needed to do was observe and think. This, and this alone, would have brought you a step closer to the gods.

ALKIMOS: But I did observe—all the time!

ATHENA: You're the exception, Alkimos. Perhaps there is something godly in you.

ALKIMOS: If that's true, then can't you help?

ATHENA: I suppose I might be able to help you.

ALKIMOS: And the billions of others like me?

ATHENA: No. It's too late for them. Even I don't have that kind of power. Now, time for you to go.

ALKIMOS: Where?

ATHENA: Back to Abdera.

ALKIMOS: Oh, dear Zeus!

ATHENA: Zeus? Oh, no, I'm afraid my father isn't home at the moment. He took the form of a swan and set off for Samos Island—he's up to no good I'm afraid.

ALKIMOS: Back to Abdera?

ATHENA: Yes. My gift to you is two thousand years of happy life.

ALKIMOS: Two thousand years? How?

ATHENA: That liquid you consumed—it was real ambrosia.

ALKIMOS: Ambrosia? Is that why I feel so ethereal?

ATHENA: Farewell, Alkimos.

ALKIMOS: Thanks you, Owl-Eyed Goddess, my dearest Pallas … Aaagghh, here I go!

The Bay of Abdera

ALKIMOS: Aaaggh!!! Oh, Bright-Eyed One! Oh, Owl Goddess! Thank you!

DEMOCRITUS: Alkimos, stop yammering!

ALKIMOS: Master! My dear, dear master!

DEMOCRITUS: Careful son. Don't knock me into the water!

ALKIMOS: I succeeded!

DEMOCRITUS: Yes, you succeeded in frightening the fish, if that's what you mean.

ALKIMOS: We're home!

DEMOCRITUS: Home? Whatever you say. You've certainly slept long enough.

ALKIMOS: Slept?

DEMOCRITUS: Yes. I had to drag you onto the boat. But then I let you sleep because I didn't want to waste the best evening hours. That's when the fish are active.

ALKIMOS: Drag me? I don't understand.

DEMOCRITUS: Hand me my knife, sleepyhead.

ALKIMOS: Sleepyhead? Have I been sleeping this whole time? You mean …?

DEMOCRITUS: What's that, son?

ALKIMOS: Nothing. Just … all we're doing is fishing?

DEMOCRITUS: What else would we do?

ALKIMOS: Oh, thank you, master!

DEMOCRITUS: Just give me the knife, Alkimos.

ALKIMOS: But it's so odd. You see, we were lying on the gravelly shore, and …

DEMOCRITUS: What are you talking about?

ALKIMOS: It was a dream. The whole thing was a dream. It must have been the wine!

DEMOCRITUS: Well, you drank the whole flask.

ALKIMOS: Me?

DEMOCRITUS: See, it's empty.

ALKIMOS: My Zeus, I guess I did.

DEMOCRITUS: Then you fell into a deep sleep—as if Athena had clunked you over the head with her spear.

ALKIMOS: Athena? But—

DEMOCRITUS: Oh, and how you snored!

ALKIMOS: Master, you wouldn't believe what I dreamed.

DEMOCRITUS: Well?

ALKIMOS: I saw the future, and I met Athena. And master, scientists discovered your atoms of life! They now know so much about them! And about the forms of living organisms, their internal systems, their development, and the history of the entire living world!

DEMOCRITUS: That's quite a remarkable dream.

ALKIMOS: Oh master, I learned and learned and learned. Hmm, but I'm not sure who I learned it from.

DEMOCRITUS: I'm sure Athena was whispering in your ear.

ALKIMOS: Yes, that's it! I met her at the end. It was terrifying.

DEMOCRITUS: What was terrifying? Her weaponry?

ALKIMOS: No. Her words.

DEMOCRITUS: I see. Well, son, you can learn a lot from dreams.

ALKIMOS: How?

DEMOCRITUS: You have to write them down.

ALKIMOS: Master, you're right! I'll write my *magnum opus*! Talaionides Alkimos—*The History of Atoms*. I'll send the parchment scrolls to the library in Alexandria.

DEMOCRITUS: Great idea.

ALKIMOS: They'll be preserved for posterity. And hundreds of years into the future the world's great minds will learn from them! And that's how we'll avoid the final ruin of the world.

DEMOCRITUS: Very noble thoughts. But I think first you'd better tell me more about this dream.

ALKIMOS: Yes, that would be best.

DEMOCRITUS: And while you're at it, bait that hook.

ALKIMOS: Yes, master. It began with us locked up in some kind of bronze box. It started shaking and then next thing we knew, we were in an unbelievable place …

DEMOCRITUS: Interesting.

ALKIMOS: I'd never seen such a place in all of Thrace.

DEMOCRITUS: Perhaps it was Attica?

ALKIMOS: No, I don't think so. I think—

DEMOCRITUS: Wait a minute, son. What's that awful noise?

ALKIMOS: Ah, the familiar clatter …

DEMOCRITUS: And that cloud of dust? It's stinging my eyes. But hmmm, it could be useful in fooling the fish …

ALKIMOS: Master, look over there!

DEMOCRITUS: What's that?!

ALKIMOS: The Deus ex Machina!

DEMOCRITUS: Oh, gods of Olympus …

ALKIMOS: Exton!

EXTON: Alkimos, Democritus, how good to see you!

ALKIMOS: Master, row towards him!

DEMOCRITUS: Yes, yes, I am.

ALKIMOS: Exton, what are you doing here?

EXTON: I decided to return to you, gentlemen. Now help me aboard. There's room for me, isn't there?

DEMOCRITUS: But of course! Give me your hand, dear Exton!

ALKIMOS: Master, you know Exton?

DEMOCRITUS: Of course. I had the same dream.

EXTON: Ah, my good friends. How I've missed your company!

ALKIMOS: The same dream?

EXTON: The main thing is we're together again.

DEMOCRITUS: Or perhaps we're still dreaming?

Appendix

A Chronology of Events Discussed in the Book

5th century B.C.: Democritus developed his theory of atoms.

5th century B.C.: Hippocrates examined chicken embryos and set forth his views on preformationism.

343 B.C.: Aristotle examined numerous embryos, argued against preformationism and stated that the embryo develops under the influence of the male seed.

c. 65: Seneca expressed his preformationist views in his tome *Quaestiones Naturales*.

1572: Paracelsus published a recipe for artificially producing homunculi.

1674: Nicolas Malebranche revived the theory of preformationism.

1651: In *Exercitationes de Generatione Animalium*, William Harvey proposed that all living creatures come from eggs.

1655: Robert Hooke described and illustrated the cells of cork in his book *Micrographia*.

1677: Antonie Van Leeuwenhoek discovered spermatozoa.

1682: In *The Anatomy of Plants*, Nehemiah Grew stated that the stamen and pistil were the male and female reproductive organs of plants.

1688: Francesco Redi's *Esperienze intorno alla generazione degl'insetti* demonstrated that flies hatched from eggs and not rotting meat.

1744: Abraham Trembley described the regenerative power of freshwater hydras.

1744: Charles Bonnet described asexual reproduction, or parthenogenesis, among aphids in *Traité d'Insectologie*.

1745: In his essay 'Vénus Physique', Maupertius refuted preformationist theories.

1759: Caspar Friedrich Wolff's treatise *Theoria Generationis* describes the gradual formation of organs from an undifferentiated state.

1765: Spallanzani's ingenious experiments disproved the theories of spontaneous generation.

1824: Jean-Louis Prévost and Jean-Baptiste Dumas proved that eggs are fertilised by sperm and described the early development of frog embryos.

1827: Karl Ernst von Baer described the mammalian ovum.

1828: Karl Ernst von Baer laid the foundations for comparative embryology in his monograph *Über Entwickelungsgeschichte der Thiere*.

1828: Friedrich Wöhler synthesised organic material (urea) from inorganic material.

1830: Giovanni Battista Amici described fertilisation in plants.

1831: Robert Brown discovered the cell nucleus.

1832: Barthélemy Dumortier described cell division in plants.

1839: Theodor Schwann laid the groundwork of cell theory in *Mikroskopische Untersuchungen über die Uebereinstimmung in der Struktur und dem Wachsthum der Thiere und Pflanzen*.

1841: Rudolf Albert von Kölliker demonstrated that spermatozoa are also cells.

1841: Robert Remak described cell division in animals.

1848: Wilhelm Hofmeister examined cell division in *Tradescantia* and observed chromosomes in the dividing cell nuclei.

1855: Robert Remak's book *Untersuchungen über die Entwicklung der Wirbeltiere* explained how animals develop from the ovum by the process of cell division.

1859: Louis Pasteur's experiments disproved the theory of spontaneous generation.

1859: In his seminal book *On the Origin of Species*, Charles Darwin revealed the common origin of all living things and expounded the theory of evolution by natural selection.

1865: Gregor Mendel's essay *Versuche über Pflanzen-Hybriden* presented his discovery of the inheritance of discrete traits, the framework of Mendelian genetics.

1866: In his book *Allgemeine Anatomie*, Mendel maintained that the cell nucleus is responsible for inheritance.

1866: Ernst Haeckel argued that ontogeny recapitulates evolution.

1869: Friedrich Miescher isolated chromatin (nuclear DNA and protein).

1873: Anton Schneider discovered chromosomes in the cell nuclei of *Mesostoma* flatworms.

1877: Hermann Fol first observed fertilisation in animals.

1882: Walther Flemming described in detail the process of mitosis in *Zellsubstanz, Kern und Zelltheilung*.

1884: Hugo de Vries showed that plant cells are enclosed in a flexible membrane that lines the rigid cell walls.

1884: Eduard Strasburger observed that in the *Orchis latifolia*, only the cell nucleus enters the seed case during fertilisation.

1885: Oscar Zinoffsky determined the atomic composition of haemoglobin.

1888: Theodor Boveri's studies of the *Ascaris* roundworm proved the permanence of chromosomes.

1889: Richard Altmann demonstrated that chromatin consists of proteins and nucleic acids.

1890: In *Zellenstudien*, Theodor Boveri described the process of meiosis.

1890: While examining sperm production in the beetle *Pyrrhocoris apterus*, Hermann Henking discovered the X sex chromosome.

1892: In *Das Keimplasma*, August Weismann argued that germ cells are distinct from soma and are the exclusive agents of inheritance.

1894: After conducting experiments on the regulative ovum, Hans Dreisch concluded that development cannot be explained by the laws of physics and chemistry.

1899: Jacques Loeb managed to induce an unfertilised sea urchin egg to develop.

1900: Carl Correns, Hugo de Vries, and Erich Tschermak rediscovered Mendel's work.

1902: Walter S. Sutton observed the chromosomes in the gametes of the grasshopper *Brachystola magna*.

1911: Thomas Hunt Morgan proved that genes are located on the chromosomes.

1912: Max von Laue discovered the ability of crystals to diffract X-rays.

1913: Alfred H. Sturtevant prepared the first genetic map.

1914: Leonard Troland posited that enzymes operate as autocatalytic molecules.

1915: Félix D'Hérelle discovered bacteriophages.

1919: Morgan and associates identified several hundred Drosophila genes through their studies of genetic mutations.

1918: Ronald Fisher showed that continuous variation could result from the combined effects of several discrete genes and was thus consistent with the laws of Mendelian genetics.

1923: Hans Spemann and Hilde Mangold investigated embryonic induction and introduced the concept of the organiser.

1926: Hermann Joseph Muller discovered that X-rays could cause mutations.

1926: James Sumner crystallised the enzyme urease.

1935: Phoebus Levene proposed a chemical structure for DNA, which he conceived of as a single strand.

1935: Wendell Meredith Stanley crystallised the tobacco mosaic virus.

1937: In his book *Genetics and the Origin of Species*, Theodosius Dobzhansky interpreted the observable variations in natural populations using the model of genetics established by Mendel and Morgan.

1940: Adolf Butenandt and Edward Tatum independently identified the substance synthesised under the direction of the vermilion gene as kynurenine.

1940: Linus Pauling and Max Delbrück proposed the phenomenon of molecular complementariness in biology.

1941: George Wells Beadle and Edward Tatum established the association between numerous genes and biochemical reactions in the bread mould *Neurospora*.

1942: Ernst Mayr expounded his concept of biological species in the book *Systematics and the Origin of Species*.

1944: In *What is Life?*, Erwin Schrödinger argued that genes are aperiodic crystals.

1944: Oswald Theodore Avery, Colin MacLeod, and Maclyn McCarty proved that the transformation of the *Pneumococcal* bacteria was caused by the transfer of DNA.

1945: Dorothy Hodgkin solved the structure of the antibiotic penicillin (27 atoms). A few years later she solved the structure of vitamin B12 (181 atoms).

1946: Joshua Lederberg discovered bacterial sex.

1949: Harvey Itano demonstrated that the structure of sickle-cell haemoglobin is different than that of healthy haemoglobin.

1950: Paul Zamecnik showed that protein synthesis takes place on the ribosomes.

1951: Frederick Sanger determined the sequence of amino acids in insulin's B chain.

1951: Linus Pauling described the basic structural element of proteins, the alpha-helix.

1952: Paul Zamecnik used rat liver extract to synthesise proteins *in vitro*.

1952: Alan Turing demonstrated that in a reaction-diffusion system, spatial patterns can arise even when the system is initially homogenous.

1952: Alfred Hershey and Martha Chase showed that during infection by a bacteriophage, only the DNA is injected into the host cell; the protein coat remains outside.

1953: Francis Crick and James Watson determined the structure of DNA.

1953: Stanley L. Miller and Harold C. Urey subjected a solution containing water, ammonia, methane and hydrogen to electric shocks, trying to simulate the circumstances of primordial Earth. This procedure resulted in the formation of several organic compounds, the building blocks of life.

1953: George Gamow proposed a model to explain how the sequence of DNA determines the incorporation of amino acids in a protein chain.

1957: Vernon Ingram demonstrated that sickle cell haemoglobin differs from healthy haemoglobin in just one amino acid.

1957: Sydney Brenner proved that the genetic code cannot consist of overlapping triplet codons.

1959: John Kendrew determined the atomic structure of the myoglobin protein.

1961: Marshall Nirenberg and Heinrich Matthaei were the first to decode a triplet codon. They found that the presence of poly U, a synthetic RNA molecule composed only of uracil, caused the production of the amino acid phenylalanine.

1961: François Jacob and Jacques Monod defined the *operon* as a unit of genetic regulation and transcription.

1962: Christian Anfinsen demonstrated that the folding of a polypeptide chain was determined by its amino acid sequence.

1964: Robert Holley sequenced a tRNA molecule consisting of 8 nucleotides.

1966: Har Gobind Khorana and Marshall Nirenberg and their research team cracked the entire genetic code.

1967: Sydney Brenner isolated the first *Caenorhabditis elegans* mutant.

1969: Lewis Wolpert formulated his theory of positional information.

1969: The first restriction endonuclease was identified.

1971: Daniel Nathans prepared the first restriction map, that of the SV49 virus.

1973: Frederick Sanger developed a new sequencing method.

1975: Walter Fiers sequenced the single-stranded RNA of the MS2 phage.

1977: Walter Gilbert and Allan Maxam devised a new sequencing method and determined the sequence of the 5243 base pair genome of the SV40 virus and the 5577 base pair genome of the G4 phage.

1979: Christiane Nüsslein-Volhard and Eric Wieschaus began screening for mutations affecting the embryonic development of the fruit fly.

1988: Wolfgang Driever and Christiane Nüsslein-Volhard described the first morphogen, the Bicoid protein in *Drosophila*.

2001: The entire sequence of the human genome was completed.

Further Reading

Arendt, D. 2003. Evolution of eyes and photoreceptor cell types. *International Journal of Developmental Biology*, 47:563–571.

Arendt, D., and Wittbrodt, J. 2001. Reconstructing the eyes of Urbilateria. *Philosophical Transactions of the Royal Society B: Biological Sciences*, 356:1545–1563.

Avery, O. T., MacLeod, C. M., and McCarty, M. 1944. Studies on the chemical nature of the substance inducing transformation of pneumococcal types. *Journal of Experimental Medicine*, 79:137–158.

Bacon, F. 1626. *The New Atlantis*. The Internet Wiretap edition. P.F. Collier & Son, New York.

Baltzer, F. 1967. *Theodor Boveri: The life of a great biologist 1862–1915*. University of California Press, Berkeley.

Bateson, W. 1900. Problems of heredity as a subject for horticultural investigation. *Journal of the Royal Horticultural Society*, 25:54–61.

Bateson, W. 1899. Hybridisation and cross-breeding as a method of scientific investigation. *Journal of the Royal Horticultural Society*, 24:59–66.

Beadle, G. W., and Tatum, E. L. 1941. Genetic control of biochemical reactions in neurospora. *Proceedings of the National Academy of Sciences USA*, 27:499–506.

Benedek, I. 1987. *Az értelem dícsérete*. Minerva, Budapest.

Benson, K. R. 2001. T. H. Morgan's resistance to the chromosome theory. *Nature Reviews Genetics*, 3:469–474.

Boveri, T. 1902. Über mehrpolige Mitosen als Mittel zur analyse des Zellkerns. *Verhandlungen der Physicalisch-Medizinischen Gessellschaft zu Würzburg*, 35:67–90.

Bowler, P. J. 2002. Climb chimborazo and see the world. Alexander von Humboldt, 1769–1859. *Science*, 298:63–64.

Bragg, W. L. 1922. The diffraction of X-rays by crystals. Nobel Lecture, from Nobelprize. org, https://www.nobelprize.org/nobel_prizes/physics/laureates/1915/wl-bragg-lecture.pdf

Brenner, S. 1957. On the impossibility of all overlapping triplet codes in information transfer from nucleic acid to proteins. *Proceeding of the National Academy of Sciences USA*, 43: 687–693.

Bridges, C. B. 1914. Direct proof through non-disjunction that the sex-linked genes of *Drosophila* are borne by the X-chromosome. *Science*, XL:107–109.

Buss, L. W. 1983. Evolution, development, and the units of selection. *Proceedings of the National Academy of Sciences USA*, 80:1387–1391.

Chargaff, E. 1978. *Heraclitean Fire. Sketches from a Life before Nature.* The Rockefeller University Press, New York.

Chargaff, E. 1979. How genetics got a chemical education. *Annals of the New York Academy of Sciences*, 325:345–360.

Cohen, L. B. 1980. *Album of Science. From Leonardo to Lavoisier 1450–1800.* Charles Scribner's Sons, New York.

Cremer, T. 1985. *Von der Zellenlehre zur Chromosomentheorie.* Springer-Verlag, Berlin Heidelberg New York Tokyo.

Crick, F. H. C., Griffith, J. S., and Orgel, L. E. 1957. Codes without commas. *Proceeding of the National Academy of Sciences USA*, 43: 416–421.

Crick, F. H. C. 1962. On the genetic code. Nobel Lecture, http://www.nobelprize. org/nobel_prizes/medicine/laureates/1962/crick-lecture.html.

Crick, F. H. C. 1974. The double helix: A personal view. *Nature*, 248:766–769.

Driever, W. 2004. The Bicoid Morpogen Papers (II): Account from Wofgang Driever. *Cell*, S116:S7–S9.

Driever, W., and Nüsslein-Volhard, C. 1988. A gradient of *bicoid* protein in drosophila embryos. *Cell*, 54:83–93.

Driever, W., and Nüsslein-Volhard, C. 1988. The *bicoid* protein determines position in the drosophila embryo in a concentration-dependent manner. *Cell*, 54:94–104.

Eisenberg, D. 1994. Max Perutz's achievements: How did he do it? *Protein Science*, 3:1625–1628.

Farley, J. 1982. *Gametes and Spores: Ideas About Sexual Reproduction 1750–1914.* Johns Hopkins University Press, Baltimore.

Ford, B. J. 1991. *The Leeuwenhoek Legacy.* Biopress and Farrand Press, Bristol, London.

Franklin, R., and Gosling, R. G. 1953. Molecular configuration in sodium thymonucleate. *Nature,* 171:740–741.

Galton, F. 1898. A diagram of heredity. *Nature*, 57:293.

Garen, A., and Gehring, W. 1972. Repair of the lethal developmental defect in deep orange embryos of drosophila by injection of normal egg cytoplasm. *Proceedings of the National Academy of Sciences USA*, 69:2982–2985.

Gasking, E. 1967. *Investigations into Generation, 1651–1828.* Johns Hopkins University Press, Baltimore.

Geison, G. L. 1995. *The Private Science of Louis Pasteur.* Princeton University.

Gilbert, S. F. 1978. The embryological origins of the gene theory. *Journal of the History of Biology,* 307–351.

Graves, R. 1985. *Görög Mítoszok.* Európa könyvkiadó, Budapest.

Harris, A. J. 1923. Galton and Mendel: Their contribution to genetics and their influence on biology. *The Scientific Monthly,* 16:247–263.

Harris, H. 1999. *The Birth of the Cell.* Yale University Press, New Haven and London.

Hershey, A. D., and Chase, M. 1952. Independent functions of viral protein and nucleic acid in growth of bacteriophage. *Journal of General Physiology,* 36:39–56.

Hunter, G. K. 2000. *Vital Forces; The Discovery of the Molecular Basis of Life.* Academic Press, London.

Jablonka, E., and Lamb M. J. 2002. The changing concept of epigenetics. *Annales of the New York Academy of Sciences,* 981:82–96.

Judson, H. 1979. *The Eighth Day of Creation: Makers of the Revolution in Biology.* Simon & Schuster, New York.

Juhász-Nagy Pál. 1992. *Változatok at ökológiai kultúra tematikájához.* Természet- és Környezetvédő Tanárok Egyesülete, Budapest.

Juhász-Nagy Pál. 1993. *Természet és ember. Kis változatok egy nagy témára.* Gondolat, Budapest.

Juhász-Nagy Pál. 1993. *Az eltűnő sokféleség (A bioszféra-kutatás egy központi kérdése).* Scientia Kiadó, Budapest.

Kohler, R. E., 1994. *Lords of the Fly. Drosophila Genetics and the Experimental Life.* The University of Chicageo Press, Chicago–London.

Laue, M. 1915. Concerning the detection of X-ray interferences. Nobel Lecture.

Loeb, J. 1899. On the nature of the process of fertilization and the artificial production of normal larvae (plutei) from the unfertilized eggs of the sea urchin. *American Journal of Physiology,* 3:135–138.

Lwoff, A., and Ullmann, A. 1979. *Origins of Molecular Biology. A tribute to Jacques Monod.* Academic Press New York San Francisco London.

Mayr, E. 1982. *The Growth of Biological Thought.* Harvard University Press, Cambridge.

Mayr, E. 2002. Interview with Ernst Mayr. *BioEssays,* 24:960–973.

Mayr, E. 2001. The philosophical foundations of darwinism. *Proceedings of the American Philosophical Society,* 145:488–495.

Mayr, E. 2004. *What Makes Biology Unique? Considerations on the Autonomy of a Scientific Discipline.* Cambridge University Press.

Mendel, G. 1865. Versuche über Pflanzenhybriden. *Verhandlungen des naturforschenden Vereines in Brünn*, Bd. IV für das Jahr 1865, 3–47.

Molnár, I. 1995. A biológiai formák evolúciója. *Természet Világa*, 1995/I különszám:32–41.

Monod, J. 1965. From enzymatic adaptation to allosteric transitions. Nobel Lecture, from Nobelprize.org, https://www.nobelprize.org/nobel_prizes/medicine/laureates/1965/monod-lecture.html

Monod, J. 1978. *Selected Papers in Molecular Biology by Jacques Monod*. szerk. Lwoff, A., and Ullmann, A (eds.), Academic Press New York, San Francisco, London.

Morgan, T. H. 1909. What are "Factors" in Mendelian explanations? *American Breeders Association Reports*, 5:365–368.

Morgan, T. H. 1919. *The Physical Basis of Heredity.* J. B. Lippincott Company, Philadelphia and London.

Morgan, T. H. 1923. The bearing of Mendelism on the origin of species. *The Scientific Monthly*, 16:237–247.

Muller, H. J. 1922. Variation due to change in the individual gene. *The American Naturalist*, 56:32–50.

Nüsslein-Volhard, C., and Wieschaus, E. 1980. Mutations affecting segment number and polarity in Drosophila. *Nature*, 287:795–780.

Nüsslein-Volhard, C. 1995. *The Identification of Genes Controlling Development in Flies and Fishes*. Nobel Lecture, from Nobelprize.org, https://www.nobelprize.org/nobel_prizes/medicine/laureates/1995/nusslein-volhard-lecture.pdf

Nüsslein-Volhard, C. 2004. The bicoid morpogen papers (I): Account from CNV. *Cell*, S116:S1–S5.

Orel, V. 1996. *Gregor Mendel: The First Geneticist*. Oxford University Press, Oxford, United Kingdom.

Pauling, L., and Delbrück, M. 1940. The nature of the inramolecular forces operative in biological processes. *Science*, 92: 77–79.

Pauling, L., Itano, H. A., Singer, S. J., and Wells, I. C. 1949. Sickle cell anemia, a molecular disease. *Science*, 110:543–548.

Pauling, L. 1951. The hemoglobin molecule in health and disease. *Proceedings of the American Philosophical Society*, 96:556–565.

Pinto-Correia, C. 1997. *The Ovary of Eve: Egg and Sperm and Preformation*. The University of Chicago Press.

Pollock, M. R. 1970. The discovery of DNA: An ironic tale of chance, prejudice and insight. *Journal of General Microbiology*, 63:1–20.

Schrödinger, E. 1944. *What is Life?* Cambridge University Press, Cambridge, United Kingdom.

Shull, G. H. 1915. Genetic definitions in the new standard dictionary. *The American Naturalist*, 49:52–59.

Slack, J. M. W. 2002. Conrad Hal Waddington: The last renaissance biologist? *Nature Review Genetics*, 3:889–895.

Spemann, H., and Mangold, H. 1923. Über Induktion von Embryonalanlagen durch Implantation artfremder Organisatoren. *Archiv für mikraskopische Anatomic und Entwicklungsmechanik*, 100:599–638.

Van Speyerbroeck, L. 2002. From epigenesis to epigenetics — the case of C. H. Waddington. *Annales of the New York Academy of Sciences*, 981:61–81.

Sturtevant, A. H. 1965. *A History of Genetics.* Cold Spring Harbor Laboratory Press, Cold Spring Harbor, New York.

Thompson, D. 1942. *On Growth and Form.* 2nd edition. Cambridge University Press, Cambridge, United Kingdom.

Tilman, D., Fargione, J., Wolff, B., D'Antonio, C., Dobson, A., Howarth, R., Schindler, D., Schlesinger, W. H., Simberloff, D., and Swackhamer, D. 2001. Forecasting agriculturally driven global environmental change. *Science*, 292:281–284.

Turing, A. M. 1952. The chemical basis of morphogenesis. *Philosophical Transactions of the Royal Society B: Biological Sciences*, 237:37–72.

Waddington, C. H. 1956. *Principles of Embryology.* George Allen & Unwin, London.

Waddington, C. H. 1975. *Evolution of an Evolutionist.* Cornell University Press.

Watson, J. D., and Crick, F. H. C. 1953. A structure for deoxyribose nucleic acid. *Nature*, 171:737.

Watson, J. D., and Crick, F. H. C. 1953. Genetical implications of the structure of deoxyribonucleic acid. *Nature,* 171:964–967.

Watson, J. D. 1968. *The double helix Watson: A Personal Account of the Discovery of the Structure of DNA.* Athenaeum, New York.

Watson, J. D. 1962. The involvement of RNA in the synthesis of proteins. Nobel Lecture.

Watson, J., Zoller, M., Gilman, M., and Witkowski, J. 1992. *Recombinant DNA.* 2nd edition. W. H. Freeman, New York.

Weismann, A. 1893. *The Germplasm: A Theory of Heredity,* Charles Scribner's Sons, New York.

Wilson, E. B. 1896. *The Cell in Development and Inheritance.* The Macmillan Company, London.

Wilson, E. B. 1905. The chromosomes in relation to the determination of sex in insects. *Science*, 22:500–502.

Wolpert, L. 1969. Positional information and the spatial pattern of cellular differentiation. *Journal of Theoretical Biology,* 25:1–47.

List of Figures

Verlagsort: Londini/London// | Erscheinungsjahr: (1686) |
Verlag: Robertus Scott & Georgius Wells.
Signatur: 13165861 Augsburg, Staats- und Stadtbibliothek —
2 Nat 97 13165861 Augsburg, Staats- und Stadtbibliothek —
2 Nat 97.
Permalink: http://www.mdz-nbn-resolving.de/urn/resolver.
pl?urn=urn:nbn:de:bvb:12-bsb11200298-0.

Index